D0867158

The Transits of Venus

"Here is all the information needed to plan for and view the great Venus transits of 2004 and 2012, wrapped in an exciting and highly readable account of four centuries of scientific adventures to remote corners of the world. In *The Transits of Venus* science history becomes science adventure."

—Dale P. Cruikshank, NASA Planetary Astronomer

"An enthralling and literate narrative of one of astronomy's most remarkable quests."

—Alan W. Hirshfeld, University of Massachusetts, Dartmouth;
Author of *Parallax: The Race to Measure the Cosmos*

"Portrays the heritage of an astronomical event so rare that it has been witnessed only five times since the invention of the telescope, and never by anyone today alive."

—Trudy E. Bell, Managing Editor, *Journal of the Antique Telescope Society*

"Brings to life the long quest for one of the holy grails of astronomy. Masterfully rendered."

—Anthony Misch, Resident Astronomer, Lick Observatory,
University of California–Santa Cruz

"A superb job of turning a rather esoteric subject into an adventure story reminiscent of Burke's *Connections* and Sobel's *Longitude.*"

—Donald C. Parker, MD, Former Director of the
Association of Lunar and Planetary Observers

"A fascinating, insightful, and eminently readable book. Sheehan and Westfall have accomplished a task almost as daunting as the exploits they chronicle."

—Gary Seronik, Associate Editor, *Sky & Telescope* magazine

"Finally the grand saga of the transits of Venus gets the book it deserves. Lucid, unpretentious, detailed, and elegantly phrased. What sheer pleasure it was to read this book."

—Michael Sims, Author of *Adam's Navel* and *Darwin's Orchestra*

"Anyone who has ever been awed by the first light of Venus should get immersed in this poetic offering from such gifted men of letters."

—Melita Wade Thorpe, Tour Operator, MWT Associates

"An ingenious history, filled with vividly told anecdotes as well as antidotes to previous misunderstandings about the transits of Venus."

—Thomas R. Williams, Editor, *Biographical Encyclopedia of Astronomers*

The Transits of Venus

WILLIAM SHEEHAN
AND JOHN WESTFALL

Prometheus Books
59 John Glenn Drive
Amherst, New York 14228-2197

Published 2004 by Prometheus Books

Inquiries should be addressed to
Prometheus Books
59 John Glenn Drive
Amherst, New York 14228-2197
VOICE: 716-691-0133, ext. 207
FAX: 716-564-2711
WWW.PROMETHEUSBOOKS.COM

08 07 06 05 04 5 4 3 2 1

Library of Congress Cataloging-in-Publication Data

Sheehan, William, 1954-
The transits of Venus / by William Sheehan and John Westfall.
p. cm.
Includes bibliographical references and index.
ISBN 1-59102-175-8
1. Venus (Planet)—Transit. I. Westfall, John Edward, 1938- II. Title.

QB509.S57 2004
523.9'2—dc22

2003022420

Every attempt has been made to trace accurate ownership of copyrighted material in this book. Errors and omissions will be corrected in subsequent editions, provided that notification is sent to the publisher.

Printed in Canada

Dedicated by William Sheehan to his Southern Hemisphere colleagues, Marilyn Head and Phil Barker, without whose enthusiastic support this book would never have become seaworthy.

Dedicated by John Westfall to those who gave their lives in the pursuit of the transits of Venus of the eighteenth and nineteenth centuries.

Such dim-conceived glories of the brain
Bring round the heart an indescribable feud;
So do these wonders a most dizzy pain,
That mingles Grecian grandeur with the rude
Wasting of old time—with a billowy main,
A sun, a shadow of a magnitude.

—John Keats, "On Seeing the Elgin Marbles for the First Time" (1817)

CONTENTS

PREFACE 13

ACKNOWLEDGMENTS 17

1. OF TIME AND NUMBER 21

The Chinese Emperor's Son 21
The Parallax Principle 23
A Classic Problem: The Size of the Earth 25
How Far to the Moon? 27
The Distances to the Planets 28

2. PLANET WATCHERS 31

The Wandering Stars 31
Celestial Alignments 35
Some Special Astronomical Alignments 36

3. VENUSIAN PRELUDES 41

Venus in Antiquity 42
Venus in the New World 45
Astronomic Models in History 47

4. A MISSED OPPORTUNITY 49

Planetary Systems 49
A Realization: The Inferior Planets Can Transit the Sun 56
Searching the Pretelescopic Archive 57
Keplerian Calculations 59
Standing Watch from an Attic in Prague 63
An Unexpected Disclosure: Sunspots 66
Kepler's Preview of the Transit of Mercury 67
The Great Gassendi's Spot-on Observation 70
Gassendi's Vain Vigil for Venus 72

5. HOMAGE TO HORROCKS 75

The Winds of War 76
"Nothing More Noble" 77
Horrocks and William Crabtree 80
Enmeshed in a Net of Calculations 81
Venus Observed 84
The Curtain Falls 87

6. A Celestial Monarchy 93

The New Astronomy and Newton 93
Turbulent Times 95
Annus Mirabilis, or What Newton Knew When 96
Newton Plays Hooky until Hooke Comes Along 97
A Glimpse at Venus in the Telescope 100
Solar Parallaxes 102
Longitude 104
Enter Halley 106
A Transit at St. Helena 108
The *Principia* 111
The Clockwork Universe 118
How the Transits Figure 123

7. "This Famed Phenomenon" 125

Halley's Grand Proposal 125
Halleyan and Cishalleyan Poles 130
The Man of Vision Lives Not to See 135
Delisle Takes Over 136
Bianchini Observes Venus 138
Anticipation Mounts 141
Instrumental Interlude 147
Globetrotters 148
Day of Fame—or Infamy 152
Unexpected Phenomena 154
Back to the Drawing Board 159

8. Cook's Tour 161

Paradise Found 161
Planning for the Great Event 162
Endeavoring to See the Transit from Tahiti 170

Tahiti Looms 173
Point Venus 175
What Is So Fair as a Transit Day in June? 180

9. PURSUING A PLANET 185

Chasing Results 185
What the Hell? 187
The "Barbarous Region" Comes into Its Own 188
Astronomer Faints during Venus Transit! 191
Chappe's Tragic Success 195
More Fortunate in Love than in Astronomy 197

10. A NOBLE TRIUMPH—SURPASSED 201

Astronomy by Numbers 201
Another World Observed 205
The Grand March of Technology 217
Eliminating the "Personality of the Eye" 220

11. FROM ENLIGHTENMENT TO PRECISION 223

A Parallax Challenged 223
Laying Plans 225
The Transit Parties 235
The "Compleat" Transit Station 249

12. UPON THE FLAME-CASED SUN 253

A Century's Wait Ends 253
The Aftermath of the Transit 263

13. 1882: The Last Hurrah 267

Eight Eventful Years: 1874–1882 267
Building on the 1874 Experience 270
The Last Great Attempt 277
David Peck Todd and Lick Observatory 287
The Closing Act of the Transits 293

14. A "Compleat" Guide to Our Transits: 2004 and 2012 297

The Grail Achieved 297
Distances to Stars and Galaxies 300
Venus Revealed 301
The 2004 Transit Approaches 304
Observing the Transits 308
Specialized Equipment 313
What to Look For 317
2012 Is Not That Far Away! 320

Epilogue 327

Appendices 333

A. Transits of Venus, –2970 to +7464 333
B. Data on Observed Transits, 1639–2012 335
C. Local Circumstances for the Transit of Venus, June 8, 2004 339
D. Local Circumstances for the Transit of Venus, June 5–6, 2012 345
E. June and December Sunshine Probability Maps 351

NOTES 355

BIBLIOGRAPHY 377

INDEX 393

PREFACE

On a starred night Prince Lucifer uprose.

. . . .

And now upon his western wing he leaned,
Now his huge bulk o'er Afric's sands careened,
Now the black planet shadowed Arctic snows.

—George Meredith, "Lucifer in Starlight" (1883)

On June 8, 2004, Venus will "transit" the sun. For the first time in over a century, it will appear as a black dot silhouetted against the solar disk—a strange sight, not witnessed by anyone now living.

We are used to thinking of the stars and planets as lights in the sky. Once we thought them to be made up of ethereal substances, but we know now they are as physical as we are. It follows that from time to time we can be caught up in their shadows. The moon interposes itself between us and the sun to produce the dazzling spectacle of a total solar eclipse. Earth passes between the sun and the moon and throws its conical shadow upon the moon, causing a lunar eclipse.

Transits of Venus are among the rarest of all predictable celestial shadow displays, occurring when that beautiful planet, the Evening Star, passes between us and the sun and throws its shadow on the earth. The last transit occurred in 1882; the next, in 2004, will be visible in its entirety from the Old World and partly visible from the United States. There will be one more in 2012, then no more until the twenty-second century.

A rare white heron exists in New Zealand. The Maori called it *kotaku,* the bird of the single flight. If you are lucky you might see it once, maybe twice, in a lifetime. A transit of Venus is an astronomical white heron, a *kotaku.*

Not only has its rarity made a transit of Venus important to astronomers; the alignment of the three bodies—the earth, Venus, and the sun—lent itself to a method of working out the elusive "solar parallax," which was the basis of measuring the distance from the earth to the sun; this *astronomical unit,* or AU, is the celestial yardstick that underpins the entire scale of astronomical distances.

The AU is equal to half the diameter of the baseline of the orbit the earth traverses once every year in its elliptical orbit around the sun. As such, this fundamental unit becomes the basis for determining the parallaxes (apparent shifts in position as seen from points one AU apart) of—and distances to—the nearby stars. From the nearby stars, astronomers can work up the ladder to farther ones, including those of the Hyades star cluster in the constellation Taurus, the Bull, whose members have been used to calibrate how a star's color (spectral class) determines its brightness. From the Hyades, astronomers have reached farther and farther out—to pulsating stars, known as Cepheids, whose periodic light variations determine their intrinsic brightnesses, and thus their distances from us, then on to the spectacularly exploding stars known as type I supernovae, which reveal distances to very remote galaxies.

Each step depends critically on the accuracy of the last. As errors and uncertainties are multiplied, the accuracy at each stage cannot be greater than that with which the previous—and ultimately, the fundamental—units are known.

Historically, the transits of Venus are important because for a long time they seemed to provide the most promising of all methods for finding out the value of the earth-sun distance. Because the method works only if the event is observed from points far separated from one another on our globe, the attempt to capture Venus's small but enthralling shadow led in the eighteenth and nineteenth centuries to scientific expeditions on an unprecedented scale. These expeditions were undertaken by astronomical Ahabs, just as monomaniacally intent on obtaining their observations as any mad whaler in harpooning his great white whale. In chasing Venus's shadow to remote places, they embarked incidentally upon some of the great explorations of our own world, reaching remote and inaccessible points—Hudson Bay, Tahiti, St. Helena, and Tierra del Fuego.

Historically, only a small segment of society was ever concerned with the transits of Venus. In the eighteenth century, those who were so concerned included many astronomers of the major European countries, along with their assistants and the captains and crews of several ships. But the majority of observers were males from continental Europe, supplemented by a few colonists living in Asia or America. In the nineteenth century, there were also a handful of observers of non-European descent and even a few women. Additionally, by then it had been discovered that anyone in the right part of the world with normal eyesight, clear skies, and a simple screen of smoked glass could watch, if not scientifically observe, a transit of Venus. Thus thousands of more or less ordinary persons, at least in those places served by the mass media of the day—newspapers—could participate in the event.

For the transits of 2004 and 2012, scientists may have lost interest—the transits are no longer the most accurate method of determining the distance from the earth to the sun. However, the thousands of transit watchers of the past will have grown to billions; many watching directly, many more through the sometimes-enlightening medium of television. They will join their numbers to the adventurers of the past who traveled to the far ends of the earth to catch *kotaku* on the wing, the rare sight of Venus projected against the sun.

This is a book that presents, as one of its overarching themes, the science of accurate measurement. Measurement requires the adoption of suitable units. We have employed the scientific ("metric") system wherever possible, followed by equivalent English units in parentheses.

We shall also be measuring small angles, thus using units smaller than the familiar degree (°). A degree is subdivided into 60 arc-minutes (symbolized ', as in 30') and 3,600 arc-seconds (e.g., 9".5). To give an idea of the size of these units, the sun and moon appear about a half-degree (0°.5) wide to us; a person standing 3½ miles (5.8 km) distant from you appears about one arc-minute high (1'.0; close to the apparent diameter of Venus when in transit); were that same distant person holding a quarter, it would appear close to one arc-second across to you.

Dates are given according to the Common Era (CE or BCE), with the Julian Calendar ("Old System," or O.S.) before 1583 CE and the Gregorian Calendar ("New System," or N.S.) thereafter. Negative years (e.g., –43) are in the astronomical system, which does have a year 0, so that, for example, –43 is the same as 44 BCE.

William Sheehan
Timaru, New Zealand, and Willmar, Minnesota

John Westfall
Antioch, California

ACKNOWLEDGMENTS

The authors acknowledge the help they received from Dr. Imad-ad-Dean Ahmad, for information about the origin of the star and crescent symbol of Islam; Dr. Michael Armstrong of the Department of Classics of Hobart and William Smith Colleges in Geneva, New York, for his help in translating some of the Latin verses of Jeremiah Horrocks; Phil Barker of the Christchurch Astronomical Society, for this enthusiasm for the project from first to last; historian of astronomy Richard Baum, for several useful illustrations; Judy Brausch, librarian of Yerkes Observatory; Rodolfo Calanca, scientific director of the Communal Astronomical Observatory of Cavezzo, for information and illustrations of the Italian expedition to Madhepur, India, in 1874; Dr. Jimena Canales of Harvard University, who shared several prepublication manuscripts on Jules Janssen and the photographic revolver; Brenda Corbin, librarian of the United States Naval Observatory; Professor Michael J. Crowe of the University of Notre Dame; Dr. Steven J. Dick of the

United States Naval Observatory, especially for his help regarding the official U.S. expeditions to observe the 1874 and 1882 transits; Thomas A. Dobbins and Karen Dobbins, for providing a number of the illustrations and much expert advice regarding instruments and observations; Dr. Audouin Dollfus of the Meudon Astrophysical Observatory and the Paris Observatory, for providing information about French transit expeditions and especially on the Janssen expeditions; Richard Dreiser of the photograpic department, Yerkes Observatory; Alan Gilmore and Pam Kilmarten at the Mt. John Observatory, New Zealand, for their help in clearing up a number of conundrums associated with past transits but especially for pointing out the correct name of Edmond Halley's first ship; Dr. Owen Gingerich of Harvard University; Marilyn Head of the Royal Astronomical Society of New Zealand, for information related to Captain Cook and the transit expeditions, including that of C. H. F. Peters, to New Zealand; Dr. John Hearnshaw of the Canterbury University Department of Physics and Astronomy in Christchurch, New Zealand, for arranging my visits there; Peter B. Hingley, librarian of the Royal Astronomical Society of London, for providing information and illustrations related to Horrocks and Crabtree; Dr. Nicholas Kollerstrom of the University of London, for sharing his work on Hooke, Halley, and Newton; Dr. Françoise Launay of the Meudon Observatory, for her help in furnishing materials on the Janssen expedition to Nagasaki in 1874; Dr. Masatsugu Minami of the Oriental Astronomical Society and his colleagues Naoya Matsumoto and Kunihau Saito, for providing much information about transit expeditions to Japan; Dr. Richard McKim of the British Astronomical Association, for providing information about his visit to astronomical sites in Prague; Anthony Misch, resident astronomer of the Lick Observatory, Mt. Hamilton, California, for pointing out the Kepler horoscope he found in the Mary Lea Shane Archives of the Lick Observatory and for his help in locating in the plate vault of the Lick Observatory of the plates obtained by David Peck Todd at the 1882 transit of Venus observed from Mt. Hamilton; Dr. Wayne Orchiston of the Carter Observatory in Wellington, New Zealand, for advice regarding the various expeditions to New Zealand and Australia; our editor at Prometheus Books, Linda Regan, whose suggestions and encouragement throughout the project were indispensable; Leif Robinson, editor emeritus of *Sky & Telescope*; Dorothy Schaumberg of the Mary Lea Shane Archives of the Lick Observatory, for assistance in locating correspondence related to the transits of Venus and for providing several of the illustrations; Dr. Bradley E. Schaefer of the University of Texas; Dr. Michael Snowden, an independent scholar/astronomer based in New Zealand and Sri Lanka, for information about Tahiti; Dr. William Tobin of Canterbury University, Christchurch, New Zealand, for providing information on Leon Foucault; Dr. Craig B. Waff, an independent scholar; Larry Webster of Mt. Wilson Observatory; Dr. Andrew T. and Louise Young, who provided invaluable comments on a draft of the manuscript and an unpublished translation of H. Kayser's great book on spectroscopy, which was very useful in providing background to the spectroscopic studies of the two nineteenth-century transits.

Above all, the authors would like to thank their wives, who made many useful comments and tolerated their absorption, at times quite absolute, in this project.

I

OF TIME AND NUMBER

It may not be wholly necessary here to enter, on behalf of authors, a protest in favor of those didactic preliminaries for which the ignorant and impatient reader have so strong a dislike. There are persons who crave sensations, yet have not patience to submit to the influences which produce them; who would fain have flowers without the seed, the child without gestation. Art, it would seem, is to accomplish what nature cannot.

—Honoré de Balzac, *The Quest of the Absolute* (1834)

THE CHINESE EMPEROR'S SON

There is a legend of ancient China, according to which the son of one of the Chinese emperors falls ill. He becomes feverish and begins to waste away. His sorrowful parents and baffled physicians fear for his life, but despite the skillful efforts of physicians and the loving ministrations of his parents, he becomes paler, more sickly, more frail.

At last the physicians learn the reason for the child's illness. He is pining away for the moon. "Unless I have it," he tells his physicians, "I shall surely die."

His physicians hold counsel with one another. One of them turns to the boy: "How large is the moon, dear prince? And what is it made of?"

The boy ponders for a moment, then looks at his thumb. "It is, I think, the size of my thumb. It is surely made of silver."

Then the physicians devise a plan. They turn to the parents, the emperor and empress. Their plan is desperate, but the emperor and empress, wan with grief, grant their approval—it must be tried. A silversmith is summoned. He is ordered to fashion a small ball of silver, a child's thumbnail in breadth, and to place it as a pendant on a silver chain.

Before the day is over, the silversmith's work is finished. The emperor and empress present the silver pendant to the sickly boy. "Here is what you have asked for," the empress tells him. "We have plucked the moon out of the sky and placed it for you on a silver chain. Here is the moon. Here is what you wished for."

The boy receives it with delight and hangs it around his neck. The ruse is successful—successful beyond expectation. From that moment the child's health begins to rally. He is soon smiling, playing, and singing, restored to his former vigor, chirruping like a small songbird.

Soon, however, a new dilemma confronts the emperor and the empress. It is now the time of the new moon, when the moon is not seen in the sky. Well enough; but what will happen when the moon returns? The ruse will be discovered; the boy will become sick again. Everyone in the palace contrives to keep dark curtains in front of all the windows. Everything is done to make sure the boy remains indoors among guardians; he is not allowed to peek out the windows at night.

For a while this works—but eventually carelessness sets in. One day the child finds himself alone. Overtaken with curiosity, he draws aside the drapes and looks out the window into the silent and lovely night. There hangs the moon, like a creambud in a neighboring cherry-tree. The emperor and empress and their physicians descend upon the scene in horror. They find the boy, standing beside the window, gazing steadily at the moon.

They fear the worst; but he is not upset. Instead, he is singularly complacent and unperturbed. The emperor and empress are baffled. The physicians are summoned. Finally the emperor asks him, "My son, why are you not upset? There is the moon around your neck, and there is the moon hanging above the tree. How can there be two moons?"

The boy laughs. "Father, don't be silly. When you lose a tooth, a new one grows in. It is the same with the moon. When you plucked the moon from the sky and put it as a pendant around my neck, the sky, too, grew a new one. It is the old moon I wear around my neck, the new moon that shines down on us through the branches of the trees."

THE PARALLAX PRINCIPLE

The emperor's ruse worked only because the Chinese prince did not know how to estimate the size of the moon. It is not an easy question. For someone with only a qualitative view of the world, it is not an answerable one; he is at a great disadvantage.

I put my thumb at arm's length and hide the moon behind my fingernail. As I run after it, the moon seems to keep pace, and for all my efforts, I cannot manage to get any closer to it. When I reach the summit of the next hill, panting, I do not seem to have gained any ground upon it. It looms as far away as ever. This experience suggests the moon must be very far away. To look so large from so great a distance, it must be much larger than the house across the way, or even than the distant mountains. But how far, and how large?

Our basic way of estimating distances is by means of *parallax*: the apparent shift of an object as seen from two different viewpoints (see fig. 1). Humans (and primates and carnivores, but not all two-eyed animals) have stereopsis. The brain carries out a series of computations, usually unconscious, making the eyes converge and

Figure 1. John Westfall illustrating the parallax principle, where the parallax is the difference in direction of his fingertip (F), as seen from his right and left eyes (R and L). (Photograph by Elizabeth Westfall, modified by John Westfall.)

causing the ocular muscles to change the thickness of the eye's crystalline lens. Because the eye-base in humans averages only about 65 millimeters, however, the eye-brain system can use this information to estimate accurately the distance out to a few tens of meters at most. Estimates of greater distances must be based on factors such as the context—the object's relationship to other things in the visual field whose distances and sizes are known.

This context-dependent method of estimating distances is not very reliable. In the case of the moon, for instance, whenever the moon stands high in the sky and is seen across "empty space," it looks much smaller than when it is seen close to the horizon next to houses and trees. The apparent difference is actually quite large, perhaps twofold or even more, though—and this is the surprising thing—the moon's actual measured diameter is slightly less when near the horizon. This is the well-known "moon illusion," which shows how hard it is to estimate sizes when there are no objects for comparison. It also proves that appearances can be deceiving.

The moon illusion underscores the need for measurements—in other words, for numbers—which brings us to the realm of the physicist and the astronomer. As Sir James Jeans once said: "Physics is an exact science because it depends on exact measurements."[1] For much of its history, and to some extent even now, astronomy existed primarily as the science devoted to determining, with increasing accuracy, the fundamental numbers of the universe. These include matters such as the positions of the moon, the planets, and the stars; their sizes and distances; and the measures of time.

A Classic Problem: The Size of the Earth

The apparent size of the moon, which we can easily measure, gives us no indication of its distance or actual dimensions. Is it the size of a fingernail, as the Chinese prince thought? Or as large as the globe which we inhabit?

To answer that, we need a much longer baseline than the distance between our two eyes. In addition, we need to make use of a fact noticed at least as long ago as the time of the Egyptian pyramid builders: that the heights of buildings such as pyramids can be measured using the lengths of the shadows they cast. The height of a pyramid is to the length of its shadow as the length of any convenient stick planted upright in the ground is to the length of its shadow. This—the law of similar triangles—can be applied to some of the most majestic of all the problems with which the human intellect has ever wrestled: the problem of the scale of the universe. With a simple stick in the ground—a *gnomon,* from the Greek word meaning "to know"—one can bootstrap one's way outward and upward, answering such questions as, How large is the earth? How far away are the moon, the sun, and the stars?

The first step in the sequence known as the cosmological distance ladder is to work out the size of the earth. More precisely, one determines the length of a small section of the earth's curved surface, say, by counting mileposts between two observing stations lying due north and south of each other. These stations lie along a *meridian,* one of the great circles passing from the poles through any given point on the earth's surface. The measured difference, on the same day, in the altitude of a celestial body when it culminates (i.e., when it reaches the point where it stands highest relative to the horizon) is equal to the difference in latitude between the two stations. Finally, the ratio of this difference to the 360 degrees in a full circle allows one to work out the circumference of the earth.

The Greek philosopher Aristotle (383–322 BCE) already knew very well that the earth was not flat but a sphere from various observations including the fact that it always casts a circular shadow during lunar eclipses. He estimated the earth's circumference at 400,000 *stadia*. Archimedes (c. 287–212 BCE) a little later revised this downward to 300,000 stadia. (Unfortunately, the length of the *stade* was not standardized. A commonly used modern conversion factor is 1 stade = 606 feet [184.7 meters].)

The most famous ancient estimate of the earth's circumference was made by Eratosthenes of Cyrene (c. 276–196 BCE), the librarian at the great library at Alexandria. By using a simple gnomon, he found that at Syene, located just above the first cataract of the Nile (near the present-day Aswan High Dam), the sun at the summer solstice cast no shadow at all: it was exactly overhead. The reason for this is that Syene is located very close to the Tropic of Cancer. At the same moment, at Alexandria, the shadow cast by the sun showed that it stood 7°.2 from vertical. This difference is equal to 1/50 of a circle. Assuming that Alexandria is due north of Syene, which is reasonably close to true, and using 5,000 stadia for the distance between them, Eratosthenes arrived at a figure for the circumference of the earth of about 250,000 stadia. Again, we don't know the exact length of the stadium he used; nevertheless, the figure he came up with was certainly of the right order and would have implied a value for the earth's radius close to 7,350 kilometers (4,600 miles), only 15 percent larger than the modern value. (It is important to point out that such phrases as "the modern value," "the accepted value," or "the actual value," useful though they may be to the reader, have little meaning in the historical context. Scientists of the past did not know what answer they were supposed to get and had nothing with which to compare their results. It is perhaps more important to take note of the confidence that was placed in the obtained values at the time by the measurers themselves or by their contemporaries. For various estimates of the radius of the earth, from Aristotle's time to our own, see Table 1).

Table 1. Estimates of the radius of the earth

Recorder	Date (astronomical system)	Value (km)	Difference from modern (%)	Source (see bibliography)
Aristotle	–340	12,200	+91.3	Smith (1997), p. 7
Archimedes	–230	9,180	+43.9	Smith (1997), p. 7
Ptolemy	150	5,513	–13.6	Van Helden (1985), p. 24
I-Sing	750	9,020	+41.4	Smith (1997), p. 13
Al-Mamum	820	6,364	–0.22	Smith (1997), p. 15
Al-Farghani	840	5,525	–13.4	Van Helden (1985), p. 170
Fernel	1528	6,338	–0.63	Smith (1997), p. 17
Snell	1617	6,369.0	–0.14	*Ency. Britannica* (1911), 25:293
Norwood	1637	6,412.2	+0.53	Clarke & Helmert (1911), p. 802
Picard	1669	6,371.2	–0.11	Clarke & Helmert (1911), p. 802
Maupertuis	1738	6,397.3	+0.30	Hooijberg (1997), p. 37
Cassini	1740	6,387.3	+0.14	Todhunter (1962), 1:127
Fr. Acad. Sci.	1743	6,376.6	–0.02	Vanícek & Krakiwsky (1986), p. 113
La Caille	1753	6,383.2	+0.08	Todhunter (1962), 1:127
IAU*	1976	6,378.14	—	U.S. Naval Observatory (2004), p. K6

*International Astronomical Union

How Far to the Moon?

Once the earth's radius is known, the earth itself can be used as a baseline for determining still greater distances—the distance to the moon and even, though this would take some finessing, to the sun. If we measure, for instance, the moon's position relative to the stars from, say, Stockholm (at 59°.27 N) and Cape Town (at 33°.94 S), which lie within half a degree of being on the same meridian, the shift in the moon's position relative to the stars amounts to well over a degree, more than twice its own apparent diameter. This observed parallax, combined with the known radius of the earth, gives the distance to the moon: on the average, some 384,000 kilometers (238,000 miles).

Of course, the ancients did not have the luxury of traveling such great distances across the earth's surface; consequently, they had to rely on less direct methods. One of them was based on the fact that during a lunar eclipse, the moon enters the shadow cone cast by the earth. By assuming that the distance of the sun is much greater than that of the moon—a fact Aristarchus of Samos (c. 310–230 BCE) already knew—it becomes possible to work out the earth-moon distance indirectly from the geometry of eclipses. Using this method, Hipparchus of Rhodes (fl. 140 BCE) worked out that the distance of the moon was 59 earth radii. It's a good approximation—within 1½ or 2 earth radii of the modern value—and showed that the moon's diameter is about a quarter of the earth's.

THE DISTANCES TO THE PLANETS

The distances to the other planets or the sun are, of course, much greater than the distance to the moon. Whereas the parallax-shift of the moon, when viewed from opposite sides of the earth, can amount to as much as 2 degrees, that of Venus, the planet that approaches nearest of all to the earth, is never more than a hundredth as much, or 0°.02. This is a small angle, equal to the apparent size of a quarter at a distance of 73 meters (240 feet). This underscores the overriding theme of all celestial parallax work: the measurement of tiny angles as accurately as possible, pushing techniques and instruments to their limits—and sometimes beyond them.

By convention, parallaxes are not referred to as if they were measured from opposite ends of the earth. Instead, the standard adopted is the *horizontal equatorial parallax*. This is defined as the parallax—the angular shift in position relative to that seen by an imaginary viewer at the earth's center—of an object measured by an observer on the earth's equator when the object viewed is at the horizon. This assumes no distortion of its direction by refraction in our atmosphere, which causes an object to appear higher in the sky than if there were no atmosphere.

The tiny angles involved with planetary and solar parallaxes were too small to be measured with even the best instruments the ancients had at their disposal. These instruments consisted, essentially, of open-barrel sighting tubes through which the observers gazed with the naked eye and read off angles on a graduated circle. The smallest angles anyone could measure with such instruments were apparently on the order of ten or twenty arc-minutes, so that the parallax of Venus, even when it is nearest to the earth, would have been beyond their reach.

The last of the great Greek astronomers, Ptolemy, lived in Alexandria, Egypt, in the second century CE; he was active during the reign of the Roman emperor Hadrian. He regarded both Venus and Mercury as moving in circular *epicycles*, the centers of which themselves moved in larger circles called *deferents*, the latter centered on, or near, the earth. He assumed, but could not prove, that these planets were situated above the sphere of the moon but below that of the sun. However, he considered that they were too far to show any discernible parallax, while that of the sun is smaller still—in reality, less than ten arc-minutes, though we will not give away the exact value now, since that would destroy the suspense. The determination of this value would become the object of one of astronomy's grandest quests, the virtual obsession of the long line of number-obsessed personalities. They began perhaps with the Akkadian-Assyrian god Ninarta, who held the "inch scale" with which he surveyed the universe, as well as such historical figures as the Greek philosophers Pythagoras and Plato. Combining the solar parallax with the size of the earth produces the earth's distance to the sun—one of the most fundamental values in astronomy; it is even called the *astronomical unit*.

The quest for the solar parallax would become the driving force behind the great eighteenth- and nineteenth-century expeditions to observe the transits of Venus. This quest eventually furnished one of the keys needed to unlock man's eternal dream: to find the distance to the roof of heaven—the height, over our heads, of the stars.[2]

2

PLANET WATCHERS

When thy folding-star arising shows
His paly circlet, at his warning lamp
The fragrant Hours, and elves
Who slept in flowers the day,
And many a nymph who wraiths her brows with sedge,
And sheds the freshening dew, and, lovelier still,
The pensive Pleasures sweet,
Prepare thy shadowy car.

—William Collins, "Ode to Evening" (1746)

THE WANDERING STARS

No one can say just when human beings first noted five bright "stars" moving among the rest of the starry host. These were indeed the *planets* (from the Greek for "wanderers"). Usually they traveled in an orderly manner, progressing in stately and majestic fashion from the west to the east; sometimes they moved more oddly, almost capriciously, pirouetting and making grand loop-the-loops among the fields of night. What were these moving lights? What purpose did they serve?

As long as humans followed a nomadic lifestyle, they seem to have paid only passing attention to the planets, indeed to the phenomena of the starry skies. This is still true of modern nomadic peoples; for instance, the !Kung—one of the Bushman groups of southern Africa—neither worship the moon or stars nor pray to them for rain or success in the hunt. "They regard [all these] as distant 'things of the sky,' beyond man's knowledge."[1] To the !Kung, the sun represents searing heat, thirst, hunger, exhaustion; it is not beneficent, but regarded as malevolent—a thing of death. It is just as well that it is far away, or it would burn people to cinders. It is rain, not the sun, that brings life to the earth.

The true rise of astronomical science is associated with the rise of agricultural civilizations, or at least of the first settled communities. Nomadic peoples generally don't stay put long enough to set up towers, pillars, plinths, and gnomons, which can serve as reference points for taking precise soundings of the positions of the celestial bodies. They may feel the wind in their face or be throbbingly alive to the romance of the stars as they gallop nightly across the plains—but this is poetry, not science.

The sharp difference between the nomadic viewpoint and that of the settled agriculturalist is nowhere better attested than in the difference between the !Kung, for whom the sun is regarded as a burden, and the agricultural society of the Egyptians. The latter had many gods; among them was the sun, which was worshiped as the god Ra. Under the Pharaoh Akhenaten, who ruled between 1379 and 1362 BCE (see fig. 2), sun-worship was raised to a level not to be seen again until the heyday of the Mayans and the Aztecs of Mesoamerica two millennia later. Akhenaten—the "heretic Pharaoh"—broke from timeless Egyptian tradition by proclaiming the sun, in the form of the solar disk (the "Aten"), as the one supreme being. (Figure 3, a representation from the treasure of Tutankhamen, shows

Figure 2. Artist "sketch" of Pharaoh Akhenaten, circa 1372 BCE. The Egyptian Museum, Cairo. (Photograph by William Sheehan.)

Tutankhamen and his wife worshiping the sun represented as the Aten, or "sun disk.") Akhenaten moved the capital from Thebes to a new site, Akhetaten (now known as Tell el-Amarna), and in a stele marking the boundary of the new city announced that it was being built on virgin land because "it belonged to no god or goddess and no lord or mistress." The Pharaohs of the Old Kingdom, the builders of the pyramids, had placed the monuments at Giza on the west bank of the Nile on the grounds that the sun set in the west, the land of the dead. The later Pharaohs were interred at Thebes, in the Valley of the Kings; they also rested on the west bank of the Nile. But Akhenaten put his city on the *east* bank, the place of the sun's rising. In a remote place where the Aten had "shut in for himself with mountains and in its midst set a plain," he, his beautiful wife, Nefertiti, and

Figure 3. Detail from Tutankhamen's golden throne. The king is anointed by the queen with perfume. The Egyptian Museum, Cairo. (Photograph by William Sheehan.)

their daughters, guarded by Asiatic and Nubian mercenaries, lived an isolated existence—evidently as the Pharaoh wished it. At Amarna, he wrote his sublime "Hymn to the Aten" (before 1362 BCE):[2]

> Thou dost appear beautiful on the horizon of heaven,
> O living Aten, thou who wast the first to live.
> When thou hast risen on the eastern horizon,
> Thou hast filled every land with beauty.
>
> . . . .
>
> When thou dost set on the western horizon,
> The earth is in darkness, resembling death.
>
> . . . .
>
> At daybreak, when thou dost rise on the horizon,
> Dost shine as Aten by day,
> Thou dost dispel the darkness
> And shed thy rays.
> The people of the Two Lands celebrate thee daily,
> Awake, and standing on their feet,
> For thou hast raised them up.
>
> (Translation from James H. Breasted, *A History of Egypt from the Earliest Times to the Persian Conquest,* 1909)

This, of course, seems to describe religion, not astronomy; but at first, the two were not distinct. Among nomadic peoples the lunar calendar has been—and still is—universal. Among agriculturalists, it becomes convenient to adopt a solar calendar, based on the sun's apparent pathway around the sky. This pathway—the *via solis,* or ecliptic—is also, of course, that traversed by the moon and the planets.

It is fair to ask whether the Egyptian priest-astronomers, as they monitored the sun from their rooftops, pyramids, or towers, keeping up their calendars, might not now and then have glimpsed the odd sunspot. They would certainly have enjoyed favorable conditions for such an observation. Even under the clearest conditions, the sun can be studied without discomfort or danger when it is near the horizon, but when it rises from or sets into mists, the comfortable zone is increased and the safe viewing period prolonged. This often occurs along the banks of the Nile. One of us (W.S.) well remembers having seen a large sunspot while on a cruise from Aswan to Karnak; it was strikingly prominent as the sun set into the mists over the west bank of the Nile.

What would they have made of such a sighting? And might they not even have captured, by chance, the odd transit of a planet—a small, dark, perfectly round spot—across the disk of the sun, without being aware of what it was?

Celestial Alignments

Transits of Venus are just one example of a class of phenomena—celestial alignments—that have been of interest to sky watchers from earliest times. Before proceeding further, it will be useful for us to briefly describe the significance of the various cases in which celestial alignments occur.

One can always draw a line between any two objects. Since celestial bodies are in constant motion, it is not surprising that sometimes three of them can line up. Exact alignments, in the mathematical sense, are exceedingly rare. However, the fact that celestial bodies are not mathematical points but objects having finite size means that even relatively close alignments produce interesting, and at times astounding, spectacles.

The various forms of alignment are referred to as conjunctions, eclipses, occultations, and transits. A *conjunction* is merely the close approach of two celestial objects in the sky—for instance, the crescent moon and the planet Venus. An *eclipse* involves the obscuration of one celestial body as it passes through the shadow cast by another. An *occultation* involves a process whereby one body, such as a star or planet, is concealed behind a nearer body, for instance the moon. A *transit* occurs when one celestial body passes in front of another. The most notable examples are the passages of Mercury or Venus in front of the sun, but the term is also used to refer to the passage of other bodies, such as the Galilean satellites of Jupiter, in front of that planet.

There is no blanket term to describe all of these phenomena, though "eclipse phenomena" comes nearest.[3] It would be impossible to list all the variations on the theme. A partial list is given in Table 2.

Table 2. Examples of eclipse phenomena

Celestial body (in order along line of alignment)					
First ("left")	Second ("middle")	Third ("right")	Name of event	Location of observer	Notes
Sun	Moon	Earth	Solar eclipse	Earth	
Sun	Moon	Earth	Eclipse of earth?	Moon	Shadow transit?
Sun	Earth	Moon	Lunar eclipse	Earth	
Sun	Earth	Moon	Solar eclipse	Moon	Surveyor 3 spacecraft
Sun	Mercury	Earth	Transit of Mercury	Earth	
Sun	Venus	Earth	Transit of Venus	Earth	
Sun	Jupiter	Satellite	Satellite eclipse	Earth	(A Galilean satellite)
Sun	Satellite	Jupiter	Shadow transit	Earth	(A Galilean satellite)
Sun	Satellite	Satellite	Mutual eclipse	Earth	(Two Galilean satellites)
Planet	Moon	Earth	Planetary occultation	Earth	
Satellite	Jupiter	Earth	Satellite occultation	Earth	(A Galilean satellite)
Jupiter	Satellite	Earth	Satellite transit	Earth	(A Galilean satellite)
Satellite	Satellite	Earth	Mutual occultation	Earth	(Two Galilean satellites)
Star	Moon	Earth	Stellar occultation	Earth	
Star	Planet	Earth	Stellar occultation	Earth	
Star	Asteroid	Earth	Stellar occultation	Earth	
Star	Star	Earth	Eclipsing variable star	Earth	Binary star
Star	Planet	Earth	Transit	Earth	Extrasolar planet

Under eclipses, we must also define the terms *total, partial,* and *annular,* depending on the extent to which one body is obscured by the shadow cast by the other. Each eclipse, occultation, or transit creates three internested conical zones in space in which conditions of totality, partiality, or annularity apply (see fig. 4).

SOME SPECIAL ASTRONOMICAL ALIGNMENTS

Conjunctions of the moon and one or more planets, or of several planets among themselves, can be spectacular and attracted the attention of ancient astrologers. Their alignments were thought to be significant in forecasting events in human affairs (e.g., the "Star

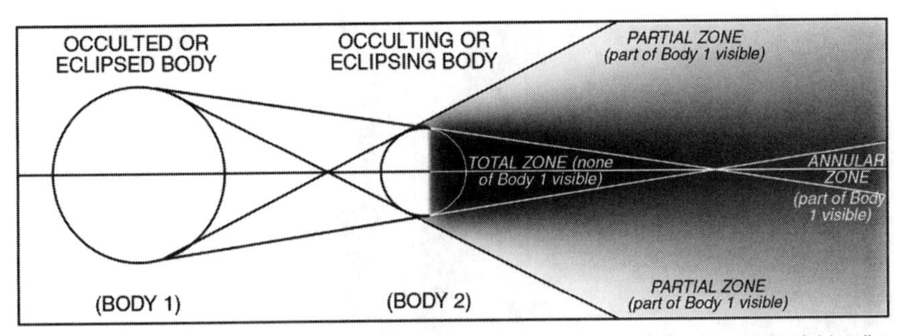

Figure 4. The eclipse-phenomenon zones that are created whenever two celestial bodies are aligned, whether for an eclipse, occultation, or transit. In the case of an eclipse (where Body 1 is self-luminous and Body 2 is not), the total zone is called the *umbra*, and the partial zone, the *penumbra*. (Diagram by John Westfall.)

of Bethlehem," thought by some to have been a "triple conjunction" of Jupiter and Saturn, which occurred in the years 7 and 6 BCE). Alignments are still avidly observed by astronomers, as they provide information not available under ordinary circumstances. (Some of these are described below, with the notable exception of the transits of Mercury and Venus—the subject of this work—which will be described in detail later.)

Solar eclipses. By a remarkable coincidence, the distances and sizes of the earth, moon, and sun dovetail in such a way that all three possible cases—partial, annular, or total eclipses—can be seen from the earth's surface. Whenever a total or annular solar eclipse occurs, the eclipse track is surrounded by a wider zone where a partial eclipse can be seen. Sometimes the central line of a solar eclipse misses the earth entirely but still passes close enough so that one of the earth's polar regions sees a partial eclipse.

To illustrate their frequency, between 2001 and 2010 the earth will experience an average of two solar eclipses per year, twenty in all: seven total solar eclipses, eight annulars, and four partials—and even one eclipse (April 8, 2005) that changes from annular to total and back again.

A total solar eclipse is the most dramatic predictable celestial event, which draws many people from long distances, at great expense, to go and experience the adrenaline rush of the spectacle. His-

Figure 5. (*a*) "Diamond Ring" at the beginning of the total phase of the July 11, 1991, total solar eclipse, as photographed at La Paz, Baja California Sur, Mexico, by William Sheehan with a 4-inch (10 cm) reflecting telescope. (*b*) Same eclipse as above, during the total phase, showing the pearly white streamers of the corona.

torically, total solar eclipses were the only way to see the sun's outer atmosphere, the *corona*. Ancient solar eclipse descriptions serve to clock the earth's varying rotation rate, and modern observations pinpoint the moon in its orbit and help to study possible fluctuations in the sun's diameter. Figures 5 and 6 show examples of total and annular solar eclipses. Note that the same categories apply to occultations, whereas a transit is analogous to an annular eclipse.

Lunar eclipses. Just as the earth passes into the moon's shadow at a solar eclipse, the exact opposite can also occur, when the moon passes through the earth's shadow, producing a lunar eclipse (see fig. 7). The major difference between the circumstances leading to solar and lunar eclipses is brought about by the earth's diameter being 3.7 times that of the moon. At the mean distances of the moon and the

Figure 6. Annular solar eclipse, with the moon's apparent disk smaller than that of the sun's. Photographed from El Paso, Texas, May 10, 1994, with a 9 cm (3.5 in) Maksutov telescope by John Westfall.

sun, the earth's penumbral shadow (the "partial zone") is 16,400 kilometers (10,200 miles) across; even the earth's umbral shadow (the "total zone"), slightly enlarged by refraction in its atmosphere, has a diameter of about 9,400 kilometers (5,800 miles), or 2.7 times that of the moon. Thus the moon never experiences an annular solar eclipse, because as seen from the moon, the earth's apparent size far exceeds that of the sun.

When the moon passes entirely within the earth's umbral shadow, it undergoes a total lunar eclipse. If, even at maximum eclipse, some part of the moon lies outside the umbra, the eclipse is partial. Finally, if the moon enters only the earth's penumbral shadow, we call the event a penumbral lunar eclipse.

In the long term, lunar eclipses are slightly more frequent than solar eclipses. Between 2001 and 2010, there occur twenty-three lunar eclipses: eight are total, six are partial, and nine are penumbral.

Obviously, a lunar eclipse as seen from the earth becomes a solar eclipse as seen from the moon—the sun is then blocked by the earth.

Eclipses of planetary satellites. Every major planet ex-

Figure 7. The partial eclipse of the moon on March 24, 1997, shown near mid-eclipse, with the northern limb of the moon (upper left) in the penumbral shadow and the remainder of the moon in the umbra. Photographed from San Francisco, California, with a 36 cm (14 in) Schmidt-Cassegrain telescope by John Westfall.

cept Mercury and Venus has one or more natural satellites. These bodies sometimes pass through their planet's shadow and become eclipsed, just as does the earth's moon.

The most common such events that are easily observed are those involving the four large satellites of Jupiter, discovered by Galileo in 1610. The three innermost of these undergo eclipses every time they orbit the planet, at intervals of 42 hours for the innermost, Io; 3.44 days for Europa (see fig. 8), and about a week for Ganymede. The fourth of the large satellites, Callisto, is so far from Jupiter that it passes north or south of Jupiter's shadow for three-year stretches. Each uneclipsed period brackets about three years of eclipses, which occur at intervals of 17 days.

Historically, eclipses of Jupiter's satellites, which for Io can be timed visually with small telescopes to twenty or thirty seconds' accuracy, provided one of the very few methods of comparing one's local time to the time at a standard location, such as Paris or London, and thus of finding one's longitude on the earth.

Eclipses of the sun and moon draw the attention of laypersons and astronomers alike. Those of planetary satellites are less enthralling but also have their uses. Yet these are only three of many different forms of celestial alignments, including the infrequent but famous transits of planets across the sun. Those involving our nearest planet, Venus, are particularly rare and serve as the centerpiece of this book.

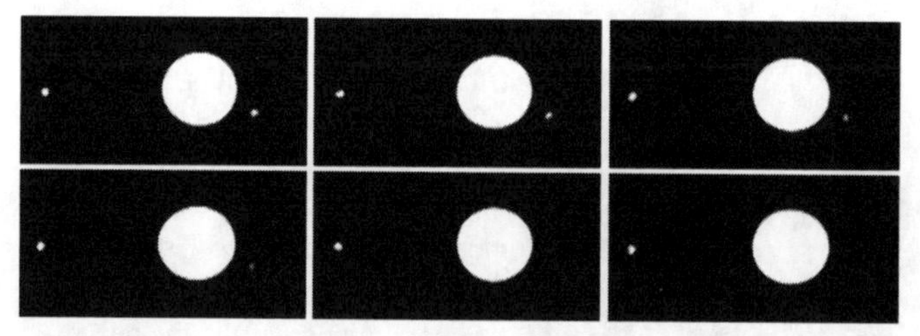

Figure 8. Jupiter's Galilean satellite Europa, to the lower right of the overexposed planet, disappears into Jupiter's shadow on April 7, 1994. The satellite Io is visible to the left. The sequence of CCD (charge-coupled device; see chap. 14) images is top row left to right, then bottom row. North is up. Imaged with a 25 cm (10 in) Cassegrain telescope by John Westfall.

3

VENUSIAN PRELUDES

Venus near her! Smiling downward at this earthlier earth of ours,
Closer on the Sun, perhaps a world of never fading flowers.

. . . .

Might we not in glancing heavenward on a star so silver-fair,
Yearn, and clasp the hands and murmur, 'Would to God that we were there'?

—Alfred, Lord Tennyson,
"Locksley Hall Sixty Years After" (1886)

The planet Venus has been monitored closely by astronomers for thousands of years, partly because of its spectacular brilliance and unsurpassed beauty, partly because of its close relationship with the sun. It tags behind the sun in the evening; at these times, the Egyptians sometimes called it *Uati,* the first star of the night. It pops up before it in the morning, when it was known as *Tiû-nûtiri,* the harbinger of the sun.

There were other names for Venus, corresponding to other guises. It was *Benin*—the heron. The bird, still common along the banks of the Nile, disappears under the river, only to rise again. In the same way, Venus, the celestial heron, disappears for periods of time—sometimes for months—only to return always and ever again to take its place as the leader of the stars of the night.

Venus in Antiquity

Unlike the Egyptians, the people of the other great agricultural civilization of antiquity, Sumer-Akkad, which developed in Mesopotamia (modern Iraq), never adopted a solar calendar. The Sumer-Akkadians remained strict lunarians, whose careful records of the moon's intricate motions remained for centuries a centerpiece of their astronomy. In their observations of the first appearance of the crescent moon, which in practical terms marks the beginning of the lunar month as it still does among Muslims, they were evidently struck—as people throughout history have been struck—by the brilliant planet that often appeared in their field of view. Indeed, the almost-new moon standing together with Venus in the sky is one of the most beautiful and awe-inspiring sights of naked-eye astronomy.

The Sumerian name for Venus was *Inanna*. She was originally a rain-deity and fertility goddess, bride of the god Dumuzi-Amaushumgalana, who represented the growth and fecundity of the date palm; hence she was sometimes known as the Lady of the Date Clusters. She set her heart on ruling the underworld and attempted to depose her sister Ereshkigal, Lady of the Greater Earth; but the attempt failed, she was killed, and in the underworld she was turned into a piece of rotting meat. Eventually, Enki, the Lord of Sweet Waters in the earth, managed to bring her back to the earth, but only on the condition that she offer a substitute in her place. She chose her husband Dumuzi when she found him feasting instead of mourning her absence. In the end, Dumuzi and his sister Gesthtinanna were allowed to alternate as her substitute; each spent half a year in the underworld, half a year above it.

What must have started out as a fertility myth became entwined with the planet Venus's alternate appearances above the horizon as the Morning Star (where it reigns for 260 days, close to the period of human gestation); followed by its disappearance (long period of invisibility, or retreat into the underworld); followed again by its triumphal reappearance as the Evening Star. Inanna became an astral deity, forming with Shamash, the sun-god, and Sin, the moon-god, the great celestial triad that was worshiped throughout the ancient Middle East. She was also identified with the Queen of Heaven, the Evening Star, who followed the sun into Kur, the underworld.

In addition to its bobbing motions east and west of the sun, which bring it alternately into the evening and morning sky, Venus also pursues the sun as it moves north and south with the seasons. It runs ahead of it along the zodiac by as much as almost two months when it is in the evening sky and falls behind it by as much when it is in the morning sky.

Thus Venus's actual motion consists of a series of intricate dance steps that trace out complex but characteristic figures (see fig. 9). Each star figure, if we may so call it, is completed in a period of 19 months, 17 days (or, more precisely, 583.9169 days), the interval between successive appearances of the planet in the morning or evening sky.

By a remarkable coincidence, every eight years Venus moves through five completed star figures and again comes to conjunction with the sun in the same part of the zodiac. In musical terms, there is a five-eight beat or polyrhythm between the orbital motions of Venus and the earth. The eight-year Venus-solar period is not quite exact, however; it falls short by a little more than 2½ days. This means there will be a slight slippage out of alignment. To return to our musical analogy, the frequencies of the planets merge, then they go off from one another just like tuning forks not quite in perfect tune.

The Sumer-Akkadian astronomers tracked all of this as they peered out into the famously clear skies of the Iraqi desert, from the terraced towers of their ziggurat observatories (one of them was dimly remembered as the biblical tower of Babel). As they pored over their account books of the heavens, they made discoveries foreshadowing by centuries the famous pronouncement of the Ionian Greek

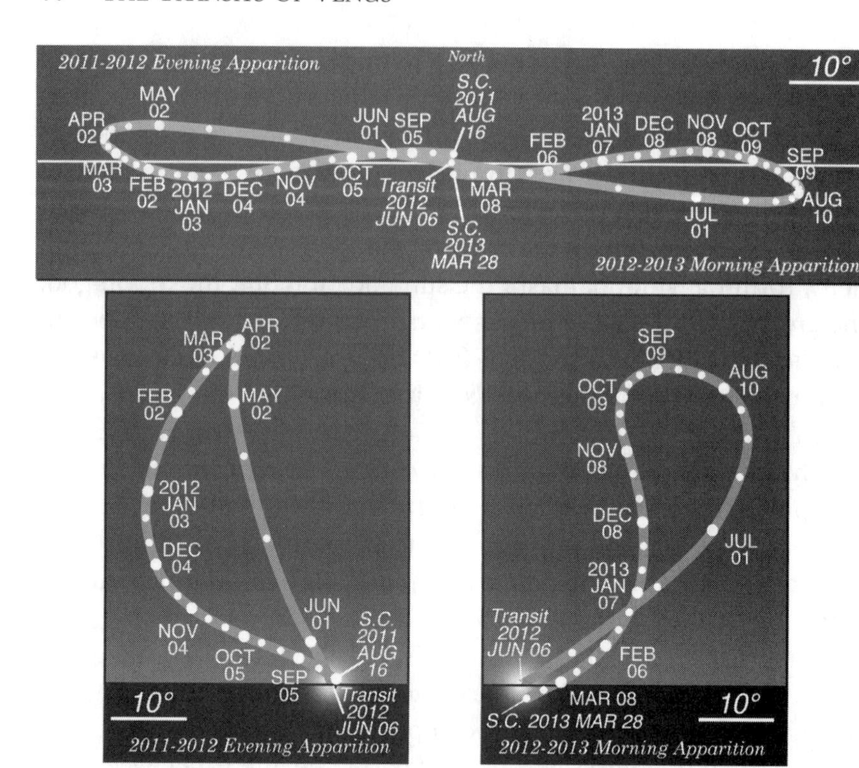

Figure 9. Venus's motion relative to the sun at successive apparitions; in this case, the 2011–12 evening apparition before and the 2012–13 morning apparition after the June 6, 2012, transit. The upper view shows Venus's motion along the ecliptic (horizontal line), while the lower views show the planet's motion relative to the setting (left) or rising (right) sun as seen from latitude 40° north. *Note:* S.C. = superior conjunction. (Diagram by John Westfall.)

philosopher Pythagoras: "All is number."[1] It was once suggested by the French popular-astronomy writer Camille Flammarion that one of their observations might refer to a transit of Venus.[2] It is recorded in a collection of ancient Venus lore from a tablet, now in the British Museum, dating to the reign of King Ammisaduqua (−1648 to −1628), the next-to-last king of the first Babylonian dynasty. We have recalculated Venus's movements for that distant era and find that a transit did occur during Ammisaduqua's reign—on May 21, −1641—but unfortunately very little of it, if any, was visible from Babylon, so regrettably Flammarion's idea has had to be discarded.

The Akkadians, whose language and cuneiform script eventually supplanted those of the Sumerians in the Fertile Crescent, used a different name for Venus—*Ishtar,* for the great goddess of the East, who invoked the power of the dawn. In Egypt and Ugarit, also among the Hittites and in Canaan, she became known as *Astarte,* Queen of Heaven. The eight-year Venus-solar period might even have inspired Akkadians' representation of Ishtar as a star with eight rays within a circle. An image of the goddess with a crescent-tipped staff was found among the ruins of Nineveh.

To the Greeks, she was *Phosphoros* and *Hesperos,* as well as *Aphrodite;* among the Romans, *Lucifer* and *Vesper.* Homer called Hesperos "the most beautiful of all the stars." Five hundred years later Sappho, the lovelorn poetess of Lesbos, penned the lines,

> All that the dawn, O Hesperos, sends away—
> Sheep, goats, and youths—regather at close of day.
> (Trans. William Sheehan)

Venus in the New World

The greatest Venus watchers of all time may have been neither the Egyptians nor the Sumerians but the Mayans of Central America, who flourished between 300 and 900 CE. They were convinced that they owed their very existence to *Kukulkan,* the brother of the sun.[3] The Mayan priest-astronomers kept careful watch over the planet from observatories like the Caracol in Chichen Itza, a two-storied cylindrical structure that in its ruined state looks remarkably like the dome of a modern observatory. Others are at Uxmal and Copan. These structures were carefully oriented toward the key points in the Venus-watching cycle; at Chichen Itza, the priest-astronomers were able to peer through a series of small apertures in the great circular tower and observe (as one can still observe) the exact positions where Venus set on the horizon. They recorded the different star figures that the planet traced in its path near the sunset or sunrise horizon and had a different name for each of them. They adopted the 260-day appearance-disappearance period as the basis of their calendar, which is still in use today in parts of Guatemala. Like the ancient Sumerians, they even believed they could correlate the planet's appearances and disappearances at the horizon with the dry season or the coming of the rain.

At about the same period, certainly by 1000 CE, those consummate oceanic navigators, the Polynesians, used various clues to help find their way across thousands of miles of trackless ocean—wave patterns, winds, seeds, and bird sightings. They also relied on the stars but had the problem that there was no usable pole star for their voyages south of the equator. In such case, they took to heart this advice, expressed in the Maori language:

Kia pai te takoto
Ote ihu o te waka
I runga ia kōpu i te pō
I te awatea kai whai i
Muri i a tama-nui-te-ra.

(During the night, on Venus keep
The vessel's fro-ward prow;
During the day, you'll go aright
If you follow the sun, I trow!)

(Trans. William Sheehan)

The reasons for the decline of the Mayan civilization, around 1000 CE, have never been clearly defined, but their military power, at least, was rivaled by yet another Venus-intoxicated people, the Aztecs of Mexico. Among the Aztecs the planet was variously known as *Tlahuizcalpantecuhti* (the lord of the dawn) and *Xolotl* (the "twin," i.e., the evening star), also as *Quetzalcoatl*, or *Quetza-xolotl*, the "feathered serpent." In this form the distant planet Venus was destined to play a critical role in the destruction of the earthly Aztec empire.

According to the story, Quetzalcoatl—who was supposed to have been tall in stature, with white skin, long dark hair, and a flowing beard—had been compelled to abandon the country. As he passed through Mexico, he stopped at Cholula, where a temple was dedicated to him, and on reaching the shores of the Gulf of Mexico, took leave of his followers, promising them that he would one day return. He then boarded a skiff made of serpents' skins and set out for the fabled land of Tlapallan.

In 1519, Hernando Cortés, the Spanish conquistador—a man with light skin, dark flowing hair, and a beard—arrived in Mexico. Among the Mexicans, it was as if Cortés and his small band of invaders "had dropped from some distant planet."[4] The superstition of Quetzalcoatl was enough to soften the resistance of the forces of the great empire of the Indian ruler Montezuma. Cortés, who combined in his person "what is most rare, singular coolness and constancy of purpose, with a spirit of enterprise that one might well call romantic,"[5] won Montezuma's confidence, even acquiescence; the result was that the Spaniard succeeded in penetrating to the very heart of the empire. William Hickling Prescott evocatively recalls the final days of Montezuma:

> With shouts of triumph the Christians tore the uncouth Huitzilopotchili [the Aztec war-god] from his niche, and tumbled him, in the presence of the horror-struck Aztecs, down the steps of the teocalli. . . . [Meanwhile], the Indian monarch had rapidly declined . . . quite as much under the anguish of a wounded spirit, as under disease. He continued in [a] moody state of insensibility; holding little communication with those around him, deaf to consolation, obstinately rejecting all medical remedies as well as nourishment. On the 30th of June 1520, he expired in the arms of some of his own nobles. . . . "Thus," exclaims a native historian, one of his enemies, a Tlascalan, "died the unfortunate Montezuma, who had swayed the sceptre with such consummate policy and wisdom; and who was held in greater reverence and awe than any other prince of his lineage, or any, indeed, that ever sat on a throne in this Western World."[6]

As Montezuma lay dying in Mexico City, Venus was the Morning Star, dropping back toward the sun but still heralding, by some two hours, its rising. As he and his empire entered upon their sudden decline, Europe, an ocean away, stood on the verge of an intellectual rebirth.

Astronomic Models in History

Ever since the Greeks—and we have temporarily skipped ahead of all that, as we have pursued our Venus-obsessed mad dash through history—Venus, the rest of the planets, and the sun were proclaimed to be in motion around the earth. Cycles such as those charted since the time of the Babylonians had been adopted by the Greek master mathematicians in working out the beats of the planets moving in circles and circles-on-circles. Theirs was a scheme that, we may assert, admittedly with necessary oversimplification, represented in geometric form the wonderful Babylonian numerological discoveries.

The diagrams worked out by the Greeks—most notably, by Claudius Ptolemy—were like orreries, those complex mechanical devices that mimic the movements of the planets, intricate and in their way exquisitely beautiful. They were elaborate calculating devices for phenomena such as the progressive elongations and conjunctions of the inner planets relative to the sun and, for the outer planets, for their conjunctions and oppositions, those points where they stand in line with, or opposite to, the sun in the sky. They explained admirably the spiraling planetary convolutions that the planets seem to trace around the centrally fixed and recumbent earth.

The sun itself—together, of course, with Mercury and Venus, oddly following in train with it—still moved around the earth in the heads of all the learned scholars in Europe—that is, all the learned heads except one. The exception was Nicolaus Copernicus, an obscure canon at the cathedral of Frauenburg. Copernicus had worked out as early as 1512—seven years before Cortés set foot in the New World—how the whole planetary system would look if the sun instead of the earth were placed at its center. In Copernicus's diagrams and thoughts, the planets began to dance with rhythms different than those known to the ancients. His scheme had implications that even he did not fully grasp, among which the possibility that Mercury and Venus could be projected, at certain points and on rare occasions, in front of the sun.

With this, our survey of the long era of astronomic prehistory is over. The incidental stage music, from the stately and majestic march of old Dumuzi-Amaushumgalana to the Siegfried-like dirge of Montezuma, comes to an end. Our preparations are complete and the curtain rises. The hushed anticipation gives way to excitement, like the fervor felt as a dazzling planet emerges out of the dawn: our proper subject, the transits of Venus, can finally take center stage. Our theater is the cosmos, our cockpit stage nothing less than the radiant circle of the solar disk itself.[7]

4

A Missed Opportunity

Not what We did shall be the test
When Act and Will are done,
But what Our Lord infers We would—
Had We diviner been.

—Emily Dickinson, untitled (poem J. 823, composed ca. 1864)

Planetary Systems

In 1543 Copernicus published *De Revolutionibus Orbium Caelestium* (On the revolutions of the celestial orbs), his magnum opus, a fully elaborated account of the planetary motions based on the assumption that the sun was at the center of the planetary system. In it, the earth was relegated to being an ordinary planet orbiting around the sun. It is said in Ecclesiastes that there is nothing new under the sun; even the Copernican system—or at least the heliocentric (sun-centered) system—did not originate with Copernicus. In ancient times, some of the Greek astronomers had speculated that either the inner planets, Mercury and Venus, orbited the sun—a partial Copernicanism—or even that the whole system of planets did.

One of the Greek astronomers was Heraclides of Pontus (c. 388–310 BCE), a pupil of Plato (c. 428–347 BCE), who pointed out how much more reasonable it would be if Mercury and Venus orbited the sun instead of the earth. He also seems to have suggested that the apparent diurnal rotation of the heavens might be owing to the axial spin of the terrestrial globe. For all this one would expect posterity to have been grateful. However, he seems to have let his successes go to his head. When his city Heraclea was beset with famine, he traveled to the oracle at Delphi in order to bribe the Pythia, the priestess of Apollo, to promise help, but only on condition the city award him a gold crown during his lifetime and proclaim him a hero after his death. According to the German historian Jacob Burckhardt: "This turned out wretchedly for all concerned. During the coronation ceremony in the theater, Heraclides suffered a stroke, and at the same moment the Pythia was bitten by a snake in the *adyton* [inner chamber] at Delphi. On the point of death, Heraclides ordered that his corpse should be smuggled away and a snake laid in his bed, as if he had gone to join the gods. This trick too came to nothing, and Heraclides was proved to be not a hero but a fool."[1]

Heraclides' scheme became known as the "Egyptian system," possibly because, given the Egyptians' well-known worship of the sun, it was thought to have originated there. Heraclides' scheme received less attention than it deserved. Eventually, it was taken up by a much greater man, Aristarchus of Samos (ca. 250 BCE), who in fact took the "Egyptian system" to its logical conclusion. Aristarchus was the first to propose workable methods to estimate the distances of the moon and the sun. His method of tackling the distance to the moon has already been considered (see chap. 1). We now consider the other part: his method of estimating the distance to the sun.

Aristarchus put forward a geometrical argument, based on determining the sun-earth-moon angle at the time the moon's phase is exactly half (see fig. 10). For this angle, which is actually 89°.86, Aristarchus used 87°; the disagreement is more significant than it might appear because the critical quantity is the *difference* between the angle and 90°. From the discrepancy we can guess that either Aristarchus did not make any direct measurements himself or was misled by the faintness of the lunar surface under oblique illumination. The latter makes the apparent phase less than that calculated; even with the crudest of instruments—a simple gnomon, or stick in the ground—he ought to have obtained a better measure of this angle than 87°, which would have led to a better value of the sun's distance.[2] Even so, he realized that the sun was at least 8 million kilometers (5 million miles) from the earth. It followed that it had to be many times larger than the moon, or even than the earth. It certainly could not be, as an earlier Greek philosopher, Anaxagoras (500–428 BCE), friend and scientific teacher of Pericles of Athens, had suggested, a hot stone only a little bigger than the Greek Peloponnesian peninsula.

It may be that Aristarchus's insight into the sun's huge dimensions led him to consider it a more likely and majestic center for the planets to revolve around than the earth. Unfortunately, since none of Aristarchus's original writings on the heliocentric system have actually survived, we do not know just what his arguments may have been. Certainly, as soon as he strove to perform the thought experi-

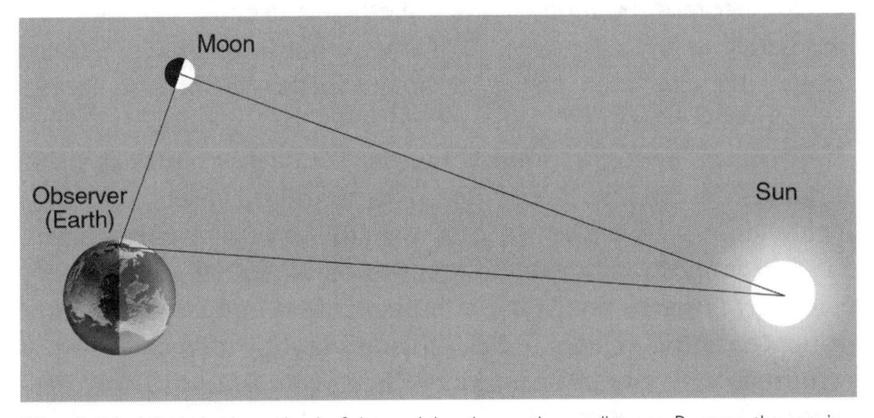

Figure 10. Aristarchus's method of determining the earth-sun distance. Because the sun is at a finite distance, the moon at half-phase is slightly less than ninety degrees from the sun. (Diagram by John Westfall.)

ment of replacing the earth with the sun, he would not only have realized the pleasing symmetry of Heraclides' arrangement for the inner planets but would also have grasped that the retrograde movements of the outer planets—their apparent backtracking or looping movement at the time of opposition to the sun—were a mere reflection of the earth's orbital motion relative to the slower-moving superior planet (see fig. 11).

There is every reason to believe that the heliocentric system proposed by Aristarchus was primitive—his book on the distances of the sun and the moon shows he did not sweat the details. Probably it was more qualitative than quantitative; but even so, it would have posed the possibility, a fleeting glimpse, of a tantalizing rationality in the motions of the planets.

But Aristarchus's system was never adopted by later astronomers, at least those whose ideas formed the reigning paradigm of the middle- and late-classical worlds. Apollonius of Perga (fl. 3rd century BCE), Hipparchus of Rhodes, and Ptolemy of Alexandria were all geocentricists, believing that the earth was the center of the universe. Since most of the ancient texts failed to survive, we can't say for certain that there wasn't an Aristarchian underground, but we do know that his ideas were rejected by mainstream astronomers on religious grounds. The planets, along with the rest of the heavenly bodies, were perfect, moving in uniform circles around the earth. Aristarchus's ideas were rejected because the motion of the earth was deemed self-evidently impossible, even "ridiculous," as Ptolemy put it. Finally, they were rejected because, even using Aristarchus's scheme, there remained small irregularities in the planetary motions that it could not explain.

The later Greek astronomers—at least the ones we know about—adopted the epicycle theory: the planets moved in small circles, "epicycles," which in turn were carried around larger circles, "deferents," around the earth. Initially, it was required that the deferents be exactly centered on the earth; in the end, when that didn't quite work, "eccentric" circles—circles placed slightly off-center—were permitted. The whole business of circles and epicycles was perfected—if that's the word—by Claudius Ptolemy, who lived in the gloaming of the once-brilliant classical world.

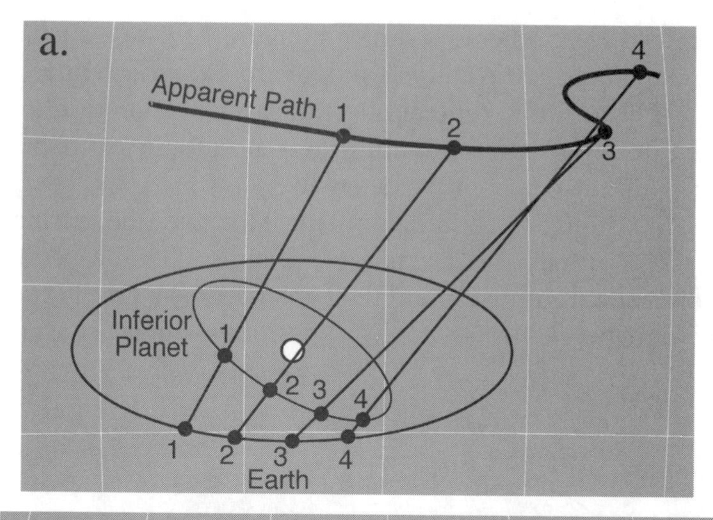

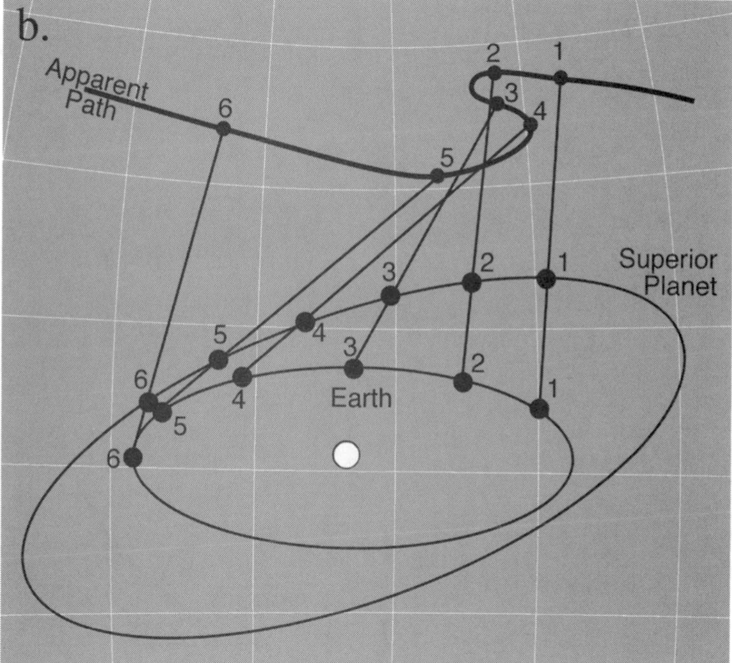

Figure 11. Apparent motions of planets, projected onto the sky, in the heliocentric system. (*a*) Path of an inferior planet (orbiting between the sun and the earth). (*b*) Path of superior (outer) planet. (Diagram by John Westfall.)

Ptolemy's basic scheme is shown in figure 12. The planet moves around the epicycle; the epicycle moves around an eccentric circle, the deferent, not quite centered on the earth. The motion around the eccentric circle is uniform, not with respect to the earth, but to a point beyond the center of the deferent.

It was all clumsy and artificial-looking. Moreover, the scheme contained some curious and inexplicable features: (1) the periods of Mars, Jupiter, and Saturn around their epicycles were exactly equal to the periods between their successive oppositions with the sun; and (2)

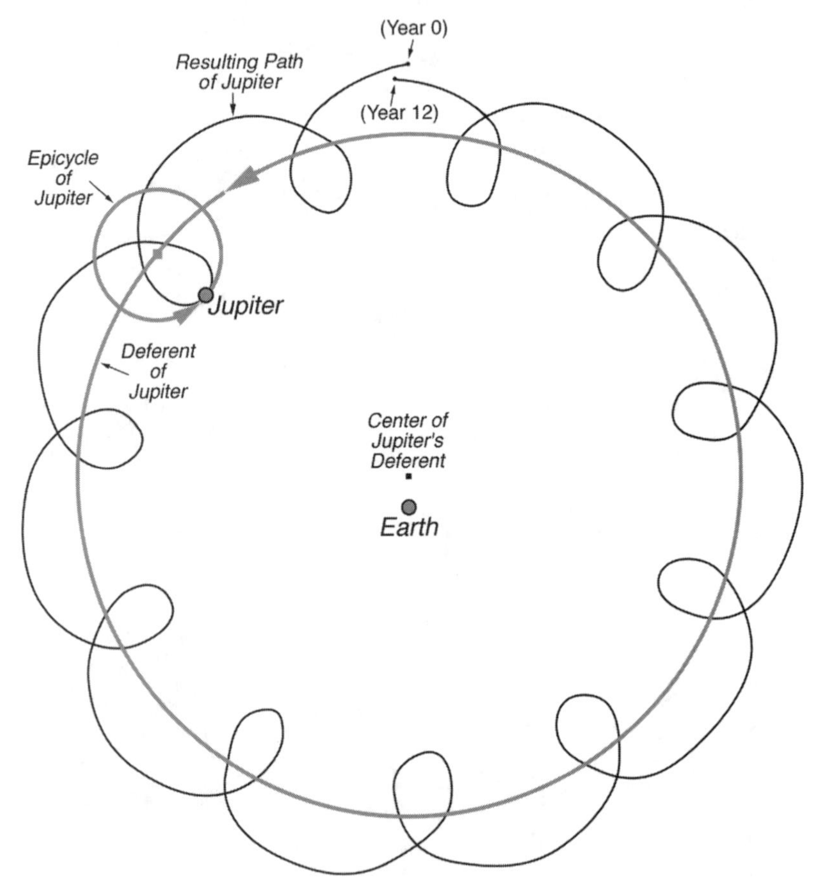

Figure 12. Jupiter as an example of the Ptolemaic (geocentric) system of epicycles and deferents. Jupiter's deferent is a circle centered near (but not at) the earth; Jupiter rotates upon its epicycle at a regular rate, while the epicycle rotates upon the deferent. (Diagram by John Westfall.)

the one-year periods of Mercury and Venus in their deferents were equal to the period of the sun in its supposed orbit around the earth (as required, of course, to keep them close to the sun in the sky).

The obvious question was: why, in a geocentric system, should the motions of the planets have had anything to do with the sun at all?

Ptolemy did not bother himself, evidently, with such questions. He was a virtuoso of geometrical diagrams—neither the first nor the last to stress the form of a theory over its content. We know that he was familiar with the general features of Aristarchus's system. In book 1 of his great treatise, the *Mathemetike syntaxis* (or *Almagest*), he says that if one concerned oneself only with the "appearances," there would be nothing to prevent things from being in accordance with this "simpler," i.e., Aristarchian, conjecture. In book 12, he shows how to convert the geocentric constructions of his system into heliocentric constructions. Ptolemy saw no reason to depart from the geocentric framework; in that sense, he was a traditionalist, not a radical experimenter. But in terms of the parameters he inherited or chose to work with, he was a master, a virtuoso who had created a delicately adjusted clockwork of the planetary notions.

With the advantage of hindsight—knowing, for instance, the simplifying beauty of Kepler's ellipses—Ptolemy's mechanisms appear to the modern mind as complicated and cumbersome. Accordingly, it has been fashionable for centuries to denigrate Ptolemy as a fastidious but rather unimaginative tinkerer. Harvard historian Owen Gingerich argues, however, that Ptolemy, "just like astronomers of today, undoubtedly built his edifice on a great array of traditional materials, rejecting, adjusting, or incorporating them as he saw fit, and molding them into a new theoretical framework of geometric planetary models."[3] Seen from that perspective, he becomes a more sympathetic figure; the *Almagest* can be appreciated for what it is—a grand entertainment of the human mind.

Ptolemy's consummate virtuosity in his own medium—as a composer of planetary baroque, including the complex counterpoint of epicycle and deferent—is proved by his remarkable durability. He established a tradition in astronomy for the next millennium.

When his grand system finally came to be challenged, it was by a man who—though he gave a new turn to the word *revolutionary*—was in at least one respect more of a traditionalist than Ptolemy had been. One of Copernicus's chief objections to Ptolemy was that he had departed from the idea that only uniform circular motion should be allowed in the heavens; the introduction of the equant made Ptolemy's system, Copernicus said, "neither sufficiently absolute nor sufficiently pleasing to the mind."

Whatever the case, Ptolemy's geocentric Ferris wheel maze reigned supreme for over a millennium. The heliocentric system was forgotten until Copernicus rediscovered it and added his own refinements.

What led Copernicus into the long-lost heliocentric fold was his sense that Ptolemy, by introducing nonuniform circular motions into his system, had made it "neither sufficiently absolute nor sufficiently pleasing to the mind."[4] He added his conviction—it was a question, finally, of aesthetics—that the sun's position in the universe ought to be fully commensurate with its nobility as a world. In *De Revolutionibus* he writes, "In the midst of [all the other planets] sits the Sun. For who could in this most beautiful temple place this lamp in another or better place than that from which it can at the same time illuminate the [whole] world? Which some not unfittingly call the Light of the World, others the Soul or Ruler. [Hermes] Trismegistus calls it the Visible God, the Electra of Sophocles the All-Seeing. So indeed the Sun, sitting on a royal throne, directs the moving family of planets."[5]

A REALIZATION: THE INFERIOR PLANETS CAN TRANSIT THE SUN

An immediate consequence of the Copernican system is that Mercury and Venus must sometimes pass between the earth and the sun. Note that in the earth-centered Ptolemaic system, the exact whereabouts of these planets on their epicycles, as they shuttle back and forth on either side of the sun, remains ambiguous—they might lie in front of the sun, they might lie behind it. Ptolemy did, in fact, assume that Mercury and Venus orbited below the sun. In a book not as well known as the *Almagest*, the *Planetary Hypotheses*, he added that the lack of observed transits across the sun did not furnish conclusive evidence against the idea, since this might just as well be a result of these planets' small diameters projected against the glare of the sun. The Copernican system, however, allowed no ambiguity as to the positions of the planets' orbits; they were precisely located in interplanetary space, so that these planets could sometimes move between the earth and the sun.

As rare as transits are, astronomers, going back to the first astronomical records of Venus we possess—those of the "Venus tablet"—have calculated that there have been no less than forty-five of them in historical times. There have been five transits of Venus definitely observed, beginning with that of 1639. What about the rest?

SEARCHING THE PRETELESCOPIC ARCHIVE

In pretelescopic times, the only chance of such an observation having been made would have been if the planet had accidentally been recorded as an ordinary sunspot, visible to the naked eye through fog or light clouds.

It was once believed that an Aztec record found near the lake at Chichen Itza—a grim place used for human sacrifices—recorded a transit of Venus. The inscription, which no longer exists, is said to have recorded a square sun rising above the horizon. Unfortunately, on the date in question, December 15, 1145, no transit actually occurred.[6]

There are extensive Chinese records of sunspots, going all the way back to the second century BCE. They have been thoroughly examined, so far without success. We have examined those in the *Catalog of Naked-Eye Sunspot Observations and Large Sunspots*, compiled by Axel D. Willmann.[7] It contains observations from 164 BCE on and furnishes only one candidate of a naked-eye observation of a transit of Venus. Undoubtedly, that one was a transit, but it hardly counts as an independent observation, being made during the transit of Venus on December 9, 1874, which was visible in its entirety in China, when many European observers were stationed there. We have also consulted a catalog of specifically far-eastern observations, but again, we come up empty.[8]

Most of the ancient Chinese sunspot observations were made quite by chance, when astronomers were attempting to determine the date of the new moon for calendrical purposes. At the moment, we know of no records of a pretelescopic transit, but in the spirit of Dickens's Mr. Micawber, with his unflagging hope that something may yet turn up, we include in appendix A, for readers interested in such matters, a list of the transits between 3000 BCE and 7500 CE, with their dates and approximate times.

In general, Europeans of the Middle Ages kept few records of astronomical phenomena. This included naked-eye sunspots, probably on the grounds that Aristotle, the influential Greek philosopher, had famously maintained that the sun and other heavenly bodies were perfect, without stain or blemish. Thus there was no reason to look for sunspots or even to keep records of them when they were seen.

One of the few records of sunspots in medieval Europe was made during the reign of the Frankish emperor Charlemagne. The chronicler, Einhard, was puzzled by so unexpected an appearance and hazarded the guess that it might be the planet Mercury seen in front of the sun: "On April 15 [807 CE] the star of Mercury appeared on the sun as a small black spot . . . which was seen by us for eight days." Though long buried in obscurity, that observation later claimed the attention of the great German astronomer Johannes Kepler. Kepler knew that Mercury would never remain in front of the sun for more than a few hours, but he assumed that the chronicler had made a slip of the pen and written *octo dies*, "eight days," for *octies*, "eight times." The explanation was never very convincing, but it was better than nothing. We now know there was no transit of Mercury in 807. There was rumor of an eleventh-century observation of a transit of Mercury by the Arab astronomer Averrhöes (Abu al-Walid Muhammad ibn Ahmad ibn Rushd, frequently known as ibn Rushd), who observed something blackish in the sun at a time when his tables indicated that Mercury should have been in conjunction.

Again, there is no reason to believe it was anything other than an ordinary sunspot.

Keplerian Calculations

Johannes Kepler (1571–1630) is, of course, one of the giants of modern science (see fig. 13). He now enters our story in a leading role, as the first astronomer to predict transits of Mercury and Venus successfully.

Kepler had been born in 1571 at Weil der Stadt, a village in Swabia, in southern Germany. Until 1599, he was a professor of mathematics at the Protestant Seminary in Graz, Austria. That was the year the Habsburg archduke, Ferdinand, adopted a policy of re-Catholicizing inner Austria by force. Kepler, a confirmed Protestant, seeking a refuge, set out to join the famed observational astronomer Tycho Brahe (1546–1601), who had settled at Prague, the capital of Bohemia.

By then Tycho himself was in decline. He had once enjoyed the patronage of the king of Denmark and had been accorded his own island, Hven, in the Baltic—Tycho always insisted the name meant the "island of Venus"—where he set up an observatory, equipped with the most accurate instruments of the day. The quality of his observations is legendary. But he proved to be a better astronomer than a landlord and eventually was ousted from Denmark. For a while he marked time in Germany but was lured to Prague by the Holy Roman emperor, Rudolf II, who created for him the

Figure 13. Portrait of Johannes Kepler (1571–1630), who predicted the transits of Mercury and Venus of 1631. (From Robert F. Ball, *Great Astronomers* [London: Ibister, 1895], p. 103.)

position of imperial mathematician and offered him a munificent salary of three thousand florins a year. (Kepler's salary at Graz was only two hundred florins at the time.) Admittedly, Tycho did not find it easy to collect his salary: the financial administration of European states at the time was primitive, and Rudolf, whose talents were not along administrative lines, was chronically short of funds. The struggle to get the money owed to him would become even more of a problem for Kepler; indeed, it would become a major theme of his life. Tycho's family and entourage were large and noisy, and at first Kepler was thoroughly miserable in Prague. He became depressed and was often physically ill. He quarreled with Tycho and admitted that he, Kepler, behaved at times like a "mad dog." At last, he could stand it no more. He decided to return to Graz. When he arrived there he found the Protestant Seminary was being closed down. With nowhere else to go, he returned to Prague.

Under the circumstances, it was probably just as well that Tycho died soon afterward, in October 1601. Kepler succeeded him as imperial mathematician and, after a struggle with Tycho's heirs, eventually took charge of the records of Tycho's observations.

Kepler's patron, Rudolf II, cut a curious figure. He was a Habsburg, which meant that insanity—as well as fanatical religious orthodoxy and the famous "Habsburg lip"—ran in the family. He was the great-grandson of the Habsburg emperor Charles V, the last emperor to attempt to realize the fading medieval dream of a spiritually and politically unified Christendom. But he, after many triumphs and some defeats, had abdicated his throne and gone to a monastery at Yuste, in western Spain, where he prepared himself for a "happy death."

Charles had fought against the tide of his time but was unsuccessful. Despite his efforts at suppression, the Protestant Reformation, inaugurated in 1517 when Martin Luther nailed his Ninety-five Theses to the church door in Wittenberg, was now in full swing. Meanwhile, the papacy, which had so long opposed Luther but also the Catholic Charles V, finally reformed itself. The forces of the Roman Catholic Counterreformation, unleashed at the Council of Trent in 1545, were spearheaded by the newly established Society of

Jesus, whose first priority was to settle the religious ferment in Germany, still the trouble spot. Jesuit settlements were established at Cologne, Vienna, Ingolstadt, and Prague, all of which were still straddling the religious fence. The new emperor, Ferdinand I, educated at the court of Spain, proved to be fiercely Catholic, and "hated compromise with Lutherans." His son, Maximilian II—Rudolf's father—would, on the other hand, be suspected of leaning toward Protestantism.

This was the situation into which Rudolf—hardly the man for the moment—was thrown when he succeeded to the throne of the Holy Roman Empire in 1576. Rudolf was frightfully shy, anxious, and unsociable, traits that only increased with age. He never married and lived alone in his castle in Prague. He preferred intellectual pursuits to the cares of statecraft. At times, he was simply unable to face the world and shut himself for days at a time in his "treasure room," where he could devote himself without interruption to his collections—paintings and sculptures, textiles, gems and coins, mechanical clocks and toys, alchemical apparatus, even wax figures and monsters. Only architecture was neglected.

It was perhaps an extension of this odd, compulsive character that he collected not only things but some of the great men of Europe, as many as he could; sometimes, as in Tycho's case, he attracted them by offering them exorbitant sums, which invariably he found impossible to pay. He was drawn to alchemy—he had a cadre of alchemists working for him in his redoubt in the castle of Prague—and also to astrology, an area in which Kepler, who was not opposed to drawing up a horoscope now and then (see fig. 14), recognized Rudolf's vulnerability. He worried that a clever and unscrupulous astrologer could easily do great harm to Rudolf by exploiting his credulity; he thus resolved to do his best to see that this did not happen. Though Rudolf himself was a devout Roman Catholic, he did not ask the scholars who came to him to profess their faith. Apparently he was not interested; late in his reign he even published a letter of majesty granting all Bohemians protection against the forceful imposition of religion.

Down the hill from the castle in which Rudolf immersed him-

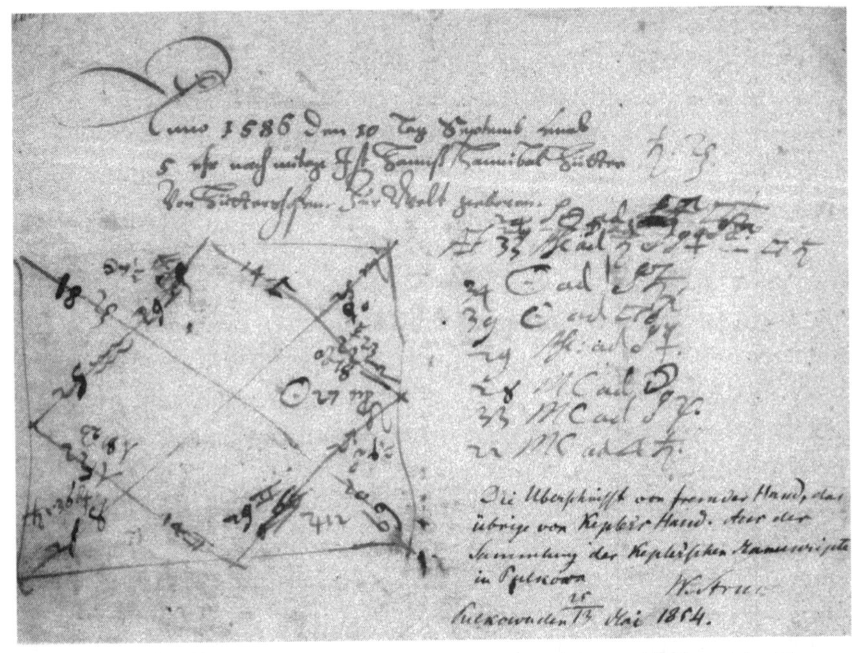

Figure 14. Horoscope drawn up by Kepler. (Courtesy of Anthony Misch and the Mary Lea Shane Archives of the Lick Observatory.)

self in his solitary passions was the house where Kepler lived—and calculated. Even today, Prague remains wonderfully alive with reminders of Tychonic and Keplerian days. One recent visitor, Dr. Richard McKim, describes his favorite tram ride, "up the hill to the ancient streets above Prague castle. If you get out at the right stop, you face a large statue of Tycho Brahe and Johannes Kepler. Just behind is the modern Jan Kepler Gymnasium, Prague's top school—the foundations of Kepler's house are to be found beneath—on the hill below the castle. . . . Tycho is buried in the Tyn Church in the Old Town Square, a stone's throw from the famous astronomical clock of the Old Town Hall."[9]

Here Kepler squeezed from Tycho's observations the elliptical form of the planetary orbits. Even Copernicus—despite his insight that the sun, rather than the earth, must be the true center of this system of planetary orbs—had been forced to exercise considerable ingenuity in introducing epicycles and deferents to satisfy the observed motions of the planets in terms of his theory, because the apparent motions of the planets are not uniform. Kepler convinced himself that it would never be possible to account for them on the basis of circular motions, however ingeniously compounded. After a prolonged series of backbreaking investigations, using Tycho's data and his own mathematical ingenuity, he gradually tightened the noose around the motion of the planet Mars. At last, in the spring of 1605, he formulated his great law: the orbits of the planets are ellipses, with the sun located in one of the foci.

He did not publish his result immediately—Rudolf's purse could ill afford it. However, Kepler at once began dreaming of putting to work his grand insight into the true form of the planetary orbits by calculating tables of their motions more accurate than any that had existed hitherto. Almost as soon as he started these backbreaking calculations, he realized that Mercury would arrive at inferior conjunction close to one of the nodes of its orbit on May 29, 1607. There was a good chance that a transit would occur, so Kepler set out to watch for it.

What would Mercury in transit against the sun look like? No one really knew, since there were no reliable measures of the diameters of the planets. All the measures available had been made with instruments designed for use by naked-eye observers, such as those perfected by Tycho, and were subject to considerable errors. As it turned out, they considerably overstated the apparent size of the planets.

Standing Watch from an Attic in Prague

Kepler's method of observation was to let the sun's light in through a crack in the shingles of his attic onto a sheet of paper, creating a pinhole camera. Because of uncertainties in the orbit of this most difficult to observe of the planets known at the time, he began his vigil the day before the predicted date of the transit. This precaution was taken by other transit watchers of that pioneering century; it would be one of the marks of progress made in astronomy that by the eighteenth century it would no longer be necessary for astronomers to cast the nets of their vigilance so broadly.

Thin, scraggly clouds drifted over Castle Hill that day. Through intermittent breaks, Kepler obtained glimpses of a spot on the sun a twentieth of the sun's diameter. He seized immediately upon this apparent fulfillment of his expectation. He had seen Mercury! Filled with excitement, he ran to the top of Castle Hill to report it to Rudolf II.

Another two years passed before Kepler published a booklet—*Phaenomenon Singulare seu Mercurius in Sole* (A Singular Phenomenon, or Mercury on the Sun)—describing his observation. At that time, he was still convinced that he had seen Mercury in transit across the sun. By now the telescope had been invented. From the first, Kepler himself had taken a keen interest in the new invention. Although his eyesight was poor and he did not make telescopic observations himself, he corresponded with Galileo about his observations of the phases of Venus and the companions of Jupiter.

This revolutionary new instrument, which provided an astounding extension of the sense of sight, provides an example where theory—the ability to predict transits to a high degree of accuracy—dovetailed with the technology to observe them. The earliest firm evidence for the existence of a telescope is the patent by a Dutchman, Hans Lippershey, in 1608, but it is possible others had used telescopes years, or even decades, earlier. Also, the camera obscura, a camera-like device employing a single lens to project an image on a screen, seems already to have been in use. In any case, by 1609 astronomers such as Englishman Thomas Harriot (c. 1560–1621) and Italian Galileo Galilei (1564–1642) had turned the new device to the sky and found wonders (see fig. 15).

Although the first telescopes revealed discoveries such as the craters of the moon and the major satellites of Jupiter, they must have been frustrating to use. With a single-element lens, typically an inch or two in diameter with a focal length of at most a few feet, they used a negative ("Galilean") eyepiece to achieve a magnification about that of a modern pair of handheld binoculars. They had three serious drawbacks: First, the single-element lens unavoidably suf-

Figure 15. Statue of Galileo Galilei (1564–1642), located outside the Uffizi Gallery in Florence, Italy. (Photograph by William Sheehan.)

fered from chromatic aberration. No two colors came to the same focus, so everything observed—bright objects especially—were irritatingly fringed with color. Second, there was also the problem of spherical aberration; the rays of light that passed through the periphery of the lens did not come to the same focus as rays passing through the center. Finally, the Galilean eyepiece, though giving a convenient upright image, furnished a tiny field of view. The larger of Galileo's two extant telescopes had a field of view only a quarter of a lunar diameter across.

Kepler proposed to solve the tunnel-vision problem by using a positive eyepiece lens instead of a negative one. Admittedly, the Keplerian eyepiece inverted the image, but observers accepted this minor inconvenience with gratitude in exchange for the enlarged field of view. Later, the new design was found to permit the use of crosswires and even, eventually, an eyepiece micrometer, which used movable wires to measure angles. Kepler proposed his form of eyepiece as early as 1611, but he did not actually make one; however, Christoph Scheiner (1575–1650) and Francesco Fontana (c. 1585–1656) were using it later that decade to observe sunspots. It is likely the transit observers in the 1630s used the Keplerian form of eyepiece.

It would be many years before the chromatic- and spherical-aberration problems were solved. Moreover, the higher the magnifications used, the worse the aberrations. On the other hand, going to longer focal ratios (the ratio of focal length divided by lens aperture) reduced these effects. Telescopes soon grew to absurdly awkward lengths; by the second half of the seventeenth century, 12-foot telescopes were commonplace, while several instruments of over 100-foot focus were actually used—all this long before equatorial mountings and clock drives!

AN UNEXPECTED DISCLOSURE: SUNSPOTS

As early as 1611, several astronomers, including Galileo, Scheiner, Harriot, and Johann Fabricius (christened David Goldschmidt, 1564–1617) had recorded sunspots. Scheiner, an Austrian Jesuit at Ingolstadt, entered into a bitter dispute with Galileo over the question of priority in this discovery—according to Stillman Drake, "this dispute appears to have had a great deal to do with Galileo's ultimate trouble with the Church."[10] Ironically, neither man was the first to see sunspots—as noted, they had already been seen by naked-eye observers long before—but Fabricius was the first to publish on the subject.

Scheiner was still influenced by the Aristotelian view of the sun as an unblemished globe and declared that the spots he saw in the sun must be "small stars"—planets, in other words—in transit between the earth and the sun. Another who held this view was Jean Tarde (1561/62–1636), canon of Sarlat, who confirmed Scheiner's observations and even gave the putative planets a name, the "Bourbonian Stars," in honor of the French house of Bourbon.

Galileo thought otherwise. In his *Letters on Sunspots* of 1613 he argued that, since the sunspots seemed to move along with the solid body of the sun, they were not planets but vast dark clouds in the atmosphere of the sun itself, since "of all the things found with us, only clouds are vast and immense, are produced and dissolved in brief times, endure for long or short periods, expand and contract, easily change shape, and are more dense and opaque in some places and less so in others," thus behaving much like sunspots. What allowed him to settle the question once and for all was the invention, by his pupil Benedetto Castelli, of a safe and effective method of observing the sun. At first, Galileo had been able to observe the sun with his telescope only occasionally, when it was near the horizon. As a result, his observations were scattered in time, and he had been unable to determine the nature of sunspots or how they moved. However, Castelli's method involved projecting the sun onto a white screen instead of looking directly at it. This allowed Galileo to observe the sun daily, in comfort and without injury to his eyes, for a full month, except for a few days when it was completely cloudy.

Kepler was among the first to concede Galileo's point. He graciously acknowledged that in 1607 he had seen not Mercury at all but an ordinary sunspot. "O lucky me," he sighed, "the first in the century to see a sunspot!" (As a matter of fact, Mercury's inferior conjunction occurred on June 1, 1607, instead of May 29 as Kepler had calculated; moreover, it did not actually transit, passing 2°.7 south of the sun.) He also reasoned that the observation from Charlemagne's time was also a sunspot, which was indeed present on the sun for eight days: "Had this discovery been made several years ago," Galileo noted, "it would have saved Kepler the trouble of . . . altering the text and emending the reported times."[11]

Kepler's Preview of the Transit of Mercury

All the while, Kepler was continuing to refine his work on the orbits of the planets. But it was never easy, not least because of the constant harassments in his personal life. As ever, Rudolf lacked the funds to wage his earthly wars—much less Kepler's planetary campaign! As he grew older, Rudolf became more and more unbalanced. Gloomy and unhappy, subject to fits of severe depression, he was increasingly reclusive and more difficult to reach. Once, an admiring Kepler had remarked how Rudolf "sits in Prague, understands nothing about the military profession, but yet, without authority . . . accomplishes wonders."[12] But the violent times were getting the better of this unwarlike, perplexed, even childlike man. His more forceful, militaristic, and fervent brother, Matthias, was depriving him piecemeal of his territory, taking over Hungary, Austria, and Moravia. At last Rudolf promised him the succession of Bohemia as well. Inevitably, as he retreated into his own world and events closed in upon him, he sought refuge in his collections and craved assurance from the distant and unfeeling stars. He was virtually a prisoner in his citadel after his cousin Leopold entered the city and occupied part of Prague, which by 1611 was ravaged by civil wars and epidemics. Kepler was directly affected. His first wife, who suffered from epilepsy, died, and his three children became afflicted with smallpox; his favorite child succumbed to the disease.

Rudolf himself died in 1612. By then, Prague was becoming a battleground. The misery that was soon to sweep over Bohemia and Germany provided the backdrop of Kepler's final years. He left Prague and settled in Linz, Austria. But he could not escape the ensuing chaos and disruption, the end of which he would never see.

What would become known in Germany as the "Great War" and in European history as the Thirty Years' War began in Prague. Since neither Matthias nor his surviving brothers had legitimate heirs, the Habsburg succession was due to pass to their cousin, Ferdinand of Styria, who had been brought up by the Jesuits and whose policies in Austria had driven Kepler out of Graz. He was hardly a figure of compromise. The Spanish Habsburg Philip III relinquished his claims and offered Ferdinand his support in exchange for cession of the Austrian territories in Alsace and Ortenau, which would join the Spanish Netherlands with the Spanish Franche-Comté and the routes to Italy. The Habsburgs of Spain and Austria still aspired—as Charles V had—to encircle France. France, led by the crafty Cardinal Richelieu, did its best to wreck the plans of the Habsburgs, as Sweden and Denmark waited in the wings. The stage was being set for a long, drawn-out struggle of the outside powers for predominance in central Europe. When, in 1618, the Protestant Bohemian estates rose up and refused to accept Ferdinand II's succession, none foresaw that the Four Horsemen of the Apocalypse were about to ride

Though religion unleashed them, the dogs of war, once let loose, were not so easily confined again. The finances of the states at the time were still primitive; the armies, poorly funded, turned into mercenary bands. Rogues and marauders roamed the countryside, pillaging and burning as they went. Few soldiers died in battle; most of the depredations were against civilians. The armies plundered not only to feed themselves but also to create wastelands where the enemy could not be supported and in which hunger, homicide, and pestilence flourished.

By 1626, Linz itself was under siege, and Kepler was again forced to flee, eventually finding refuge at the court of Ferdinand II's opportunistic and increasingly dilatory general, Albrecht von Wallenstein, at his newly formed Duchy of Sagan, in Silesia. In Ulm, Kepler at last published the long-awaited tables of the motions of the planets: the Rudolphine Tables, named in honor of the long-dead and (but not for Kepler) forgotten patron. In this work, Kepler predicted that a transit of Mercury would occur at midday in Europe on November 7, 1631, followed, within a month, by a much rarer transit of Venus on December 6, 1631. In letters, he urged astronomers to observe these rare celestial events. Mercury's orbit remained uncertain, in part because of difficulty in observing the planet—it is visible only when low in the sky or when swamped by the glare of the sun. He suggested, therefore, that astronomers keep watch from November 6 to 8. The transit of Venus would not, according to his calculations, be visible at all from Europe. Nevertheless, he urged European astronomers to remain on the qui vive anyway, just in case.

Meanwhile, Kepler's worldly struggles continued. His wages still badly in arrears, he set out in frustration, in October 1630, from Sagan to Regensburg, in hopes of conferring with the Diet of the Emperor about yet another possible residence for himself and his family (he had by this time remarried). Weakened by the strain of overwork and chronic ill health, the trip proved too much for him. After a short illness, he died, on November 15, 1630. He would not live to see either of the transits he had predicted. Kepler wrote his own epitaph:

> I, who once measured the heavens;
> Now I measure the earth's shadows.
> Mind came from the heavens.
> Body's shadow has fallen.[13]

Measuring the shadows of the inner planets' transits would become the preoccupation of future generations of astronomers, beginning with a number of his contemporaries, who made plans to witness the transits he had forecast.

Kepler guessed that Venus in transit should have an apparent diameter of 7 arc-minutes—almost a quarter of the sun's apparent diameter. He did not make a specific prediction regarding Mercury, though similar reasoning would have produced a figure around 2½ arc-minutes. Inevitably, many observers were betrayed into expecting—as Kepler himself had done in 1607—that Mercury in transit would appear as a very large dark spot easily observable by means of simple devices, such as pinhole projections or cameras obscura.

Using modern orbital data, we now know that Mercury and the sun were above the horizon for all of Europe during most of the transit. Unfortunately, though, on the day of the transit, heavy clouds hung over much of Europe. Even where the skies were clear, observers who equipped themselves with pinholes or cameras obscura saw—nothing. The planet's image projected upon the sun proved to be much smaller than anyone had surmised; as a result, only three astronomers, who had taken the precaution of arming themselves with telescopes, are known to have succeeded in catching a glimpse of Mercury as a black spot against the sun: Remus Quietanus (Johannes Remus) at Rouffach, Alsace; the Jesuit Father Cysat (Cysatus, 1588–1657) at Innsbruck, Austria; and Pierre Gassendi at Paris.

THE GREAT GASSENDI'S SPOT-ON OBSERVATION

The most thorough observations were made by the great Pierre Gassendi (1592–1655), canon of the parish church at Digne, in his native Provence, but who then lived in Paris (see fig. 16). For many years he had been an enthusiastic observer of sunspots, using Castelli's projection method. He adopted the same method for the transit of Mercury. On the day of the projected transit, he admitted the sun's light into his apartment through a small aperture in the window, behind which he placed his telescope. The sun's image,

Figure 16. Portrait of Pierre Gassendi (1592–1655), the French astronomer who observed the November 1631 transit of Mercury but, despite maintaining a careful vigil, failed to see that of Venus in December (which actually took place at night in Paris). (Courtesy of Yerkes Observatory.)

some nine or ten inches across, was received onto a white screen, on the surface of which he marked off a circle the same diameter as the sun's image (see fig. 17). The circle was divided into a grid to allow the planet's progress across the sun to be charted.

Gassendi's friend and mentor from Digne, Nicolas-Claude Fabri de Peiresc (Peirescius, 1580–1637), warmly applauded his success in capturing Mercury against the sun and regarded this observation as the most important in many centuries. Flushed with his achievement, Gassendi now planned for an even greater triumph. He resolved to stand on the lookout for the even rarer and more spectac-

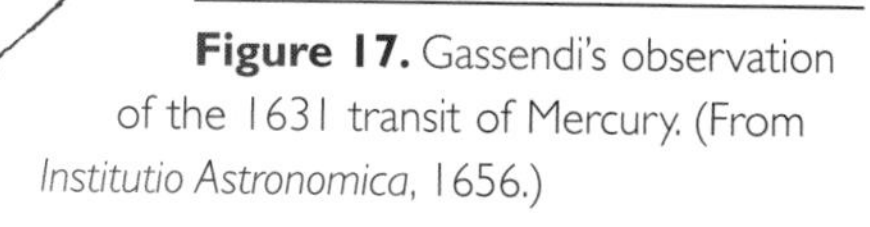

Figure 17. Gassendi's observation of the 1631 transit of Mercury. (From *Institutio Astronomica*, 1656.)

ular transit of Venus, which Kepler had predicted for December 6, 1631.

GASSENDI'S VAIN VIGIL FOR VENUS

Kepler had predicted that the transit of Venus would occur when it was night in Europe. Still, uncertainties existed, and as with the transit of Mercury, Gassendi put wide margins around his watch. He began to observe on December 4. Unfortunately, storms raged that day and the next. Only on December 6—the day for which Kepler had predicted the transit—did Gassendi manage to obtain furtive glimpses of the sun through the clouds. He kept up his vigil until 3 o'clock that afternoon, but in vain: no spot of darkness appeared within the white circle of his screen. On December 7 he continued his watch through the whole forenoon. Again, nothing. Gassendi, who a month earlier had captured the always elusive Mercury against the sun, failed to inveigle Venus into the same net. Fate had overruled.

We now know that the transit of Venus indeed took place, just as Kepler had foretold, when it was night in Europe, December 6–7. It would have been visible from the still-uncharted lands of New Zealand and Australia, from Asia and Africa, and even—the very end of it—from eastern Europe. The fourth contact of Venus with the sun would have been visible just at sunrise along a swath that included Danzig, Olmütz, Ingolstadt, Innsbruck, Bologna, Florence, Rome, and Naples. But from Paris and points farther west, the transit was already over before sunrise.

If only Kepler had lived another year, he might have seen it himself.

As far as we know there were no observers of the transit. Apart from Gassendi, no one even attempted to view it. We can blame this failure partly on the fact that mails at the time were disrupted; Kepler's entreaty to his fellow astronomers probably did not reach many of them. Even the personal safety of observers could not be vouchsafed in much of war-torn central Europe. In Italy, conditions were more settled, but anyone who took inspiration from the Copernican idea that the sun was the center of the solar system was silenced by the 1616 decision of the Roman Inquisition to put *De Revolutionibus* on the index of forbidden books and, later, by Galileo's arrest.

The forces of the Reformation and the Counterreformation attacked and counterattacked across Germany. Until two months before the transit, Catholic forces were in control of most of the region. The Catholic general Tilly, who routed the Bohemian Protestants near Prague at the Battle of the White Mountain in November 1620, laid siege to Magdeburg on the Elbe, hoping to use it as a base from which to thwart the advancing expeditionary force of Gustavus II Adolphus of Sweden. Magdeburg fell; as it did, it went up in flames, depriving Tilly of his base. Tilly became known as the "butcher of Magdeburg." Gustavus Adolphus pushed on and in September 1631, at the Battle of Breitenfeld, destroyed Tilly. In all probability—so the historians say—this victory enabled Protestantism to survive in Germany.

Still, the contest was far from over. The titanic forces of destruction would continue to grapple with one another until they were spent. The moment of peace was unforseeable by the dim prophetic powers of the human imagination. In any event, the Thirty Years' War would drag on for another seventeen years. In that era of turbulence, when terrestrial matters seemed so unsteady and unpredictable, Kepler's forecast that Venus would pass in front of the sun on such-and-such a date must have been hard, even for astronomers, to anticipate. How could there be certainty about the events in the heavens when there was so much uncertainty about the events on the earth? Even if there were such certainty, other things must have seemed more urgent to the preoccupied minds of men than watching for Venus's passage in front of the sun.

The motions of the planets were uncertain. Mercury, because of its proximity to the garish sun, had always been elusive for astronomers. Now Venus was being downright coy. After the December 1631 transit of Venus, which Kepler had predicted, there would not be another until 1761, a date that must have seemed then as remote as that of the Last Judgment.

But this time Kepler had made a mistake.[14]

5

Homage to Horrocks

I weep for Adonais—he is dead!
O weep for Adonais! though our tears
Thaw not the frost which binds so dear a head!
And thou, sad Hour, selected from all years
To mourn our loss, rouse thy obscure compeers,
And teach them thine own sorrow, say: with me
Died Adonais; till the Future dares
Forget the Past, his fate and fame shall be
An echo and a light unto eternity!

—Percy Bysshe Shelley, *Adonais: An Elegy on the Death of John Keats* (1816)

THE WINDS OF WAR

The Thirty Years' War continued, like a soaking rain, taking its heavy toll. Tilly and Gustavus Adolphus were killed in battle; Wallenstein—for whom Kepler had cast a horoscope predicting a long life—was murdered by his general staff at the age of fifty. Many of the participants in the early years of the war were swept violently from the scene. At last the armies seemed to have grappled to a vast and confused standstill. With the signing of the "peace" of Prague in 1635, the Germans attempted to end the convulsions within their own borders; but the most brutal, destructive, and homicidal phase of the Thirty Years' War followed, as the peripheral European powers, chiefly Sweden and France, struggled to gain control.

Only at the end of this frightful struggle did it become clear that the scales were finally tipping toward France's king Louis XIII, vindicating the policies of its hard-working, conscientious, and order-obsessed chief minister, Cardinal Richelieu. Spain was the overall loser. At that moment, England and France also moved closer together, as Charles, the Prince of Wales, married the French princess, Henrietta Marie, daughter of Louis XIII. A year later, after the death of his father James I, the prince ascended the throne of England as Charles I.

In the ardor of his youth, in compensation for a stammer and paralysis he had suffered as an infant, Charles determined to prove himself a man of action. He declared war on Spain (the attack on Cadiz collapsed) and attempted to assist the Protestant Huguenots of La Rochelle (his troops failed to take the citadel). These unsuccessful actions "lowered the prestige of monarchy in England."[1] By the time the Fourth Parliament sat, in 1629, it was clear that Charles and Parliament could not work together. Charles dissolved Parliament, and the era of "personal rule" began.

Meanwhile, in the Americas, the Dutch West India Company was in control of the Atlantic Coast from Chesapeake Bay to Newfoundland. English Pilgrims, landing from the *Mayflower* at New Plymouth in 1620, gained a foothold in Massachusetts. By 1635, the first secondary school in North America, English High and Latin School in Boston, was founded; at New Towne (now Cambridge, Massachusetts), Harvard College began, endowed by John Harvard, which became its name from 1639 onward.

"Nothing More Noble"

During these generally placid thirties, this "age of ease and tranquillity,"[2] the first of England's great astronomers, Jeremiah Horrocks (sometimes Horrox, 1619–41), launched his precocious and brilliant, if tragically shortened, career in astronomy. Horrocks will always be remembered as the first man to observe Venus in transit across the sun.

Horrocks was born in 1619 near Toxteth (see fig. 18), then a small village three miles from Liverpool—now swallowed up in its suburbs. Both his father's family, the Horrockses—his father was either a poor farmer named William or a watchmaker named James—and his mother's family, the Aspinwalls, seem to have been

Figure 18. Jeremiah Horrocks's birthplace, at Otterspool, near Toxteth, near Liverpool. (From *The Observatory*, 1883.)

established for sometime in Toxteth, where "watchmaking and similar mathematically based handicrafts were becoming important local trades."[3]

There were Puritans in Toxteth. Richard Mather, the well-known Puritan divine, himself had once been a tutor at the school there. Jeremiah was presumably quite often in the home of the Aspinwalls, who are said to have been strict Puritans. At school he received the usual smattering of instruction in Latin and Greek. Already by the time he was thirteen, he had made enough of an impression to be sent to Emmanuel College, Cambridge. We do not know how he got there or who recommended him; however, we do know his position there. He was a sizar, a position squarely at the bottom rung of the Cambridge social structure; this was only to be expected, given his meager family means. As sizar, he "would have been expected to run errands for his tutor, fetch his provisions from the buttery and probably wait on the fellows at the high table and dine on what was left over."[4]

Though he began inauspiciously, Horrocks made the most of his opportunities. He mastered the few subjects then in the standard curriculum, including Latin versification, at which he became an adept. He also acquainted himself with several other young men interested in mathematics—most notably, John Wallis (1616–1703), later Savilian professor of geometry at Oxford. Wallis's mathematics of infinitesimals (arbitrarily small quantities) would later serve as one of the textbooks from which Isaac Newton learned his mathematics. Wallis entered Emmanuel the same year as Horrocks, 1632, and John Worthington (1618–71), subsequently Master of Jesus College, a native of Manchester, who may have been responsible for putting Horrocks in touch with William Crabtree (1610?–1645). Wallis and Worthington would remember Horrocks after he had achieved his fame, but it does not seem they were companions as undergraduates.

Preoccupied and serious about the subjects that fascinated him, Horrocks otherwise attracted little attention. He may have presented a solemn, puritanical exterior. If so, it masked the enthusiast.

Astronomy soon became his ruling passion. "It seemed to me," he later wrote, "that nothing could be more noble than to contemplate the manifold wisdom of my Creator, as displayed amidst such glorious works; nothing more delightful than to view them no longer with the gaze of vulgar admiration, but with a desire to know their causes, and to feed upon their beauty by a more careful examination of their mechanism."[5] He was virtually alone at Cambridge in such interests. Perhaps he was even lonely; after all, Cambridge did not even have formal instruction in mathematics or physical science. At the university he was probably no better off with respect to instructors in mathematics and astronomy than if he had remained in Lancashire. By his own admission, his ambition to obtain astronomical information proved at first difficult:

> The abstruse nature of the study, my inexperience, and want of means dispirited me. I was much pained not to have any one to whom I could look for guidance, or indeed for the sympathy of companionship in my endeavours, and I was assailed by the languour and weariness which are inseparable from every great undertaking. . . . And yet to complain of philosophy on account of its difficulties would be foolish and unworthy. I determined therefore that the tediousness of study should be overcome by industry; my poverty (failing a better method) by patience; and that instead of a master I would use astronomical books. Armed with these weapons I would contend successfully.[6]

At first Horrocks suffered the usual fate of the untutored, straggling into blind alleys. The first book he turned to was a tract by Henry Gellibrand (1597–1636), professor of astronomy at Gresham College, London, praising Philippe van Lansberg (1561–1632), a Flemish astronomer, who had pursued a rather literalist Copernican program of epicycles and eccentrics to achieve—or so he pompously claimed—the "perfection of astronomy." Horrocks devoted much time and effort to working out the motions of the moon and planets from Lansberg's theory, time he later regarded as wasted.

In 1635, he left Cambridge without taking a degree and returned to his parents' home in Toxteth. The circumstances of his departure are shrouded in mystery; perhaps he had a personal crisis of some kind, although there is no evidence of this. In any case his departure "was not unusual in the seventeenth century, where many young men entered the university for a few years to broaden their knowledge, rather than to gain an explicit qualification."[7] Probably he left for no other reason than that, as a poor sizar, he could not afford to pay the college the sum that would have been required for him to receive a degree.

HORROCKS AND WILLIAM CRABTREE

Once reestablished in Toxteth, Horrocks began making his own astronomical observations. By June 1636, he had made contact with William Crabtree, a linen draper and amateur astronomer from Manchester. Despite the depiction of Crabtree as an elderly man in Ford Madox Brown's famous mural in the town hall in Manchester, he was still a very young man himself, only twenty-six at the time, and proved a much more encouraging and useful ally than anyone Horrocks had encountered at Cambridge. Equipped with a telescope, he had become a systematic observer of the sunspots first recorded two decades earlier by Scheiner, Galileo, and Fabricius. He was knowledgeable about mathematics—Horrocks, indeed, described him as having "few superiors"—and was reasonably independent-minded and distrustful of authority. Before long, both Crabtree and Horrocks had proved, by their own observations, the unreliability of Lansberg's tables. Searching for clearer and more reliable inspirations, they closed their Lansberg and opened their Kepler.

Kepler's reputation stood high on the Continent, but so far he was scarcely known in England. His works, together with those of Tycho Brahe, were a revelation to Horrocks. As he afterward recalled, "It was a pleasure to me to meditate upon the fame of these great masters of science, and to emulate them in my aspirations."[8]

The clouds having now dissipated, the light broke in. Horrocks perused with delight Kepler's beautiful demonstration of the elliptical form of the planetary orbits. He also came to grasp the importance of the so-called harmonic law, which had been formulated by Kepler at Linz in May 1618, scarcely a week before the horrors of the

Thirty Years' War had begun to unfold in Prague. It was a precious discovery—the culmination of his life's work: "The squares of the periods of revolution [of the planets around the sun] are proportional to the cubes of their distances."

With this law, Kepler revealed in a single formula the harmony of the planetary system, the grail that he had sought since the beginning of his career. The discovery seemed to the discoverer radiant and numinous. "I am now free," Kepler wrote in the *Harmonice Mundi* (1619), "to give myself up to the sacred madness. . . . If you forgive me, I rejoice; if you are angry, I can endure it. The die is cast; I am writing the book—whether to be read by present-day or by future readers, what does it matter? It may wait a hundred years for its reader, since God himself has been waiting for six thousand years for one to penetrate to the heart of his work."

The harmonic law would assume majestic importance in the chase after the transits of Venus. It expressed the order of the planetary movements, the rationality of their system. It also carried an immensely practical implication: it allowed Kepler, as well as later astronomers, to map the solar system. It gave the relative positions and distances of the orbits of the planets compared to that of the earth. It furnished a complete scale map—provided one could just somehow work out the scale. If only one grasped the distance between any two bodies in the solar system—the earth to Venus or the earth to the sun—the measures of all the others were automatically determined. Find a single cranny or foothold, in other words, and one stood in view of the lofty summit. The harmonic law opened up astounding vistas.

Enmeshed in a Net of Calculations

Inspired, Horrocks began to unfold an ambitious program of research. Kepler's tables of planetary motion—the Rudolphine Tables—henceforth occupied the foreground of his thought. He fully appreciated their value. But he also appreciated that even the accuracy of Kepler's tables could be further improved by modifying some of the numbers, by refining the values the German astronomer had used. This now became his preoccupation. He began—it was a fateful decision—by reconsidering Kepler's theory of the motion of Venus. He pursued this work in fits and starts. Inevitably investigations imposed themselves and exerted their fascination: above all, the complicated motion of the moon.

In June 1638, Horrocks visited Crabtree in Broughton and returned to Toxteth to resume work on Venus. It was at almost the very moment when John Milton (1608–74), the great poet and future foreign secretary under Cromwell, arrived in Italy to begin his "Grand Tour." While there he met Galileo, then under house arrest at Arcetri. Milton would later, as author of the epic poem *Paradise Lost,* capture perfectly the cumbersome artificiality of pre-Keplerian astronomy (both Ptolemaic and Copernican), with their "Centric and Eccentric scribbl'd o'er, / Cycle and Epicycle, orb in orb."[9] If Milton had ever read Kepler, he would have known that his work had moved astronomy decisively and irretrievably beyond the epicycle. Keplerian—and now Horrocksian—astronomy swept aside the awkward epicycle-machinery of Ptolemy, Copernicus, and other tinkerers in those unwieldly whale-bone and hoop-skirt intellectual fashions once and for all.

In June 1639, Horrocks moved from Toxteth to Hoole, a small village eight miles southwest of Preston, in Lancashire: a rather romantic, if desolate, site. It is bordered by a morass on the east and by Marton Mere and the Douglas River on the south, two miles north of the point where the Douglas meets the Ribble. "Though doubtless an open situation for an astronomer," writes Horrocks's worshipful nineteenth-century biographer, the Reverend Arundell Blount Whatton, "it could not have been a very agreeable residence." Horrocks was attached to the Anglican Church of St. Michael's and by tradition lived at Carr House, about a half-mile south of the church and a mile north of the river Douglas, "a quaint and substantial structure of brick, with projecting porch and gable facing south."[10] The Reverend R. Brickel, Rector of Hoole in the nineteenth century and a worshipful Horrocksian, supposed that the center room on the first floor, which contained a window over the porch, was where Horrocks observed the transit in 1639; there is no reason to suppose otherwise. Both St. Michael's and Carr House are still standing today—the latter has been converted into a doll museum! What Horrocks's exact role at St. Michael's might have been has never been clear. He was too young to have been ordained, though he may have taken minor orders, in which case he might have served as a priest's assistant or a lay reader. Another possibility is that he was employed as a schoolmaster. Whatever he was, he was most assuredly not, as often stated, a "Reverend" or "Curate of Hoole." And it was apparently a meager living: writing nearly a century later, Thomas Hearne described the Hoole Church as a "very poor pittance" even then.[11]

Uninterested in the avocational pursuits that absorbed his neighbors—hunting and hawking—and his responsibilities at the church being not, apparently, so very great, he made a fresh assault on the motion of Venus in early October 1639. The timing was providential: before the month was out, he had discovered that at the planet's next crossing of the node of its orbit—one of the two intersection points of its orbital plane with that of the earth—on November 24 (O.S. [Julian]; December 4, N.S. [Gregorian]), the planet would probably transit the sun. Though Kepler had in part disentangled the complicated cycle that produces transits at intervals of approximately 120 years, he had failed to recognize that transits actually usually occur in pairs eight years apart. Thus Kepler had correctly predicted the transit of 1631 but had missed this one.

On repeating the calculations, Horrocks realized that a transit would take place at about 3 PM local time on the date in question. The moment was fast approaching, so Horrocks had time only to alert his younger brother Jonas Horrocks, then living in Liverpool, and, more importantly, Crabtree, to whom he wrote on October 26, 1639: "My reason for now writing is to advise you of a remarkable conjunction of the sun and Venus on the 24th of November, when there will be a transit. As such a thing has not happened for many years past, and will not occur again in this century, I earnestly entreat you to watch attentively with your telescope, in order to observe it as well as you can."[12] He was especially eager that his friend attempt to make the observation, because he was worried that—with Venus, Mercury, and Jupiter all lying in close proximity to the sun—it seemed likely that "in many places the sky will be cloudy." Though skeptical of astrology in general, he was convinced, "from experience," that the positions of the planets directly affected the weather! He advised Crabtree to forward the information in turn to Samuel Foster, of Gresham College, London, of the transit. There was, however, insufficient time.

VENUS OBSERVED

Horrocks prepared himself for the momentous event, never to be repeated in his lifetime. From the beginning of his astronomical career, he had been troubled with the want of satisfactory instruments, but in 1638 he had obtained a half-crown telescope. Apparently this was the instrument he used to observe the transit.

He used the well-proven projection method. Horrocks set up his small telescope to project the sun onto a sheet of paper so as to form an image six inches across. Though he would have preferred a larger image size, this was the maximum that the smallness of his rooms would allow. On the circle corresponding to the sun's image, he marked the degrees along the circumference and divided its diameter into thirty equal parts. Despite the fact that his calculations indicated that the ingress of Venus onto the sun, the beginning of the transit, would not take place until the afternoon of November 24, he began his vigil on the twenty-third, "lest through overconfidence I might endanger the observation."[13] It remained cloudy most of the day. On the twenty-fourth, a Sunday, there were patches of blue, and he observed the sun from sunrise until nine o'clock, then from a little before ten until noon, and again at one in the afternoon. During the interruptions, he was "called away by business of the highest importance which, for these ornamental pursuits, I could not with propriety ignore." (What business could possibly take priority over an event that would not occur again for 122 years? It has generally been assumed, without direct evidence, that from nine until ten he was called away for matins, from one until three in the afternoon prayers or catechizing, with sermon. Admittedly, his was a religion-obsessed age, and attendance at divine services was regarded as mandatory. But if it was matins that called him away, why didn't Horrocks simply say so?) Finally, at quarter past three in the afternoon, with the sun far gone in the west, Horrocks returned to the telescope: "The clouds, as if by divine interposition, were entirely dispersed, and I was once more invited to the grateful task of repeating my observations. I then beheld a most agreeable spectacle, the object of my sanguine wishes, a spot of unusual magnitude and of a perfectly circular shape, which had already fully entered upon the Sun's disk on the left, so that the limbs of the Sun and Venus precisely coincided, forming an angle of contact. Not doubting that this was really the shadow of the planet, I immediately applied myself to sedulously observe it."[14]

It was a tantalizingly brief spectacle. The strange sight of the planet silhouetted against the sun was visible for only half an hour before the sun set. Horrocks made the most of the occasion. He had prepared himself well and made the best of the occasion (see fig. 19). The most important single result was the

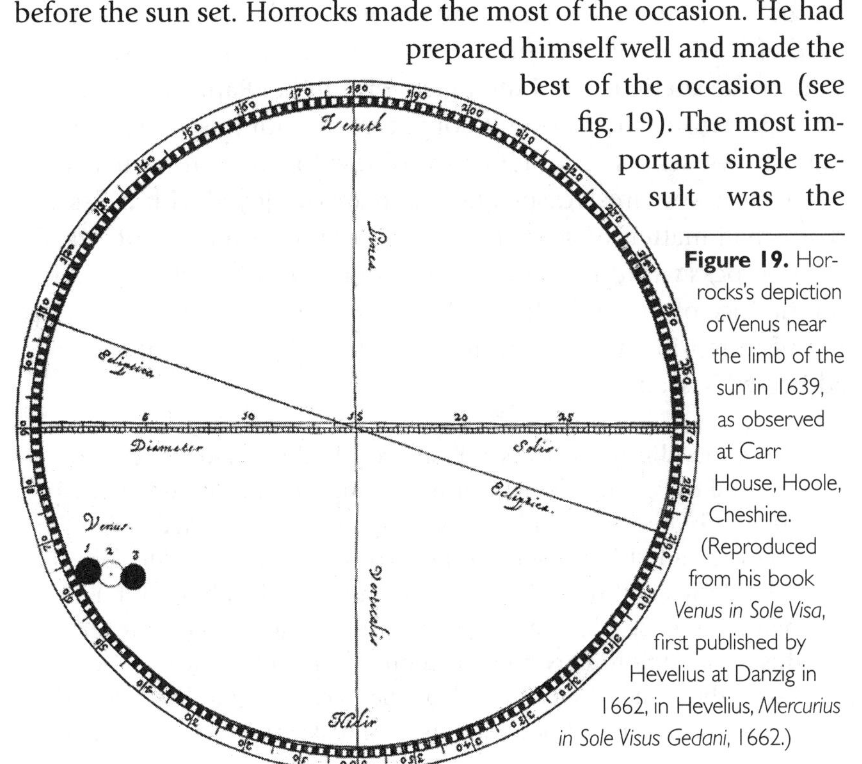

Figure 19. Horrocks's depiction of Venus near the limb of the sun in 1639, as observed at Carr House, Hoole, Cheshire. (Reproduced from his book *Venus in Sole Visa*, first published by Hevelius at Danzig in 1662, in Hevelius, *Mercurius in Sole Visus Gedani*, 1662.)

determination of the diameter of Venus: Horrocks set it at 1' 16", ten times smaller than expected. In addition, he obtained at once improved orbital elements for the planet—in particular, the exact position of one of the nodes—and corrected the ratio of the mean distances of Venus's orbit to that of the earth from Kepler's 0.72414 to 0.72333 (the latter is essentially the value still used in astronomical tables today). This was, of course, only a ratio—the absolute distance of the earth to the sun was still unknown. Horrocks took a wild guess at it by assuming that Venus was about the same diameter as the earth, and though the assumption was arbitrary, it turned out to be curiously near the truth. In that case, the sun's distance from the earth would be about 125 million kilometers (78 million miles, about five-sixths of its actual distance).

Though Horrocks's brother Jonas was clouded out, Crabtree succeeded in catching a fleeting glimpse of the transit from Manchester. The sky, which was cloudy all day, mercifully opened up just a few minutes before sunset, gratifying Crabtree with "the pleasing spectacle of Venus upon the sun's disk." Horrocks afterwards learned of Crabtree's experience and identified with him. "Rapt in contemplation," he wrote, "he stood for some time motionless, scarcely trusting his own senses, through excess of joy; for we astronomers have as it were a womanish disposition, and are overjoyed with trifles and such small matters as scarcely make an impression upon others."[15]

Horrocks wrote an account of the event, *Venus in Sole Visa* (Venus on the Face of the Sun). It is a charming work; unfortunately, it has not been reprinted for a century and a half. At one point, Horrocks addresses Gassendi:

> Thou, too, illustrious Gassendi, above all others, hail! thou who, first and only, didst depict Hermes' changeful orb in hidden congress with the Sun. Well hast thou restored the fallen credit of our ancestors, and triumphed o'er the inconstant Wanderer [Mercury]. Behold thyself, thrice celebrated man! associated with me, if I may venture so to speak, in a like good fortune. Contemplate, I repeat, this most extraordinary phenomenon, never in our time to be seen again! the planet Venus drawn from her seclusion, modestly delineating on the Sun, without disguise, her real magnitude, whilst her

> disk, at other times so lovely, is here obscured in melancholy gloom; in short, constrained to reveal to us those important truths, which Mercury, on a former occasion, confided to thee.[16]

It did not escape Horrocks's notice that all the other astronomers of Europe, including Gassendi, had missed the transit. (And what was the "illustrious Gassendi" up to during the transit of 1639? He was once more in Digne, in the throes of a depression he had suffered after his friend Peiresc had died two years earlier. The only therapy he allowed himself, apparently, was writing his friend's biography. Perhaps this accounts for his lack of interest in the goings-on in the heavens.)

In France, the transit would have been visible slightly later than in England, but still before sunset, had anyone looked. Only in the Americas had the entire transit been visible, but there it had gone unobserved. Horrocks could not resist a poetic gibe at what he considered a still-barbarous region. In using the term "barbarous," probably he was thinking of the Spanish possessions, or perhaps—as a Puritan zealot—of the heathen natives. But the young English colonies in America were also hardly ready for the luxury of astronomy. The newly founded Harvard College was as much a backwater of astronomical science as Emmanuel had been, with only the Ptolemaic system being taught. Horrocks wrote mockingly,

> Venus was visible in the Sun throughout nearly the whole of Italy, France, and Spain; but in none of those countries during the entire continuance of the transit.
>
> But America! . . . Venus! what riches dost thou squander on unworthy regions.[17]

The Curtain Falls

The observation of the transit of Venus represented the pinnacle of Horrocks's short career. Moreover, the transit occurred at a cusp of world history. On the Continent, the Thirty Years' War was still raging; in England, Charles I's era of personal rule was crumbling. It all came down to money; Charles needed it, and the nation—in the form of John Hampden, a former member of the dissolved Parliament—refused to pay. Eventually the Crown prevailed, but public opinion had been stirred against the king.

Admittedly, the reasons for the nation's discontent were complicated, but until 1638, there had been no armed resistance. At that moment, the Scots—Presbyterians increasingly restive under the efforts of Charles's heavy-handed Anglican archbishop William Laud to impose the English prayer book on the Scottish church—decided to revolt. While Charles, in the interests of frugality (and not having to recall Parliament) had studiously avoided foreign entanglements, many a Scot had fought in the armies of Gustavus Adolphus and the Protestant princes on the Continent. When they returned they were "camped . . . advantageously on Dunse Law, ready to dispute against Charles the passage of the Tweed."[18] When the Scottish General Assembly refused to dissolve upon the demand of the king's commissioner, Charles, confident of his ability to control the House of Commons, summoned Parliament to vote the taxes needed to suppress the rebellion. The Short Parliament, in which John Pym and John Hampden emerged as the leading figures, was dismissed after only a few days; the Scots crossed the Tweed; and Charles was forced to summon Parliament a second time in November 1640.

Thus began the famous Long Parliament (November 1640 to August 1641), more accurately the House of Commons, dominated by the Puritans, in revolt against the king. "The Long Parliament," writes George Macaulay Trevelyan, "is the true turning-point in the political history of the English-speaking races. It . . . prevented the English monarchy from hardening into an absolutism of the type then becoming general in Europe," notably the type being promoted in France under Richelieu, Mazarin, and Louis XIV. The dun clouds of war gathered over England. The struggle between king and Parliament would drag on, to be decided, in the end, by money and the army of Oliver Cromwell—that hard-headed, brooding, but in the end decisive, character.

We do not know what Horrocks made of the troubling events

occurring at this time. He remained at Hoole for a year, then returned to Toxteth. In October 1640—just before the Scots crossed the Tweed and the Long Parliament began to sit—he wrote to Crabtree of the "uncertainty of his affairs." We do not know what the note of foreboding may have been; we may never know. In December he wrote to Crabtree again and noted his inability to make any observations for the previous three months. Neither could he do as he wished and visit Crabtree at Broughton, because of "a great necessity, by which I am either unwillingly detained at home or compelled to journeys less pleasing" (a comment reminiscent of the one he made on the day of the transit). Still, he hoped that he would be able to make a visit around Christmas. In his last letter to Crabtree—dated December 19—he set a date for his proposed visit: January 4, 1641. Perhaps he had a premonition, for he added an ominous note: "unless prevented by something unexpected."[19]

On the morning of January 3, 1641, Jeremiah Horrocks died suddenly of unknown causes; he was not yet twenty-two. One wonders if the interruptions of Horrocks's viewing of the 1639 transit were due, at least in part, to health problems. Crabtree noted his passing "on the day before he had arranged to come to me. Thus God puts an end to all worldly affairs. I have lost alas! my most dear Horrocks. *Hinc illae lacrimae* [thus fall the tears]. Irreparable loss!"[20]

Horrocks was buried, in an unmarked spot, in the chapel at Toxteth. Two and a half centuries later, on December 9, 1874—the date of another transit of Venus—a marble scroll was placed on the pedestal of the monument of John Conduitt, Isaac Newton's nephew, in Westminster Abbey (see fig. 20). In view of Newton's own tomb, it bore an inscription com-

Figure 20. Stained-glass Horrocks Window at Hoole Church, Lancashire. (Photograph by William Sheehan, 1993.)

memorating Horrocks's "prediction from his own observations of the transit of Venus."

This violent time was not one of long or happy lives. Apparently, Crabtree did not long outlive his friend. He is said to have died relatively young, at only thirty-four, in 1644 (according to Oxford historian Allan Chapman, the last testament of Crabtree, which turned up early in the twentieth century, is dated July 1644; we are unable to find any source for Richard Proctor's claim that Crabtree perished at the Cromwellian battle of Naseby, June 14, 1645). Another promising scientific talent who died much younger than most contemporary scientists was William Gascoigne, the inventor of the filar micrometer, an ingenious instrument that would lead to much more accurate measures of the sun, moon, and planets. Gascoigne began corresponding with Crabtree about invention in late 1640, and Crabtree paid him a visit at his home in Leeds. But Gascoigne, too, was swept into the violence of the English Civil War: he perished at Marston Moor (July 1644), where a combination of Scottish Presbyterians and Cromwell's East Anglian Sectaries broke the Royalist power in the North. Cromwell later wrote of that checkered day of alternating rain and sunshine, charges and countercharges, "We drove the entire cavalry . . . off the field. God made them as stubble to our swords."[21] Among that stubble was William Gascoigne.

Historian of astronomy Robert Grant writes, "Amid the angry din of political commotion, the name of Horrocks was completely forgotten."[22] But Horrocks's papers survived: some fell into the hands of the Royal Society, others have turned up only recently. A copy of *Venus in Sole Visa* was taken to the Continent by the Dutch astronomer Christiaan Huygens (1629–95) and was published by Hevelius at Danzig in 1662. Manuscripts of his lunar theory were later studied by John Flamsteed and Isaac Newton.

Though he passed from the scene with tragic prematurity, his youthful days "scarce complete" at only twenty-one, he had achieved great work. Has anyone so young ever reached anything like his place in astronomy? His accomplishments are all the more remarkable in that he did them in the almost total intellectual isolation of the Lancashire countryside. He has been aptly called an "astronomical Keats"—although Keats lived to twenty-five. It is likely that if Horrocks had lived to an even moderately advanced age, "his fame would have surpassed that of all his predecessors." His theory of the planets, and especially the moon, whose motions he modeled so well that even Isaac Newton found it difficult to improve upon it, must be regarded as the last important steps taken before the establishment of the theory of gravitation.

In the popular mind he will always be best remembered for having observed "that which had never been observed since the World began . . . the most agreeable spectacle" of Venus silhouetted against the disk of the sun.

The sight would not be seen again until the eighteenth century. By then, how many turns of the world would there be, what social and political revolutions? What scenes would pass in that phantasmagoric pageant of events whose motto has ever been *sic transit gloria mundi* (thus passes the glory of this world)? In 1639 no one could glimpse, through the still-shut fenestrations of time, the revolutions of the world between then and when Venus would next appear against the sun.[23]

6

A Celestial Monarchy

Camille is so intelligent that she will grasp everything you tell her straight away. In fact, one day she was even able to understand the inverse ratio of the square of the distances.

—Honoré de Balzac, *Lost Illusions* (1837–43)

The New Astronomy and Newton

Jeremiah Horrocks died in January 1641. He was the first in England to be touched by the Astronomical Revolution that is identified with the achievements of Copernicus, Tycho Brahe, Kepler, and Galileo. In 1642, Galileo, aged and blind, also passed from the scene.

That same year, "in the eye of" what some have called "the greatest political storm in English history,"[1] Isaac Newton (1642 O.S./1643 N.S.–1727) was born at Woolsthorpe, near Grantham, Lincolnshire. Newton, because of his far-reaching theory of gravitation, which allowed the movements of the planets to be reduced virtually to clockwork regularity, and because of his closeness to Edmond Halley, is a background looming figure in our story of the transit of Venus expeditions.

Newton was born after his father had died and, after his mother's remarriage when he was three, lived on at Woolsthorpe Manor with his maternal grandparents. Later he was sent to school in Grantham, where he lived with an apothecary known to history only as "Mr." Clark. Soon he filled Clark's house with sundials: "his own room, other rooms, the entry, whenever the sun came. He drove pegs into the walls to mark the hours, half-hours, and even quarter-hours and tied strings with running balls to them to measure the shadows on successive days."[2] He was reported never to have played with the boys at his school, while they in turn were "not very affectionate toward him."[3] Though little mathematics was taught there, already some of Newton's other passions were becoming evident. Besides his tinkering with sundials, he entered extensive alphabetical lists of words under various headings, which his biographer, Richard S. Westfall (no relation to the coauthor of this book) has suggested "were the example of a prominent characteristic of Newton, his desire, perhaps even compulsion, to organize and categorize information."[4] This was the first of many notebooks Newton kept. Later, he amassed innumerable notebooks at Cambridge and hoarded them obsessively. Not the least part of his genius was his "ability to organize what he learned so that he could retrieve it."[5]

When Newton was seventeen, his mother recalled him to Woolsthorpe to manage the estate. The experiment proved to be a disaster. Instead of watching the cattle, he spent his time either reading or making "strange inventions," including a model mill with a mouse on a treadmill as his miller.

The mistake in recalling him to Woolsthorpe was soon recognized. He was sent back to the Grantham school and recommended for the university. In June 1661, he left what one of his biographers has called "the idiocy of rural life"[6] for Trinity College, Cambridge.

Turbulent Times

When Newton arrived at Trinity, England was in a somewhat more settled state than it had been for twenty years. The unrest of the civil wars, whose violence had during Newton's infancy veered unsettlingly close to Woolsthorpe itself—the beheading of Charles I, an unsuccessful experiment of rule by the House of Commons, followed by the era of the strongman, Oliver Cromwell, as Lord Protector—all this had led, at last, to the army-backed restoration of Charles Stuart, an Anglican with a Roman Catholic wife and Catholic proclivities. Recalled from France in May 1660, Charles Stuart, now Charles II, returned, amid much rejoicing, to London, "to enjoy his own [realm] again."

The second Charles was not like the first. He once said, "God will never damn a man for allowing himself a little pleasure."[7] He was able to alter with the winds, when necessary; above all he was a survivor. He did not, however, start out very well as king. He had been put on a short leash financially by Parliament and began amassing debts. His first priority was to revive the British Navy, a project already started under Cromwell, which brought the new king into conflict with Holland.

In general, England had the better of these wars with Holland. However, even as the terms of a settlement were being negotiated, a Dutch fleet, sailing up the Thames and the Medway, burned the finest warships of the British Navy as they lay at anchor at Chatham. The Medway disaster was followed, in quick succession, by an outbreak of the bubonic plague in London, killing forty-seven thousand people in August and September 1665. Finally, much of London burned to the ground in the great fire of 1666, which at least probably brought the plague under control.

Even these mighty events hardly seem to have penetrated Isaac Newton's deep isolation. He was so preoccupied and self-absorbed that even his chamber-fellow, John Wickins, hardly knew him; he was a man who rarely ate with others, or even in his chamber (as a result his cat grew fat). He neglected his sleep and sat up so long in the year 1664 observing a comet that he found himself "much disordered." (That comet became associated, in the popular mind, with the plague, even as another that appeared in 1665 became associated with the great fire.)

ANNUS MIRABILIS, OR WHAT NEWTON KNEW WHEN

But though Newton was seemingly inwardly unperturbed by the troubles of King Charles's realm, his outward circumstances changed. As the plague spread to Cambridge, the university closed, and Newton returned to Woolsthorpe Manor. The plague years, 1665–66, have often been telescoped for convenience into one remarkable "year" (Newton's *annus mirabilis*). Newton's own descriptions of his accomplishments, written in old age, are the basis of these accounts, in which he claimed to have developed his method of fluxions (the calculus), to have carried out his investigations with the prism into the nature of colors, and begun

> to think of gravity extending to ye orb of the Moon & (having found out how to estimate the force with w[hi]ch [a] globe revolving within a sphere presses the surface of the sphere) from Kepler's rule of the periodical times of the planets being in sesquialterate proportion of their distances from the centers about w[hi]ch they revolve . . . compared the force requisite to keep the Moon in her Orb with the force of gravity at the surface of the earth, & found them answer pretty nearly. . . . In those days I was in the prime of my age for invention & minded Mathematicks & Philosophy more than at any time since.[8]

This is clearly an oversimplification. Richard Westfall, who generally accepts Newton's heroic view of himself, admits, "When we examine Newton's grandiose adventure minutely, it turns out to be a mixture of discrete pieces rather than a homogeneous mélange. His career was episodic."[9]

In particular, the suggestion is questionable that Newton at that time combined what we would now call the law of centripetal force—a force drawing objects toward the center, like the tension on a string when a stone is whirled rapidly around—with Kepler's harmonic law to derive an inverse-square law of gravitational force; the claim is unsupported by any datable document. The first Newtonian manuscript to demonstrate the inverse-square law dates only to 1684.

Newton's actual notes from this period contain his speculation that gravity might be caused by the descent of a subtle invisible matter striking all bodies and pushing them down. He even pondered designs for perpetual-motion engines to harness this flow!

During this period, Newton believed that Kepler's harmonic law was exact only for the outer planets. He was still under the influence of the French philosopher and mathematician René Descartes (1596–1650), who postulated that the planets were carried around the sun by the whirling vortices of a rarified substance (aether). According to the vortex theory, there was no reason to believe that Kepler's period-distance relation would hold more than approximately, and Newton's own computations seemed to indicate that it broke down for Mercury and Venus, the two planets closest to the sun. But if so, we must discard the idea many of us learned in school—that Newton had grasped the law of universal gravitation by 1666. The question why there was a twenty-year delay from the time he had supposedly made this discovery and its full publication in the *Principia* is answered. As yet, according to D. T. Whiteside, "he had nothing of the sort to communicate."[10]

Newton Plays Hooky until Hooke Comes Along

During most of these twenty years of silence, Newton was massively absorbed in matters other than motion and gravitation. He was obsessed with biblical chronology, with the prophecies of the books of Daniel and Revelation, and with attempts to divine the correct number of persons in the Godhead. Next to biblical prophecies and chronology, his other preoccupation during these years was alchemy. His sizar Humphrey Newton (no relation) remembered him during these years as a remote, self-absorbed man, with little need for contact with other people:

> I never knew him to take any recreation or pastime either in riding out to take the air, walking, bowling, or any other exercise whatever, thinking all hours lost that was not spent in his studies, to which he kept so close that he seldom left his chamber unless at term time, when he read in the schools as being Lucasianus Professor, where so few went to hear him, and fewer understood him, that ofttimes he did in a manner, for want of hearers, read to the walls. . . . So intent, so serious upon his studies that he ate very sparingly. . . . He very rarely went to bed till two or three of the clock, sometimes not until five or six, lying about four or five hours, especially at spring or fall of the leaf, at which times he used to employ about six weeks in his laboratory, the fire scarcely going out either night or day; he was sitting up . . . till he had finished his chemical experiments, in the performances of which he was the most accurate, strict, exact. What his aim might be I was not able to penetrate into.[11]

Newton's inaugural lectures as Lucasian professor concerned his theory of colors, an offshoot of his meditations on light that finally drew him out of himself and, as it were, into the light. In his researches on light, he came to recognize the problems with refracting telescopes of the day, which used simple lenses that bend the colors of the spectrum unequally, producing hazes of unfocused color around all bright objects. He constructed a different kind of telescope, using a mirror instead of a lens; this was the first reflecting telescope, employing a one-inch mirror he himself had ground in his chamber at Trinity (see fig. 21). In 1671 he presented it to the Royal Society for the Improvement of Natural Knowledge, where it created such a furor that within the year he was elected a fellow of the Royal Society.

Through the Royal Society, Newton first came into contact with others seriously interested in experimental science and in light, colors, and the motions of the moon and the planets. For a long while he participated only intermittently in the activities of the Royal Society, remaining as if hermetically sealed in his chambers at Trinity, pursuing his biblical and alchemical work.

Even without Newton's regular participation, the Royal Society was a hugely gifted group. Its members included Christopher Wren

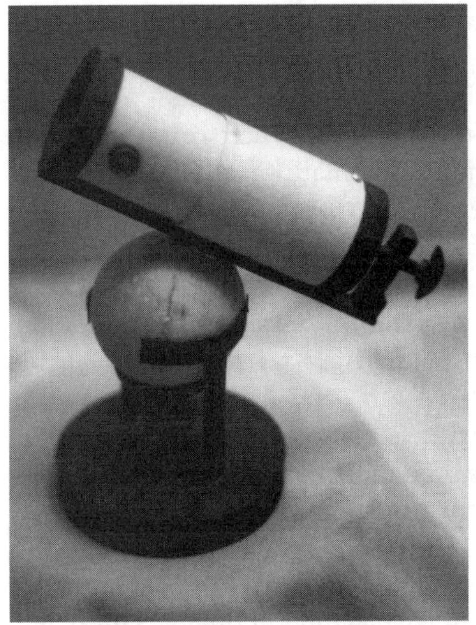

Figure 21. Model of Newton's reflecting telescope, with a mirror 1 inch in diameter, on display at the Adler Planetarium in Chicago. (Photograph by William Sheehan, 1978.)

(1632–1723), Savilian professor of astronomy at Oxford and the architect who would play such an important role in rebuilding the churches of London after the great fire, best remembered for his design of St. Paul's Cathedral. Also, there was Robert Hooke (1635–1703), a physically crooked little man who was bursting with ideas, gregarious, and a frequenter of coffee-shops.

Hooke had started out as a subsizar at Oxford, later became professor of geometry at Gresham College, London, and was a versatile experimentalist and a pioneer of both the telescope and the microscope. In his 1665 book, *Micrographia,* he describes his telescopic observations of a spot on Jupiter—probably the Great Red Spot, of which he was the first observer—and also his microscopic observations of the compound eyes of flies and the cells of plants—it was he who coined the term "cell" in biology. He was also an architect: with Wren he drew up plans for the rebuilding of London and was responsible for the design of the Bedlam Hospital for the insane.

The Royal Society had originally been known as the "Invisible College for Promoting of Physico-Mathematical Experimental Learning" and had become "Royal" when Charles II gave it his charter. Until this time, Newton had hardly had any contact with the outside world, except through books. But he had also experienced no discord. At first Newton looked to the Royal Society as a haven; but there were obvious dangers as well for a man of his psychological make-up. He was sensitive to criticism and suffered from the same extreme jealousy and possessiveness for ideas that some other men felt when threatened with the dispossession of a woman they loved.

Hooke baited him over his theory that white light was composed of the other colors of the spectrum. Their correspondence over the next several years provides a fascinating study. At first, it was civil enough, even courtier-like.

Though Newton's later hatreds would become legendary, Hooke, though much more sociable than Newton, also had his share of controversies. He feuded with Newton and the Astronomer Royal, John Flamsteed (1646–1720), for most of his life. Says one scholar, "He resented both of them, they disliked him, and they themselves were eventually on very bad terms with one another."[12] The reasons for Hooke's "great antagonisms lies probably in one of his characteristics as a scientist. He had a very lively mind and a fertile imagination, so that he was forever putting out ideas and plans, but he was not good at applying himself to working them out in any detail."[13] In any case, it would be Hooke who would draw Newton out on gravitation and fully awaken both Newton the genius—and the monster.

A GLIMPSE AT VENUS IN THE TELESCOPE

Meanwhile, as England had been preoccupied with civil wars and restoring its monarchy, France had been steadily advancing its arms and influence across a continent in which it met little opposition. Spain was enfeebled and in long decline; Germany had been exhausted by the seemingly interminable ordeal of the Thirty Years' War; Italy and the popes were caught in the never-ending dynastic struggles between the Habsburgs and the Bourbons; and Austria was preoccupied with defending the approaches of Vienna from the Turks.

The "Sun King," Louis XIV, had seized on the dream of his minister, Cardinal Richelieu, to make the "king absolute . . . in order to establish therein order and rule."[14] Nearing the zenith of his power and magnificence, Louis was incessantly building new and expensive projects like the palace of Versailles, which would swallow up a fortune, eventually drowning France in its own magnificence.

A more useful building project was the Royal Observatory at Paris, established in 1667 and one of the great scientific institutions of the age. Funded by Jean-Baptiste Colbert, Louis's wizard minister of finance, the designs for it were submitted by Claude Perrault, who had been architect of the new facade of the Louvre. Louis managed to entice to the new observatory Giovanni Domenico Cassini (1625–1712), Italy's greatest observational astronomer (see fig. 22). Cassini had been making tantalizing progress in working out tables of the positions of the satellites of Jupiter, tables showing their positions at regular intervals. This research he directed toward producing a workable method of solving the all-important problem of longitude at sea (in this, he was elaborating on a method first suggested by Galileo). He also used the unwieldy telescopes of the day to discover spots on the disks of the planets, the rotation periods of Mars, Jupiter, and even—so it seemed—Venus.

For a long time Cassini had been frustrated by the paucity of detail on Venus, not realizing that all that can ever be seen from the earth is the bright upper surface of its stratum of pale-yellow cloud. But at

Figure 22. Portrait of Giovanni Domenico Cassini (1625–1712), first director of the Paris Observatory. (Courtesy of NASA/European Space Agency.)

last in April 1667 he seemed to make out a conspicuous bright cloud (see fig. 23). He attempted to follow its motion, but his statement, published in the *Journal des Sçavans,* was most confusing. The spot, he said, followed a long rotation—perhaps rather a moonlike wobble or libration—of 23 days; at the same time, his unclear wording seemed to say that the planet rotated in 23 hours. The latter figure, with its appealing analogy to the rotation of the earth, was eventually adopted by Cassini's own son, astronomer Jacques Cassini. It would long bias future perceptions of the planet.[15]

SOLAR PARALLAXES

After G. D. Cassini's arrival in Paris in 1669, he never again saw any spots on Venus. Perhaps in part this was owing to the observatory's design. Despite its architectural elegance, it made, Cassini said, "no sense" from the viewpoint of a working astronomer. In particular, it was not designed for the use of the long telescopes then in vogue. Cassini had to set them up in the courtyard! Nonetheless, by 1672 he was involved in a grand project to measure the solar parallax by a method that would provide the most accurate measure of the solar parallax ever produced before the transit of Venus of 1761.

Cassini took advantage of the planet Mars's close approach to the earth in early September 1672 (see fig. 24). While he remained in Paris, his colleague, Jesuit astronomer Jean Richer (1630–96), was sent to Cayenne, in the South American territory of French Guiana. (Incidentally, Cayenne is located just twenty-four miles from the modern rocket launch facility of Kourou, used by the European

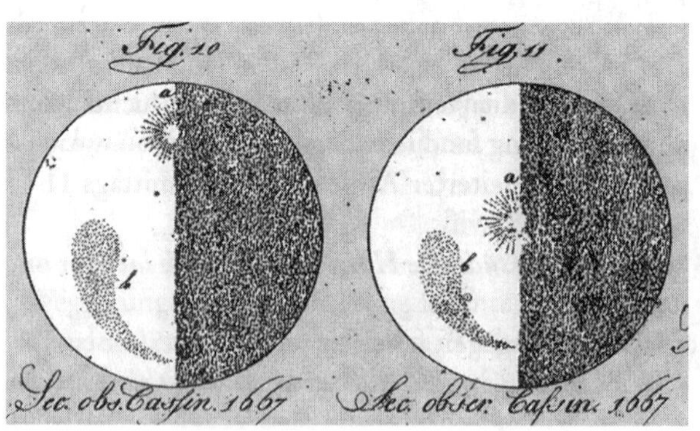

Figure 23. Two drawings of Venus, made in 1677 by G. D. Cassini. (From Camille Flammarion, *Les terres du ciel,* 1877.)

Figure 24. Observing Mars from the Paris Observatory. (Illustration from Camille Flammarion, *Les terres du ciel*, 1877.)

Space Agency.) The observing stations were far enough apart to produce a large baseline.

The principle was the same as that of the finger held before the eyes, but now the baseline was on a global scale. With Mars near opposition, Richer, at Cayenne, measured the position of the planet relative to three small telescopic stars in Aquarius, at the same time that Cassini did so from Paris. The observed displacement, due to the distance between Paris and Cayenne, could be used to work out the parallax of Mars (see fig. 25). Cassini found that the planet shifted almost 25 arc-seconds—about its own apparent diameter—between the two stations. Knowing the distance of Mars at opposition date was 0.3812 astronomical units, Cassini's value for the solar parallax came out to 9.5 arc-seconds, implying an earth-sun distance of 138 million kilometers (86 million miles; only 8 percent less than the current figure, though historians generally concede that the apparent agreement with the modern value is due more to chance than method).

Cassini and a Danish colleague at the Paris Observatory, Ole Rømer (1644–1710), tried measuring the solar parallax on three nights in September 1672 by another method that had been originally proposed by Rømer's compatriot Tycho Brahe. The positions

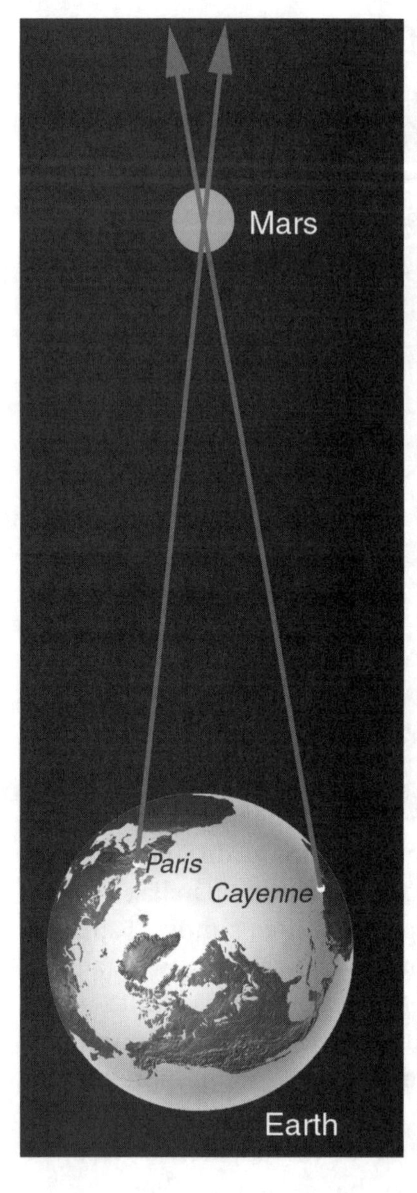

Figure 25. Cassini and Richer's method of measuring the parallax of Mars by means of its apparent shift when simultaneously observed from Paris and Cayenne. (Diagram by John Westfall.)

of Mars rising and setting relative to the background stars were measured using an eyepiece micrometer. Because of the earth's rotation, between the rising and setting of Mars the observer has effectively been carried over a baseline equal to a substantial part of the diameter of the earth. Thus the parallax of Mars is found by measurements made at a single station, with no traveling required. Using this method, Cassini and Rømer found a value of 24–27 arc-seconds for Mars's parallax, again leading to a solar parallax of 9–10 arc-seconds.

At the same apparition of Mars, John Flamsteed, at Derby, England, measured Mars's position hourly over a 6-hour, 10-minute period on the night of October 6, 1672. From this series of measures, Flamsteed determined the parallax of Mars and also deduced that the solar parallax was at most 10 arc-seconds.

LONGITUDE

Meanwhile, England's war with Holland, which had been funded in part by Louis, was bankrupting Charles and forced him once more into the arms of parliamentary control. The Dutch had, moreover, gained a startling success against the combined British and French fleets at Solebay. This underscored the need to rebuild the British Navy; as part of this effort, Charles became convinced of the need for his own observatory. In June 1675, he issued this warrant: "In order to the finding out of the longitude of places for perfecting navigation and astronomy, we have resolved to build a small observatory within our park at Greenwich." The building was designed by Wren and Hooke. John Flamsteed was named "His Majesty's Observator," later Astronomer Royal (see fig. 26); not without a little grumbling, he was forced to provide his own instruments. Hooke did not like Flamsteed; he called him Vulponi—the fox—and regarded him as arrogant and ignorant. Flamsteed, for that matter, did not care for Hooke. But he proved to be a first-rate observer, setting out to recatalog the stars of the Northern Hemisphere, especially those near the ecliptic and thus whose angular distance from the moon could be measured by navigators.

This interest in mapping of the stars for triangulation with the moon reminds us that in the seventeenth and eighteenth centuries the most vexsome and pressing problem of astronomy and navigation was that of determining longitude at sea. Latitude is easily determined, simply by measuring the height of the sun at noon or the height of the pole star above the horizon. Longitude is a different matter: since the earth

Figure 26. Portrait of John Flamsteed (1646–1720), the first Astronomer Royal. (Reproduced by permission of the National Maritime Museum, London.)

rotates on its axis, longitude gets inextricably mixed up with time, which varies regularly from place to place; a 1-hour difference in time equals a 15-degree difference in longitude. Until the development of clocks accurate enough to keep time during an ocean voyage, the most promising method of determining longitude at sea seemed to be one based on observations of celestial phenomena, including the times of the eclipses of the satellites of Jupiter or the angular distance between the moon and a fixed star worked out for a standard longitude, such as Greenwich or London. An observer at sea measuring the local time at which such an event occurred, compared with that computed in the tables, could work out the difference in longitudes. Obviously, to make the so-called method of lunars viable, several things were needed, each accurate to 1 arc-minute or better: (1) an instrument for measuring the moon-star distance, (2) a star catalog, and (3) tables giving lunar positions relative to the stars at given times. This was obviously a work in progress, and all of these prerequisites were not realized until about 1750. It was the star catalog that Flamsteed set out to provide.[16]

Flamsteed went to work at Greenwich. A hard worker, obsessive and perfectionistic as most great observers must be, he scanned the heavens "through the long, cold nights for half a century as a form of divine service."[17] But he did allow occasional intrusions into the single-mindedness of his routine, one of which came from a man who will occupy a large stage in the transit of Venus saga—Edmond Halley.

ENTER HALLEY

Edmond Halley had been born in 1656 in Hackney, then a rural borough of London. His father was a soap boiler and salter who owned fourteen houses on Winchester Street, London, some of which were destroyed in the great fire. At the time of the fire the Halleys were living at St. Giles's, Cripplegate, which was also consumed. Nevertheless, Edmond Halley Sr.'s businesses continued to thrive, thanks in part to a postplague boom for soap and an increasing demand for salted meat by the Royal Navy. The latter's voyages were becoming wider and wider ranging in competition with those of other nations, especially the Dutch. Edmond Halley Jr., whose interest in astronomy had started in his "tenderest years," was able to attend one of the best schools in England, St. Paul's, almost in the shadow of the famous cathedral.

Halley entered Queen's College, Oxford, as a "commoner," a step up from a sizar. The curriculum at Oxford emphasized grammar, antiquities, rhetoric, logic, and moral philosophy. But Halley was more interested in mathematical subjects and astronomy. He had taken with him to Oxford a 24-foot-long telescope that his father had purchased for him. He wrote to Flamsteed about errors he had discovered in the published tables of the motions of Jupiter and Saturn and in some of the star positions in the great star catalog of Tycho Brahe. Soon afterwards he explained to Flamsteed that he was "reasonably well provided in instruments in which I can confide to one minute without error by means of telescopicall sights and a skrew for the subdivision."[18] Flamsteed was impressed; in June 1675 they got together in London for a solar eclipse. Two weeks later, they observed a lunar eclipse from the Tower of London. Flamsteed later saw to it that Halley's first paper, "A Direct and Geometrical Method of Finding the Aphelia, Eccentricities, and Proportions of the Primary Planets, without Supposing Equality in Angular Motion," was published in the *Philosophical Transactions* of the Royal Society.

Thus Halley came into the "humorless band of compulsive geniuses—Hooke, Newton, Flamsteed"[19]—as a refreshingly different sort of personality. He was worldly and full-blooded, indifferent to the nice and quarrelsome matters of religious doctrine, affable, even jocular, handsome, and given to drinking brandy. He was daring, fearless, versatile, and adaptable enough to maintain close companionships with men ranging from the delicate, touchy, always-suspicious Newton, whom he managed always somehow to handle, to the young czar of Russia, Peter the Great. While visiting England and learning what he could do to open Russia to westernization, Peter stayed at Deptford, near the British shipyards, and caroused in its dives and taverns with Halley.

A TRANSIT AT ST. HELENA

Halley began to grow restless at Oxford, and so—without taking his degree—in November 1676 left rather abruptly and, with the endorsement of the Fellows of the Royal Society and Charles II, set sail for St. Helena, a small, bleak, windswept volcanic island in the Atlantic off the west coast of Africa. The island, accidentally bumped into by the Portuguese, was taken over by the British East India Company, briefly disputed by the Dutch, and has been in British hands since 1674. St. Helena then represented the southernmost point on the map then under British rule. One hundred and forty years later it would become infamous as the spot where Napoleon was exiled and died.

The twenty-one-year-old Halley would remain for a year on St. Helena, accomplishing in that year nothing less than mapping the southern skies with his 24-foot telescope, complementing Flamsteed's mapping of the northern skies from Greenwich. His self-imposed assignment was, as Carl Sagan and Ann Druyan put it in their book *Comet*, "to bring back half the sky."[20]

When he arrived at the only point suitable for landing, Jamestown, Halley found that the whole island was little more than a sparsely vegetated rock isolated in the middle of the Atlantic. Its extreme length was ten miles; the topography was sharp and rugged, with valleys running along steep cliffs to the sea from the centrally located, 2,700-foot-high Mounts Actaeon and Diana. A good share of the population at the time consisted of black slaves, and the settlers, under a governor who was unstable at best, were on the verge of mutiny. At first, Halley seems to have done what any astronomer worth his salt would do: he considered building an observatory on one of the island's high peaks. However, he found that the peaks were shrouded in clouds much of the time; in addition, the nighttime temperatures hovered too near the dew point, causing a layer of moisture to gather like a sheath around his telescope.

All in all, the weather at St. Helena—which Halley had naively expected to find first-rate—proved even worse than the notoriously bad weather of England! On the rare occasions when the sky cleared, the stars were indeed small and bright and sharp. However, the months of August and September were almost continually rainy, and ferocious winds rattled the telescope mercilessly.

Nevertheless, Halley used the scraps of time when the weather permitted to obtain observations for his star catalog, *Catalogus Stellarum Australium* (published 1679). He even introduced a new constellation—*Robur Carolinum*, after the oak tree in which Charles II had taken refuge after the Battle of Worcester in September 1651. The nineteenth-century collector of star lore, Richard Hinckley Allen (1838–1908), claims that this politic move was later a factor leading to Halley's receipt of his master's degree from Oxford, by express command of the king. Halley also measured the positions of numerous stars and nebulae hitherto unknown. He did not find a good southern pole star, however, as none exists.

His most important observation was of the transit of Mercury on November 7, 1677 (N.S.). The tables Halley brought to St. Helena were not very accurate, so for good measure Halley started watching for the transit as soon as the sun rose. The weather the night before had been cloudy and windy. Though the wind dropped, the clouds persisted on the day of the transit, and Halley had to monitor Mercury as best he could through breaks in the clouds. (See fig. 27 for photographs of a transit of Mercury.)

It was the first transit of Mercury observed in its entirety, and only the third since Gassendi's pioneering effort in November 1631. The others had been the transit of 1651, observed by Jeremiah Shakerley (1606–c.1670) from Surat, India, and the transit of 1661. The latter was recorded by, among others, the Dutch astronomer Christiaan Huygens, then in London and about to join Louis XIV's intellectual entourage in Paris.

Halley described the transit to one of his London friends, Jonas Moore: "I have . . . had the opportunity of observing the ingress and egress of [Mercury] on the [Sun], which compared with the like Observation made in England, will give a demonstration of the Sun's Parallax, which hitherto was never proved, but by probable arguments."[21]

Unfortunately, the transit had been clouded out entirely from England. In Europe, only French observer Jean Charles Gallet,

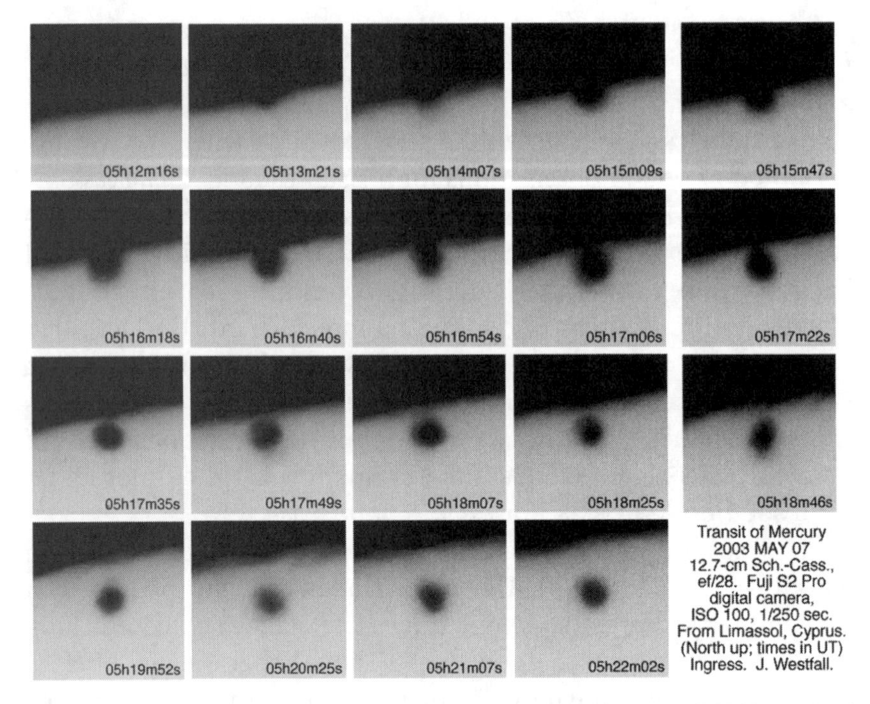

Figure 27. Digital-camera photographs of the ingress phase of the May 7, 2003, transit of Mercury, taken by John Westfall. The Universal Time is indicated within each frame.

known as the "hermophile" because of his perennial interest in observing the innermost planet, saw the planet on the solar disk. But clouds interfered until the very end of the transit, when they parted to allow him a glimpse of egress.

From the 1677 transit, Halley concluded that it ought to be possible from such observations to work out an accurate value for the solar parallax. Observations of future transits of Mercury might be good enough, but those of the much-rarer transits of Venus—because of its greater proximity to the earth—would serve far better. Thus, at the age of only twenty-one, he began to meditate upon the "Great Method," which he worked up and published in detail forty years later:

> [W]hen I was in the island of St. Helena, observing the stars about the south pole, I had an opportunity of observing, with the greatest

> diligence, Mercury passing over the disk of the Sun; and (which succeeded better than I could have hoped for) I observed, with the greatest degree of accuracy, by means of a telescope twenty-four feet long, the very moment when Mercury, entering upon the Sun, seemed to touch its limb within, and also the moment when going off it struck the limb of the Sun's disk, forming the angle of interior contact; whence I found the interval of time, during which Mercury then appeared within the Sun's disk, even without an error of one second of time. For the lucid line intercepted between the dark limb of the planet and the bright limb of the Sun, although exceedingly fine, is seen by the eye, and the little dent made on the Sun's limb, by Mercury's entering the disk, appears to vanish in a moment; and also that made by Mercury leaving the disk seems to begin in an instant.[22]

The *Principia*

All that would lie ahead in an enormously busy career. At the moment, we pick up once again the Newtonian thread, for after Halley returned to England, he found himself thrust into the grandest role of his life: as supporting actor in the unfolding drama involving the development of the theory of gravitation.

Up to this time, Newton remained immersed in his usual interests—the Bible and alchemy. In 1679, Hooke vetted him with a series of letters, in which he raised a series of questions about the motion of bodies. Newton's ideas about gravitation—such as they were—still involved forces generated within the swirling aether-vortices postulated by Descartes.

Hooke, however, in a lecture entitled "Attempt to Prove the Motion of the Earth" (1674), had proposed that all "Coelestial [*sic*] Bodies whatsoever, have an attraction or gravitating power towards their own centres . . . but that they also attract all other Coelestial bodies that are within the sphere of their activity." This comes close to being a statement of a law of universal gravitation, disqualified only by the final modifying phrase. In letters to Newton, Hooke suggested that, instead of the aether-gravity that Newton had been working with, it might be productive to assume an attractive force toward the center of the orbit. This central force would be counterbalanced not by an aethereal pressure but by the tendency of the moon or the earth to move away along a tangent-line to the orbit. Newton later said that he was "inclined" to try Hooke's mode of analysis, and that he found—by late 1679 or early 1680—the demonstration by which he afterward "examined ye Ellipsis." Still, his interest had hardly been stirred. When Hooke's letter arrived, Newton had just returned from his mother's funeral and was absorbed in a study of the book of Revelation; still, he took time to respond to Hooke's queries. It would be hard—without knowing the bitterness of their relationship later—to detect any note of uncivility in these letters. He simply told Hooke he was "almost wholly unacquainted wth what Philosphers at London or abroad have of late been imployed about. And perhaps you will ye more believe me when I tell you yt I did not before ye receipt of your last letter, so much as heare (yt I remember) of your Hypothesis of compounding ye celestial motions of ye Planets, of a direct motion by the tangt to ye curve . . . though these no doubt are well known to ye Philosophical world."[23]

Newton and Hooke exchanged a few more letters, and Newton made observations of the huge comet that appeared in the winter of 1680. It made a hairpin turn around the sun as it approached within a single solar radius at perihelion (the point of its orbit nearest the sun). According to Flamsteed, this comet had appeared in the predawn sky in December 1680 and emerged two weeks later into the evening sky with a larger tail. Newton disagreed; he thought two comets were involved. But his thinking at this stage was still dominated by the whirling aether of the solar vortex. Thus he questioned the "Royal Observator's" conclusion that the comet had passed in front of the sun at perihelion, since then it would have had to have moved in reverse direction to the tide of the aether.

Because of the comet's almost straight-line approach to the sun, Cassini proposed that this comet might be in orbit around Sirius! Newton assumed that its next return would spell the end of the world.

Nicholas Kollerstrom of the University of London believes it was not this comet but another that appeared two years later—the one now known as "Halley's Comet"—that finally prodded Newton's thinking about the theory of gravitation.[24] The comet of 1680 had followed a path sharply inclined (by 61°) to the ecliptic but moving in the same direction the planets traveled in their orbits around the sun. The comet of 1682 followed a trajectory that was less sharply inclined to the ecliptic than the comet of 1680. However, it moved in reverse, or *retrograde,* to the motion of all the planets. It was beating against the tide that swept everything else around the sun.

Up until this point, Newton had been running well behind Hooke. The comet of 1682 came as a revelation. According to Kollerstrom,

> The comet moved in between Mercury and Venus at its perihelion. A tension built up between all that Newton had said publicly on the subject and the new testimony of the heavens. At last, the skies became—for him—empty. Only then was he moved to ponder the words Hooke had earlier addressed to him [on motion]. His alchemical experiments continued, until in August 1684 his furnace was allowed to expire . . . and an entirely new phase of his life began. Claims that he accepted or formulated an inverse-square law of gravitational attraction prior to that date are mere fabrications; Newton himself was the prime author.[25]

Now things began to happen; not in Cambridge, but in London. Since the early 1680s, Wren, Hooke, and Halley had been accustomed to meet at a coffeehouse by St. Paul's. In January 1684, they met to discuss "the Curve . . . that would be described by the Planets supposing the force of attraction toward the sun to be reciprocal to the square of their distance from it."[26] By then, each of them had already deduced an inverse-square law; Wren sometime in the 1670s, Halley that very month. They had done so privately—up to this point, only Hooke had advocated it publicly.

Nevertheless, Hooke was more a qualitative than a quantitative thinker; he had difficulty relating the inverse-square law to elliptical motion. In August, Halley visited Cambridge and looked up Newton. He put to him the problem of what would be the motion of a body if they supposed an inverse-square law of attraction. Newton replied immediately that the curve would be "an Ellipsis";

when Halley asked him how he knew it, "why saith he I have calculated it."[27] He claimed he could not then find his paper; possibly it had never existed. He promised Halley he would recalculate it.

The result was a treatise, *De Motu Corporum in Gyrum* (On the Motion of Bodies in an Orbit), a first sketch of the principles that were later to be worked up into Newton's masterpiece, the *Principia—The Mathematical Principles of Natural Philosophy*. Newton did present a mathematical proof of sorts; it is not easy to follow. Derek T. Whiteside has inferred that even Halley probably did not understand it. Moreover, what Newton proved was not what he told Halley he had already done—given an inverse-square law, to demonstrate that the motion of the body must be an ellipse. Instead he proved the reverse: given the orbit is an ellipse, the force must vary according to the inverse square of the distance.

Though the sketch shows that Newton can hardly have done the calculation rapidly and then misplaced it, he had reached the point where his investigations into celestial dynamics could properly begin. At this point the Newton of the history books finally made his appearance; his definitions became clear and precise. He defines centripetal force: "that by which a body is impelled or attracted towards some point regarded as its center." Resistance: "that which is the property of a regularly impeding medium." He dropped explicit reliance on ideas about the existence of a material aether as the cause of gravitation, noting that, as far as the motions of the moon and planets are concerned, "the resistance is nil."

Then he thrusted his astounding intellect forward in a new direction: the way was clear for him to work out the mathematical implications of the law of universal gravitation. In a sense, he was following a path first trodden by Hooke, but the fact that others had wandered, however inconclusively, into the field of deep research that lay ahead of him was something that, psychologically, he could not allow and would go to any length to deny. The rest of Isaac Newton's life, apart from the *Principia*, was dominated by his ruthless attempts to obliterate all indications of his intellectual debts to anyone and to crushing his rivals, real or perceived; Hooke would be only the first to bleed.

In late 1684 and early 1685, Newton began systematically investigating centripetal forces as they determined orbital motion. His own moment of insight came, he recalled, when Charles II's brother, James Stuart, succeeded him as King James II of England in February 1685. It was at this moment that Newton grasped the generalization Hooke had earlier formulated, that of an attraction "arising from the universal nature of matter." He revised *De Motu* to reflect this bold new conception, giving not only the reason for the ellipses of the planets' orbits but also their minute but perceptible departures therefrom. From this moment he was seized by the majestic problem that Hooke vainly had earlier set before him. There followed eighteen months of the most intense concentration of which a human being is capable, by all odds the most remarkable burst of mathematical virtuosity the world has ever seen. Now in his early forties, Newton absorbed himself in the struggle to give birth to the *Principia*. It was then, if ever, that Newton bore the face of the man later eulogized by the poet Wordsworth (in *The Prelude*, book 3, 1850 ed.):

> . . . Newton with his prism and silent face,
> The marble index of a mind forever
> Voyaging through strange seas of thought, alone.

Of course, Newton had intimated that he was then only working through what he already had in his grasp as early as 1666. In doing so, he helped establish another myth—that theoretical physicists almost invariably do their best work in their twenties and are washed up by the time they are in their thirties. This was not true in Newton's own case; indeed, his best work was done in his forties. He was older when he wrote the *Principia* than Shakespeare was when he wrote *Hamlet*.

Newton's manuscripts reveal his struggles—there was no "'Let Newton be!' and all was light," in the phrase of poet Alexander Pope. To the contrary, his *Principia* papers might well be compared to the messy drafts Beethoven produced in writing one of his symphonies. At the end of all this effort, Newton emerged with a tour de force of complicated geometrical diagrams. He never would admit that Hooke had provided an indispensable starting point; he admitted only that Hooke's "'correction' of his 'fancy' [about aether-gravity] had acted as a dare, a challenge, a stimulus, a diversion, but not, he insisted, an aid. The mathematical proofs of universal gravity, not the concept or even the formula, was for Newton the crux of the problem."[28]

Halley, still a young man, had been closely identified with Newton throughout this creative ordeal and sequestration. Indeed, it is safe to say that except for Halley's intervention, Newton would have remained an eccentric recluse. Having first presented to the Royal Society Newton's solution of the problem of what form of curve would be followed by a body moving according to an inverse-square law, Halley remained loyal for the rest of his life to the man whose towering genius he recognized. That meant, of course, staying on good terms with Newton, no easy task. As Newton's friend, the philosopher John Locke, admitted, Newton was "a nice man to deal with," not meaning "nice" in the modern sense, but rather touchy and hypersensitive.

Halley was by now full-time secretary of the Royal Society. He was responsible for tackling the Royal Society's voluminous correspondence, taking minutes of its meetings, and editing the *Philosophical Transactions*. His father was now dead; he no longer enjoyed the fortune he had once depended on, and though as secretary he was granted a salary, the Royal Society was cash-strapped after having seriously overestimated demand for a book they had just published, Francis Willoughby's *History of Fishes*. Halley had to accept part of his salary in remaindered copies. Even so, he undertook to put Newton's *Principia* through the press, which he paid for out of his own pocket.

One of the most delicate matters that Halley had to deal with during the prepublication process was Hooke's claim that he was the originator of the inverse-square law. Even if there was truth to it, Halley recognized at once the dangers of asserting that claim to the precariously balanced genius for whom he was serving as midwife. He mentioned the impending cloud on the horizon in a letter to Newton: "There is one thing more that I ought to inform you of, viz, that Mr Hook has some pretensions upon the invention of ye rule of the decrese of Gravity. . . . He sais you had the notion from him."[29]

Newton reacted to Hooke's claim with rage, and like the child he was—emotionally—threatened to suppress the all-important third book of the *Principia,* the "System of the World," which contains "the breakthrough discoveries . . . about gravitation."[30] Notably, it was this part of the book that included all the celebrated demonstrations about the shape of the earth, the tides, the precession of the equinoxes, and the motions of the moon. It also provided groundbreaking diagrams showing how comets move in elongated, but closed, ellipses around the sun. Indeed, the equivalence of the comets that had appeared in 1531, 1607, and 1682 would be demonstrated by Halley himself. That comet would, as Halley expected, return again in 1758. It did so, and Halley's comet has made further returns in 1835, 1910, and 1986; it is next due back in 2061.

Eventually, Halley—who perhaps began to see himself lugging home copies of yet another remaindered edition—managed to assuage Newton's injured feelings. He fudged and assured him that no one had taken Hooke's priority claim seriously. He even added the comment that of Wren, Hooke, and Halley, Hooke had been the last to come by the inverse-square law! This was the first, but it was hardly the last, time that Halley would allow himself to become enmeshed in Newton's unsavory plots against his rivals. Eventually, Halley became a willing, perhaps even an enthusiastic, participant.

At the moment, Halley's performance was enough to make Newton relent. The *Principia,* third book and all, appeared in July 1687, introduced by a Latin poem of tribute by Halley. Though little less than rapturous, it hardly exaggerates the immensity of Newton's achievement:

> O mortal men,
> Arise! And, casting off your earthly cares,
> Learn ye the potency of heaven-born mind,
> Its thought and life far from the herd withdrawn!
> Nearer the gods no mortal may approach.
>
> (From *Sir Isaac Newton's Mathematical Principles of the World,* trans. Andrew Motte, 1729)

Halley thus introduced the view that increasingly he himself came to believe—one gets the impression he worshiped the ground Newton walked on. It was the view of Newton that would hold until the nineteenth century, when Francis Baily published *An Account of the Rev. John Flamsteed*, showing how badly Newton had behaved toward yet another rival. In the nineteenth century, this publication produced a scandal, since, as Augustus de Morgan wrote at the time, "That Newton was impeccable in every point was the national creed."

THE CLOCKWORK UNIVERSE

The *Principia's* mathematical virtuosity ensured that it would be read and appreciated by experts. It would have been harder to predict that it would also give rise to what Betty Teeter Dobbs and Margaret C. Jacob aptly termed a "Culture of Newtonianism."[31] Newton himself—with the socially adept Halley as his front man—spent much of the forty years of life that remained to him after the publication of the *Principia* perpetuating the Newtonian myth of his own single-handed and monolithic achievement. He moved to London, where he occupied an apartment where everything was crimson red, and became Master of the Mint, a sinecure which, being Newton, he took seriously. This was the work for which he was officially knighted, and he even dragged Halley into coinage matters, naming him Deputy Comptroller of the Mint in Chester, where Halley spent two miserable years until the mint closed down.

He posed frequently for portraits, usually in great wigs, assuming an arrogant, increasingly prosperous and jowly face (see fig. 28). After 1704, Newton became the autocratic president of the Royal Society, from whose presidential bench he could carry on, mostly through younger proxies, his insatiable blood feuds with his enemies, including Hooke. Newton's backhanded "compliment" to Hooke, "If I see further than others, it is because I have stood on the shoulders of giants," reeks with malice—Hooke, after all, was a hunchback. Flamsteed was another mortal enemy; as Royal Observator, he had observations of the moon that were indispensable to

Figure 28. Portrait of Isaac Newton (1643–1727) in an engraving by W. T. Fry (1789–1843).

Newton's lunar theory. Newton wanted them, but Flamsteed understood enough about observational astronomy to know that these raw observations, without being corrected for refraction, would have been useless to Newton's purpose. Newton, who made some effort to furnish the required theory of refraction—it was never good enough—eventually became frustrated. He became more demanding, and eventually pried the observations loose. Nor were Newton's enemies found only in England: on the Continent, the foremost was Gottfried Wilhelm Leibniz, who contested Newton's priority in the invention of the calculus.

Newton became increasingly a figure of the past while still alive.

He was a man of the seventeenth century, with its bloody preoccupations with religion and the turbulence of civil and religious wars. The period of his great scientific work was over; but he continued, into advanced age, to tinker with the problems of the moon's motion and the stability of the solar system. He saw the solar system in the process of running down, a kind of divinely appointed clockwork that had to be wound up every now and then by the Creator. The Creator he referred to in the *General Scholium* attached to the third book of the *Principia* was the absolute monarch of the universe: "The Being that governs all things, not as the soul of the world, but as Lord over all; and on account of his dominion he is wont to be called . . . Universal Ruler . . . a Being eternal, infinite, absolutely perfect." His disciples—the Newtonians—eventually doubted the necessity of active divine intervention. Instead they were seized by the notion of an intricately balanced, self-regulating system.

Part of the inspiration for this preoccupation with balance and a self-regulating system was the politics of the day. After Charles II's death, his Catholic brother, James II, began his troubled tenure on the throne. Things were not settled until the "Glorious Revolution" of 1688, when William of Orange, the stadtholder of the Dutch Republic, set out, with the Dutch army, for England; James fled, and the Protestant succession was assured.

Henceforth England, under William and his consort Mary—in contrast to its role under Charles II and James II—successfully opposed the designs of Louis XIV on the Continent through a series of long, ferocious, bloody wars fought at places like Landen, Malplaquet, and Blenheim. It was, moreover, a limited monarchy: "English [political] theorists described their king-in-parliament government as having been providentially ordained. . . . Not only were their representatives sitting in Parliament, but Providence watched over them from the Newtonian heavens and all was right—at least for them in their world."[32]

Even as the great powers of Europe were embroiled in wars on their own continent, they were establishing the world's first (and perhaps last) overseas empires. The reasons for the "European explosion" involved trade and plunder; religious zealotry (to put it kindly); the technological advantages of ocean-going ships, firearms, and better and better maps and navigational methods; and, not least, an unshakeable belief in their cultural superiority. Seafaring little Portugal had led the way, beginning its probing voyages down the west coast of Africa as early as 1416. A century later, the trading empire of the Portuguese extended from Brazil and Africa to the shores of the Indian Ocean and the East Indies. Spain followed; in less than fifty years from the conquest of Muslim Granada, Spanish gold-seekers found themselves in Kansas. The sprawling Spanish Empire, from the early sixteenth to the early nineteenth centuries, covered most of the tropical and subtropical Americas, with an outlier far off in the Philippines.

Unlike the Spanish, the Dutch were motivated almost purely by trade; their holdings included places as disparate as New Amsterdam (later New York City), Ceylon (Sri Lanka), Formosa (Taiwan), and scattered small islands in the Caribbean and the Indian Ocean. Growing from small trading settlements, their major asset became the Dutch East Indies—the modern Indonesia.

The two powers that dominated eighteenth-century geopolitics, Britain and France, came to empire building relatively late, but by the time of the Seven Years' War (1756–63), they were grappling with one another for control of India and North America.

The orderly universe—running according to law, like a supremely intricate clockwork, with or without the need for the occasional intervention of divine Providence, who most presumed had first wound it up—led to a preoccupation in astronomy with order (see the orrery in fig. 29). The astronomers of the eighteenth century—especially those in France, who became the most devoted adherents of the "Majestic Clockwork"—labored with single-mindedness at the project of reducing all the phenomena of the heavens to principles of order and regularity.

The French enthusiasm for Newton dated to the 1720s, when the anglophile Voltaire (1694–1778) returned to France from England with a favorable view of the openness, toleration, and freedom of publishing and politics of English society, as compared with his own. He associated this with the influence of the ideas of "Le Grand

Figure 29. Eighteenth-century orrery, with the sun and moon, the planets from Mercury through Saturn, and the major satellites of Jupiter and Saturn. (On display at the Science Museum, Kensington, London; photograph taken by William Sheehan in 1987.)

Newton." "The French," quipped Voltaire, "always come late to things, but they do come at last."

The moon and the planets—even comets, as Newton indicated in the *Principia* and as Halley proved in the case of the comet of 1682—followed the astronomical tables that were designed to calculate their progress through space and time. In the end, the astronomers' goal was to account for the minute perturbations in the movements of the moon and planets, to eliminate the errors of the theory, and thereby remove the weeds, as it were, from Newton's celestial garden.

To know the exact moment in which a planet—moving, to all intents and purposes, frictionlessly through the void—reached a given point in its sweep through space and time was, in a sense, a way of checking to make sure the clockwork was running correctly and on time. Halley thought he saw how to do this by observing Venus during one of its transits and timing when its limb came into contact with the rim of the brilliant solar disk.

By rightly choosing the moment of observation, the planetary bodies could be forced to measure their own positions and reveal their relative distances, to almost any arbitrary degree of accuracy. In addition, the same set of observations offered the possibility of detecting the exact distance from the earth to the sun. Since through the Newtonian inverse-square law the forces between mutually attracting and disturbing bodies depended on the distances between their centers of mass, there emerged the hope that ever more fine-tuned and intricate balancing of the grand clock mechanisms of the planetary motions might be attained.

How the Transits Figure

It was this project—this obsessive quest to demonstrate the divinely appointed order of the heavens—which would become the defining scientific project of the "Celestial Century," setting off a kind of space race among the Great Powers to the far corners of the globe. What they aspired to seemed at the time to be the ultimate end of science: the quest for the absolute standard of measurement.

The hope was that the transits of Venus would provide it. As nineteenth-century commentator W. Stanley Jevons wrote: "The Sun forms a kind of background on which the planet is marked, and serves as a measuring instrument free from all the errors . . . which affect human instruments."[33]

Though guided by the most exacting of predictions and armed with instruments of the most absolute attainable precision, these expeditions were to prove sometimes dangerous and always difficult. The results would fall short of expectation, and the men who set out on them were often stalked by doubt and uncertainty. Not infrequently, for all their orderly methods and discipline, these followers of Newton and his grand clockwork universe would be undone by forces unpredictable and unavoidable, in regions of the earth that, despite the long reach and attempted sway of European empires, remained largely governed by hazard, chance, and havoc.

7

"This Famed Phenomenon"

We are alone with every sailor lost at sea
Whose drowning is repeated day by day. The sound
Of bells from buoys mourning sunken ships round
Us, warning away the launch that journeys you and me
On last Cytherean trips in spring.

—Dunstan Thompson, "Largo" (1943)

Halley's Grand Proposal

Horrocks, and briefly Crabtree through a break in the clouds, had been the only seventeenth-century humans—perhaps the only humans ever—to witness Venus's silhouette pass across the sun. Observing in "pathetic isolation," they long remained the only members of an exclusive fraternity, sole beholders of what Edmond Halley called "that sight which is by far the noblest astronomy affords."

Halley himself would be chiefly responsible for assuring that the next transit of Venus would be one of the most eagerly anticipated and widely observed celestial events of all time. In Horrocks's day, the motions of the planets were still being worked out. A century later, the secret of their motion—the law of gravitation—had been intimated by Robert Hooke and fully divulged by Isaac Newton. This changed everything; as Harry Woolf has written, "The bare mechanism behind the movement of things lay satisfactorily revealed, and the urge to complete the system of the world by discovering its actual dimensions took on a new significance."[1]

The transits of Venus of 1761 and 1769 would be pursued, then, not for the sole satisfaction of witnessing Venus silhouetted on the face of the sun, curious as that sight might be, but in the ardent quest of a number defining the scale of the solar system: the solar parallax. Its value would allow not only the distance to the sun to be ascertained but also the distance of every other object in the solar system.

The preparations for this quest took place on a scale unprecedented for any scientific endeavor of the time. "It is quite likely," notes Woolf, "that no other particular scientific problem in the eighteenth century brought so many interests to a single focus as the concern for the solar distance."[2]

Halley (see fig. 30) had been meditating on his method of using the transit of Venus to measure the solar parallax ever since he observed the transit of Mercury from St. Helena in 1677. That experience taught him that the duration of the transit could be observed exactly—he thought *very* exactly. But it was four decades before he

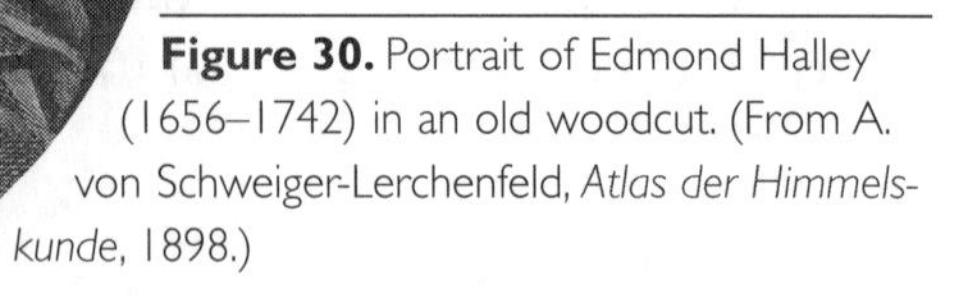

Figure 30. Portrait of Edmond Halley (1656–1742) in an old woodcut. (From A. von Schweiger-Lerchenfeld, *Atlas der Himmelskunde*, 1898.)

got around to publishing the concept in detail; until then, he had little leisure to pursue the matter. He had been the busy secretary of the Royal Society; he had guided the *Principia* through to publication; and then he had turned to calculating comet orbits. He had even sailed a large part of the world between 1698 and 1700, after the British Admiralty placed a war sloop, the *Paramour*, at his disposal. This was the first sea voyage ever undertaken for strictly scientific purposes; Halley charted a course to Spain, the Canary Islands, Africa, Brazil, and the West Indies in order to map the magnetic variations of the earth. He returned to England ahead of schedule, just as a mutiny, led by his second in command, was gathering.

Back in England, Halley assumed the position of Savilian professor of geometry at Oxford (see his home in fig. 31). He did so over the opposition of Flamsteed, who had complained, among other things, that Halley was "lewd" and had expressed himself coarsely, in the manner of a sea captain![3] Halley avenged himself by pirating and rushing into print Flamsteed's long-delayed star catalog, *Historia Coelestis Britannicae*, a deed enacted with Newton's connivance and even urging. When the pirated edition appeared in 1712, Flamsteed was horrified; he promptly burned every copy he could get his hands on. Still, Flamsteed never did finish the book. It was more than he could single-handedly accomplish. The final version was not published until after his death, by Joseph Crosthwait, one of his assistants.

Only in 1716 did Halley—at sixty, no longer a young man—finally get around to writing up his solar parallax proposal. His paper, published in the *Proceedings of the Royal Society*, sketched a procedure both marvelously straightforward and breathtakingly ingenious. There had been a hint in Horrocks's writings that he had much earlier grasped its gist: he referred to "certain and easy methods of proving the sun's distance and magnitude." No doubt he would have elaborated upon these methods, if only his life had not been cut so tragically short. The Scottish mathematician James Gregory (1638–75) had also foreshadowed Halley. In the scholium to the eighty-seventh problem of his *Optica Promota* (1663), where he describes how to determine the parallax of a planet from a close conjunction of two planets: "The problem has a very beautiful appli-

Figure 31. William Sheehan standing in front of Halley's house at No. 8, New College Lane, Oxford, where Halley devised the great method of determining the solar parallax by means of observations of the transit of Venus. (Photograph by Bernard Sheehan in 1987.)

cation, although perhaps laborious, in observations of Venus or Mercury when they obscure a small portion of the sun; for by means of such observations the parallax of the sun may be investigated."

For that matter, as soon as one makes the connection between transits and the solar parallax, the points Halley makes in his paper become almost self-evident to anyone trained to look at things from the point of view of a geometer. As Halley wrote in his paper,

> There are many things exceedingly paradoxical, and that seem quite incredible to the illiterate, which yet, by means of mathematical principles, may be easily solved. Scarcely any problem will

> appear more hard or difficult than that of determining the distance of the Sun from the Earth, very near the truth; but even this, when we are made acquainted with some exact observations, taken at places fixed upon and chosen beforehand, without much labour, be effected. And this is what I am now desirous to lay before this illustrious Society (which I foretell will continue for ages), that I may explain beforehand to young astronomers, who may perhaps live to observe these things, a method by which the immense distance of the Sun may be truly obtained to within a five-hundredth part of what it really is.[4]

Halley's method could be applied to transits of either Mercury or Venus, although those of Venus later proved the much more useful. Halley called for observers of future transits to be stationed in at least two—and preferably more—places scattered over the earth's surface and separated widely in latitude. The only prerequisite as to longitude was that the observers be located in the zone where the entire transit, from beginning to end, would be visible.

Because of the latitude difference of the observers, Venus would appear to move along chords of different length over the disk of the sun (see fig. 32). The motion of Venus being nearly uniform, the length of each chord would be proportional with the duration of the transit. Thus, observers would not actually have to *measure* anything; they would only have to *time* the transit. Fortunately, existing pendulum clocks were more than sufficiently accurate for this purpose.

The greatest strength of Halley's method was that it did not require special instruments or any other equipment than, as Halley pointed out, "common telescopes and clocks, only good of their kind; and in the observers nothing more is needful than fidelity, diligence, and a moderate skill in astronomy. . . . It is sufficient . . . if the times be well reckoned from the total ingress of Venus into the Sun's disk to the beginning of her egress from it."

The duration of the transit was determined from the exact moment when Venus's orb had just moved fully onto the sun's disk—the limbs just touching—to the corresponding moment when Venus had moved off the sun.

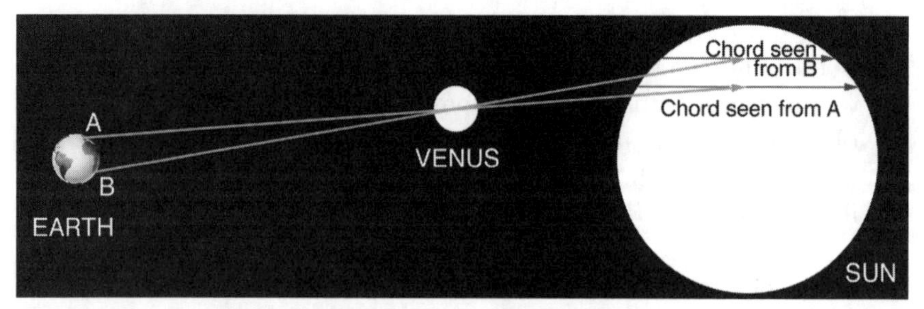

Figure 32. Halley's method for determining solar parallax: chords across the sun from two stations, A and B, widely separated by latitudes within the zone from which the transit is visible in its entirety, traverse chords of different lengths (and thus different durations) when crossing the sun's disk. (Diagram by John Westfall.)

HALLEYAN AND CISHALLEYAN POLES

To accentuate the difference in the chord lengths, hence the accuracy of the method, the observing stations needed to be situated at points where the entire transit was visible, in latitudes as far apart as possible. Ideally, one station should be at the point on the earth of maximum duration; the other should be at the point of minimum duration. These two points are called the *Halleyan Poles*. If Venus passes north of the center of the sun's disk, the Halleyan Pole of longest duration (H) lies in the Arctic or North Temperate Zone; that of shortest duration (H') lies in the Antarctic or South Temperate Zone. The opposite is true when Venus passes south of the center of the solar disk.

The Halleyan Poles themselves, being geographically so far-flung, were not always accessible. There was the added inconvenience that from one Halleyan Pole the sun would be at the horizon for both ingress and egress. Due to atmospheric distortions, clouds, and so on, this would effectively ruin the observation. From the other Halleyean Pole, the sun would actually be below the horizon! In that case, for practical purposes, the ideal location would be the point on the earth's surface closest to the ideal point where both internal contacts can be seen, which we have defined as the *Cishalleyan Pole*. Even there, the sun will be on the horizon.

In practice, observers usually tried to locate within, say, one to two thousand kilometers of a Halleyan Pole, rather than at its exact location. Areas notable for clouds and unfavorable weather were obviously to be avoided, but meteorological information was very hard to come by in the eighteenth century.

Another problem with Halley's method is that, though Halley stated that accurate longitudes were not required, serious errors in station positions throw off the distances calculated between pairs of stations. These errors in turn affect the value derived for the solar parallax.

One last circumstance must be pointed out. The difference in chord lengths is greatest for marginal transits, particularly those where the planet nearly grazes the sun's limb, where a small shift in track position creates a large change in track length. Thus off-center transits will provide the most favorable application of Halley's method. (Working against this advantage is the very slow rate of entry and exit of the planet from the sun's disk, which makes precise timing difficult; as was glaringly obvious at the "graze transit" of Mercury in 1999.) At the other extreme is a transit where the planet passes along a nearly central chord, in which case transit durations are nearly equal worldwide, making Halley's method much less accurate.

In a transit of Venus or Mercury, there occur four contacts of the planet's limb with that of the sun. Contacts I and II constitute *ingress*; contacts III and IV, *egress* (see fig. 33). The first and last contacts are called *external contacts* and are difficult to time accurately.

For an observer applying the Halleyan method, the *internal* contacts became the all-critical points. Halley, indeed, had suggested they might be determined to practically any degree of accuracy. At St. Helena, during the transit of Mercury, he found that "the interval of time, during which Mercury . . . appeared within the Sun's disk, even without an error of one second in time. For the lucid line intercepted between the dark limb of the planet and the bright limb of the Sun, although exceedingly fine, is seen by the eye, and the little dent made on the Sun's limb, by Mercury's entering the disk, appears to vanish in a moment; and also that made by Mercury leaving the [Sun's] disk seems to begin in an instant."[5]

Unfortunately, since what is obtained from the observations of a transit is the relative parallax of the two bodies, the planet and the

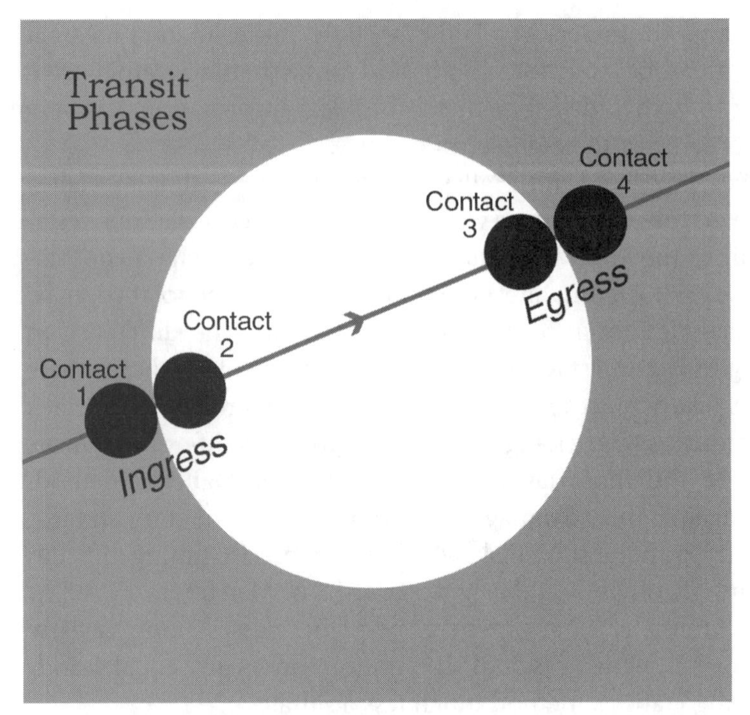

Figure 33. Sample transit (of a rather large planet), showing the two phases and the four contacts as the planet moves left to right across the sun's disk. (Diagram by John Westfall.)

sun, it turns out that transits of Mercury are not really very useful; the planet is simply too close to the sun for the relative parallax to be readily measured. Even Halley himself saw this; transits of Mercury were, he wrote, "not to be looked upon as fit for our purpose." That left the very rare transits of Venus.

The opportunity of observing a transit of Venus would not, alas, occur during Halley's own lifetime, as he realized: "On account of the very strict laws by which the motions of the planets are regulated, Venus is seldom seen within the Sun's disk; and during the course of 120 years it could not be seen once—namely, from the year 1639 (when this most pleasing sight happened to that excellent youth Horrocks, our country man, and to him only since the Creation), to the year 1761, in which year, according to the theories which we have hitherto found agreeable to the celestial motions, Venus will again pass over the Sun on May 26 [O.S.; June 6, N.S.]."[6]

Halley foresaw that the Halleyan Poles for the 1761 transit would lie in very inconvenient places, as shown in the map in figure 34. One fell in the wasteland of Siberia; the other was in the untraveled reaches of the southern Indian Ocean, one of the cloudiest and stormiest climatic zones on the whole earth! The ingress would not be visible from England at all. It would be visible, however, in the frigid Arctic, where the sun, nearing the June solstice, would not set all night: along the north coast of Norway, beyond the city of Nidrosia (now Trondheim) as far as the North Cape.

At the mid-point of the 1761 transit, the sun would stand overhead in the East Indies. At Hudson Bay Venus would enter the sun's disk just after sunrise and leave it just before sunset. Combining observations from these two locations would maximize the latitude-difference required for implementation of Halley's method. Halley concluded in his 1716 paper,

> If . . . it should happen that this transit should be properly observed by skilful persons at both these places, it is clear that its duration will be seventeen minutes longer as seen from Port Nelson [now Churchill, on Hudson Bay], than as seen from the East Indies. Nor is it of much consequence (if the English shall at that time give any attention to this affair) whether the observation be made at Fort George, commonly called Madras, or at Bencoolen, on the western shore of the island of Sumatra, near the equator. But if the French should be disposed to take any pains herein, an observer may station himself conveniently enough at Pondicherry, on the west shore of the Bay of Bengal [twelve degrees north of the equator]. As to the Dutch, their celebrated mart at Batavia will afford them a place of observation fit enough for the purpose. . . .
>
> And indeed I could wish that many observations of this famed phenomenon might be taken by different persons at separate places, both that we might arrive at a greater degree of certainty by their agreement, and also lest any single observer should be deprived by the intervention of clouds of a sight which I know not whether any man living in this or the next age will ever see again; and on which depends the certain and adequate solution of a problem the most noble, and at any other time not to be attained

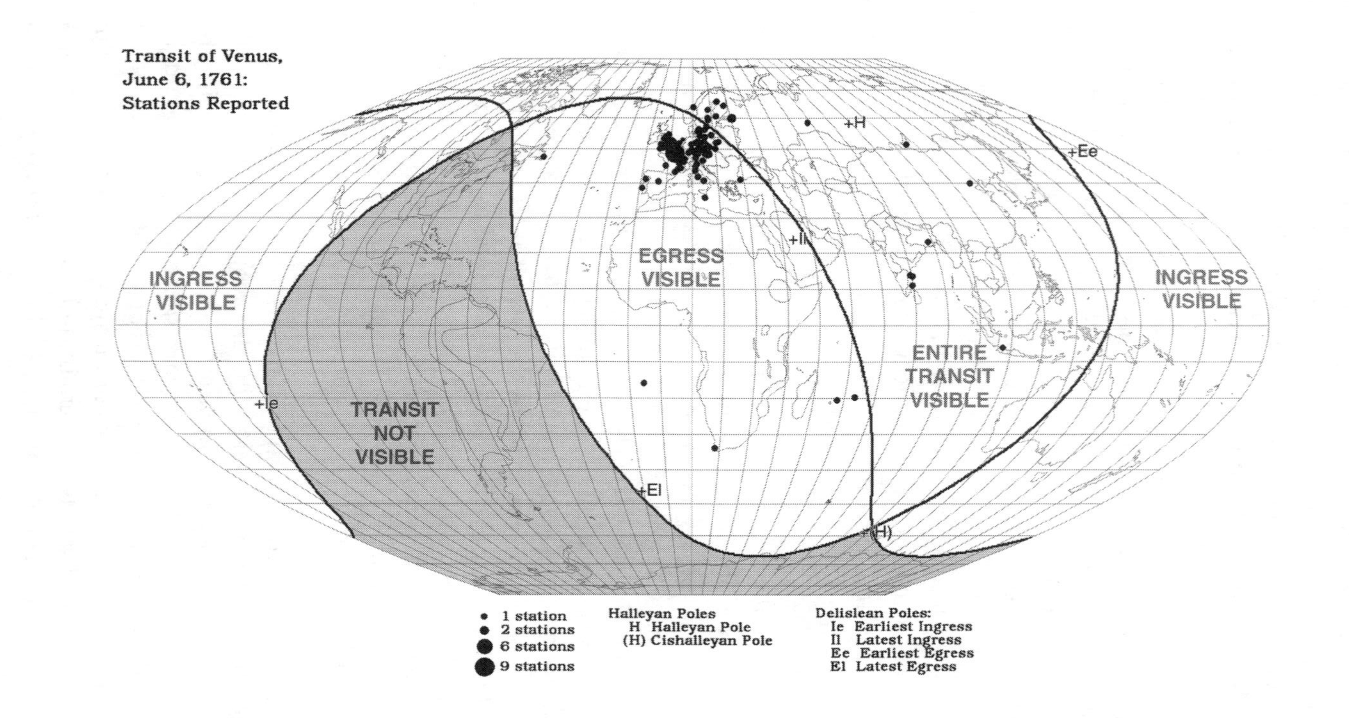

Figure 34. Visibility zones, Halleyan and Delislean Poles (see "Delisle Takes Over," below), and observing stations with published results, for the transit of Venus of June 6, 1761. (Map by John Westfall.)

> to. I recommend it therefore again and again to those curious astronomers who (when I am dead) will have an opportunity of observing these things, that they would remember this my admonition, and . . . not by the unreasonable obscurity of a cloudy sky be deprived of this most desirable sight, and then, that having ascertained with more exactness the magnitudes of the planetary orbits, it may redound to their immortal fame and glory.[7]

The Man of Vision Lives Not to See

At Oxford, Halley seems to have made no further investigation into the subject; nor did he take it up again when, three years later, he moved to Greenwich as Flamsteed's successor as Astronomer Royal.[8]

Halley's vision had always stretched farther than most men's beyond the limits of his own lifetime. He had predicted that the comet of 1682 would return in 1758; this was the one that had cleared the air so that Newton could properly begin his disquisition on the law of gravitation. "Wherefore," Halley noted, "if according to what we have . . . said it should return . . . candid posterity will not refuse to acknowledge that this was first discovered by an Englishman." His method for the transit of Venus similarly laid the groundwork for an observing program that would come to fruition, as he realized, "when I am dead."

Both of these great events transcended Halley's lifetime. He knew only too well that he could not defy the mortality tables that—yet another of his contributions—he himself had first calculated. He died at the age of eighty-six, on January 14, 1742, nineteen years before the transit of Venus and sixteen years before his comet returned. Having asked for a glass of wine, he quaffed it, and without uttering another word, slumped over in his chair. Whereas Newton, who had died in 1727 at age eighty-four, was buried amid the pomp and circumstance of Westminster Abbey, Halley was laid to rest more humbly, beside his wife Mary, in the church of St. Margaret, Lee, at Blackheath, about a mile from the Royal Greenwich Observatory. All the same, his memory will always be kept fresh, the torch of it borne tirelessly by his enduring comet.

DELISLE TAKES OVER

Halley's prestige ensured that his method for measuring the transit of Venus would be developed by other astronomers. Prominent among them was the great organizer-in-chief of French transit operations, Joseph-Nicolas Delisle (1688–1768; see fig. 35). Historically, he was connected with Halley, whom he had visited in London in 1723, shortly before Voltaire's visit to England, a time when French admiration for English ideas—including Newtonianism—was on the upswing.

Delisle belonged to a family of historians, geographers, and astronomers. His early education was provided by his father and the faculty of the College Mazarin. But Delisle traced his interest in the heavens, as many budding astronomers have done, to the experience of witnessing an eclipse of the sun, in his case that of 1706. At first he planned a career as a professional geographer (his brother was First Geographer of Louis XIV). He became skilled in producing maps, which served him well when later he was required to prepare maps of zones of visibility for the transits of Venus. However, he failed to obtain the first post he sought—Surveyor Royal of Martinique. On the other hand, the disappointment had a positive effect: it threw him into the arms of astronomy, a field closely tied to surveying and navigation.

At age twenty, Delisle began regularly to frequent the Paris Observatory. Under a virtual dynasty, the observatory's directorship had passed from Giovanni Cassini to his son, Jacques (1677–1756), and would later pass to his grandson, César François Cassini Thury (1714–

Figure 35. Portrait of Joseph-Nicolas Delisle (1688–1768). (Courtesy of Yerkes Observatory.)

84), then to his great-grandson, Jacques Dominique de Cassini (1748–1845), making four generations of astronomical Cassinis.

Delisle soon made plans for his own observatory. He received permission to use the cupola of the Palace of Luxembourg, but only if he provided his own instruments. He was evicted soon afterwards when the Duc d'Orleans' older sister, the Duchesse de Berry, took up residence in the palace. Later, Delisle was freed of financial worry when he was appointed to the chair of mathematics at the Collège Royal, the position that Gassendi had once held. A tireless letter writer, he entered into a wide correspondence with astronomers throughout Europe. He even came to the attention of Peter the Great, the czar of Russia. As part of Peter's program of westernizing Russia he hoped to found a school of astronomy and an observatory at St. Petersburg. Delisle was intrigued by Peter's offer, but before he could decide, Peter had died.

Instead, Delisle went to England, arriving shortly after the Mercury transit of 1723. That event, coupled with his meeting Halley, awakened his preoccupation for the rest of his life.

The more Delisle studied Halley's method, the more he recognized its inadequacies. Though it only required timing the duration of the transit, it could be applied only in those places where the transit was visible from beginning to end. If clouds blocked the view of either ingress or egress, the observations would be useless. Observers would have to be sent out to remote locations such as, in 1761, Bencoolen and Hudson Bay. Then they were at risk of failure unless they received the full cooperation of the weather over a period of several hours. In addition, for the six transits of Venus between 1761 and 2012, the difference of minimum and maximum duration at the Halleyan and Cishalleyan Poles varies from about 13 minutes to 39 minutes. The 1761 transit had the smallest difference and thus offered poor prospects for the Halleyan method (see Table B1 in appendix B).

These were serious disadvantages, fatal in Delisle's eyes. Instead, he envisioned a different approach: observers at a pair of stations would both time the same phase of the transit, either the beginning or the end of the transit; timing both was not necessary. Thus there was a greater likelihood the weather would cooperate. Inevitably, there was a trade-off, since now observers had to work out the exact longitude of their station. The entire success of Delisle's method, in fact, turned upon observation of the absolute time of contact (e.g., in terms of Greenwich or Paris time) at both stations. As observers could time either the ingress or the egress contacts, there were actually four Delislean Poles, which for the 1761 transit were as follows (refer to fig. 34):

- Earliest ingress (Ie): in the South Pacific, near yet-undiscovered Pitcairn Island.
- Latest ingress (Il): in the central Arabian Peninsula.
- Earliest egress (Ee): in the North Pacific, south of Attu in the Aleutian Islands.
- Latest egress (El): in the South Atlantic, roughly midway between the tips of Africa and South America.

Looking at the six locations of the Halleyan and Delislean Poles, the expedition planners were made keenly aware of the fact, only too familiar to modern eclipse chasers, that celestial alignments are in no way planned for human convenience!

There was now to be a long interlude. On returning to France, Delisle was approached again about coming to Russia, this time by Peter's second wife, the Lithuanian peasant-turned-czarina Catherine I (not to be confused with the later czarina Catherine II, called "the Great"). He accepted, perhaps not expecting to stay long—the Collège Royal granted a leave of absence of four years, during which he could still reclaim his position. As it turned out, he remained in Russia for twenty-two years.

BIANCHINI OBSERVES VENUS

While Delisle was away in Russia, in 1726–27 an Italian astronomer and priest named Francesco Bianchini (1662–1729) carried out the first careful telescopic study of Venus since the time of the first Cassini. Bianchini's telescope had been made by the Roman instrument maker Campani, and he sometimes set it up in the Farnese Gardens, on the Palatine Hill. At other times he made observations in the Alban Hills, southeast of Rome.

Until Bianchini set to work, the telescope had revealed little concerning our brilliant sister planet. Bianchini regarded even Cassini's bright spot of 1666–67 as the product of an instrumental defect. Apart from that, the record had been a blank. The telescope had readily shown the planet's phase, but otherwise its great brilliance had effectively rebuffed the attempts of observers to make out something definite in form and feature. Christiaan Huygens, who along with G. D. Cassini was one of the greatest telescopic observers of the seventeenth century, had been responsible for discovering that Saturn was surrounded by a thin, flat ring and for making out markings on Mars that pointed to its nearly 24-hour rotation period, but he was baffled by Venus. He even wondered whether it had an observable solid face at all: "What then, must Venus have no Sea, or do the Waters reflect the Light more than ours do, or their Land less? or rather (which is most probable in my opinion) is not all that Light we see reflected from an Atmosphere surrounding Venus, which . . . hinders our seeing anything of the Globe itself, and is at the same time capable of sending back the Rays that it receives from the Sun?"[9]

In the pellucid skies of Italy, Bianchini seemed at last to catch actual features on Venus. He made out various rounded patches scalloping the terminator—the line between day and night (see fig. 36). "They were similar," he wrote, "to the larger lunar ones which can be seen with the naked eye. . . . In order to see these markings more clearly it is necessary to choose days free from mist, but also to wait until twilight is more advanced."[10] Convinced of their reality, he gathered them into a map, which would serve as the foundation, he hoped, of a new branch of science—"celidography" he called it (an awkward coinage from the Greek word for "spots").[11] From the spots' movements, he attempted to deduce the rotation period of the planet. His result: 24 days, 8 hours.

Excitement over the results of Bianchini's observations rippled through the astronomical world. A Jesuit colleague, Melchior Briga,

Figure 36. Venus drawings by Francesco Bianchini (1662–1729), based on his observations using using a 2½-inch aperture, 66-foot focus refractor. (From *Hesperi et Phosphori Nova Phaenomena* [Rome, 1728].)

wrote to him: "Is it really true? Am I forced to admit that real marks have been seen by mortal eyes on the Mirror of Divine Beauty, the brightest of all bodies . . . the planet Venus? . . . But then why should not the heavenly Venus have lakes of greater or lesser extent on her surface [just like the Earth]?"[12] At Paris, Jacques Cassini, unable to make out any of the spots with his telescope, attempted to reconcile Bianchini's observations with those of his father. In doing so, he reworked the rotation to a satisfyingly terrestrial period—23 hours, 21 minutes—a result that would remain in the literature of the planet for a century and a half.

Anticipation Mounts

Meanwhile, Delisle, aged fifty-nine, returned to France in 1747. Louis XV, wanting to ensure that he remained this time, appointed him *Astronome de la Marine* (Naval Astronomer) and granted him a lifetime annuity. Delisle established as his base of operations the Hôtel de Cluny, near the Sorbonne.

His main preoccupation was now the solar parallax. Unlike Halley, he had not given up entirely on transits of Mercury as a means of establishing its value. A transit of Mercury was due in 1753, when the whole question could be put to the test. Delisle and other astronomers were also well aware that the Mercury transit could serve as a dry run for the upcoming event involving Venus. "During most of 1752," notes historian Harry Woolf, "the circulation of letters dealing with the transit among astronomers must have been very heavy indeed, if the extant material in Delisle's papers is a fair sample."[13]

By far the most ambitious plan for observing the transit of Mercury was undertaken by French abbé Nicolas-Louis de Lacaille, who received permission from the Dutch East India Company to observe from the Cape of Good Hope. In addition to observing Mercury, he remapped the stars of the southern skies and measured the length of a meridian in the attempt to ascertain the exact shape and dimensions of the earth. In the end, Lacaille's observations of the transit of Mercury did not allow a determination of the solar parallax, and henceforth even Delisle was convinced that only the rare transits of Venus would lead to that number, the long-sought holy grail of astronomy.

The director of the Paris Observatory, César François Cassini de Thury, put the matter well: "This alone could resolve absolutely our uncertainties. . . . Happy our century! to which is reserved the glory of witnessing the one event which will render it noteworthy forever in the annals of the sciences!"[14]

Between the transit of Mercury in 1753 and that of Venus in 1761, at least one other notable astronomical event occurred. As noted, Halley had predicted that the comet of 1682 would reappear in 1758. As the date approached, Delisle took a leading role among those hoping to recover it. He produced a series of charts of the sky showing the zones where he expected it to be visible and furnished them to a keen-eyed assistant, Charles Messier (1730–1817). Without much formal training and originally hired not for his astronomical expertise but for his neat handwriting and skill in drafting, Messier proved to be an enthusiast. He dutifully swept the skies for the comet with a 4½-foot Newtonian reflector housed in the cupola of the octagonal tower at the Hôtel de Cluny. When the event occurred, their plans came to naught. Delisle and Messier were forestalled by the discovery of the comet on Christmas night of 1758 by a German farmer and amateur astronomer, Johann Palitzsch (1723–88). Despite this disappointment, Messier went on to a splendid career as a discoverer of comets—he found thirteen of them, so many that Louis XV is said to have referred to him as the "ferret of comets." He achieved even greater fame as the cataloguer of diffuse objects, many of them now known as *nebulae*, named for the Latin word for 'clouds.' They were scattered around the heavens, and he took note of them simply because he did not wish to confuse them with comets.

In addition to the astronomical distraction of Halley's comet, there was also an international distraction. The measure of the solar parallax was by its very nature a project that transcended national interests. Its success depended on the ability of countries to safely and expeditiously send a diaspora of observers around the globe and to coordinate a mass of observations. Delisle had been at the center of French preparations. The Royal Society in London made similar, though more embyronic, plans. Empires on which "the sun never sets," so extensive that some portion of them always is sunlit—true of the British Empire (and later Commonwealth) since 1685, as was the case for Spain between 1570 and 1821, Holland between 1612 and 1947, Portugal between 1674 and 1822, and France from 1853 to the present—were more than national vanities when it came to planning global astronomical expeditions. This status would be very convenient for nations hoping to observe the transits of Venus from stations confined largely to their own territories.

Astronomers themselves had every intention of cooperating fully with one another. In fact, Delisle himself had written to the

Royal Society pointing out that the transit would be observable from the English colonies, and an exact observation made there, combined with an observation from the East Indies, would be of the "utmost consequence" for the calculation of the solar parallax. It nevertheless remains true that at the time of the eighteenth-century transits, as well as of those a century later, the majority of observers of whatever nationality tended to observe from their own territory, or from one of their colonies.

At this moment—when international cooperation was critical—the Seven Years' War, as it is generally known to European historians, broke out. It was in fact a series of confused and incoherent European conflicts that would draw in all the major powers of Europe. The background of the war included the attempt by the Austrian Habsburgs to regain from Prussia the fertile agricultural province of Silesia (now mostly part of Poland). In the winter of 1755–56, the Austrian statesman Count Wenzel von Kaunitz, hoping to reverse the loss, assembled a coalition of Austria, Saxony, Sweden, Russia, and France against Prussia. But before they could act, Frederick the Great of Prussia seized the initiative by invading Saxony. At the time, Britain was already fighting France in a colonial war in North America, called the French and Indian War. The two global superpowers, Britain and France, were also wrestling for supremacy in India, and Britain joined the Prussian side.

In May 1759—just as Halley's comet was as its brightest, soon to fade and not to return to visibility for another three-quarters of a century—the British Navy captured Guadeloupe, the richest sugar island of the West Indies. In July the British seized forts Ticonderoga and Niagara, and in September, Brigadier General Wolfe sailed up the St. Lawrence to begin the siege of Québec, after which the huge province of Canada exchanged hands. That vast region, favorably disposed by the circumstances of celestial mechanics for the Halleyan method of determining the solar parallax, was now opened up for exploitation by the British as a site for their possible expeditions to observe the upcoming transit. A month later, George III—a man of strong emotions but slow intellectual development—became king of Great Britain and Hanover.

Because of the war, French and British preparations for the transit were forced to go their separate ways. Instead of being carried out in cooperation, the expeditions would be sent out in a spirit of intense rivalry, and once hostilities were declared, the planning for them became infinitely more complicated and hazardous.

The Royal Society proposed one major expedition for observing the transit of Venus: Greenwich Observatory astronomer (later Astronomer Royal) Nevil Maskelyne (1732–1811) and his assistant Charles Mason (1730–87) were to travel to St. Helena. Meanwhile, the French— thanks largely to Delisle's early strategic vision and persistent efforts—made more ambitious plans. Delisle carried out extensive calculations and drew up a *mappemonde* (map of the world) showing where the various contact points of the different phases of the transit would be visible (see fig. 37). The map was earmarked for international distribution. It was intended to direct a fleet of observers to points on the globe far-flung in latitude in order to cover the contact points from as many locations as practicable.

Unfortunately, Delisle did not finish his calculations until August 1760. Then he found that Halley's predicted track for Venus was in error. By that point, though, the transit was only ten months away.

Before the British committed to St. Helena, the French considered observing from one of the numerous Portuguese and Dutch trading colonies located between Angola and the Gold Coast in south-

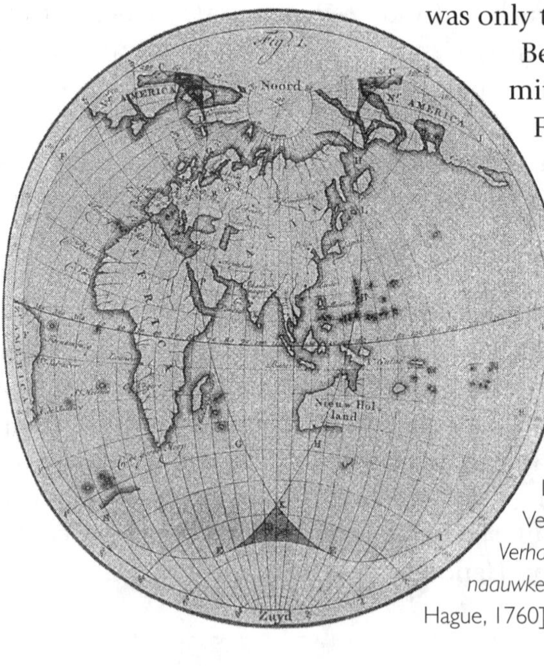

Figure 37. *Mappemonde* of Delisle for the 1761 transit of Venus. (From Dirk Klinkenberg, *Verhandeling, beneffens de naauwkeurige algemeene* ... [The Hague, 1760].)

western Africa. They also eyed Prince's Island and St. Thomas Island, in the Gulf of Guinea, and the East Indies, at one extreme end of the transit's zone of visibility.

Eventually, the African coast was abandoned, either because of unfavorable weather prospects or failure to secure the cooperation of the Portuguese and the Dutch. The French opted for the East Indies, and so a smattering of astronomers were duly dispatched there. One of them was Alexandre-Gui Pingré (1711–96), a member of the Order of Saint Genevieve and a professor of theology in Paris. He got into trouble with the church hierarchy because of his unorthodox religious views; in retaliation he was relegated to an elementary school, where he was assigned to teach Latin grammar. There, too, his doctrine was regarded as suspect, and he received five lettres de cachet in five years. He was rehabilitated at the age of thirty-eight, when he became astronomer for the Academy of Sciences at Rouen. After a few years, he was elected a member of the Academy of Sciences in Paris, named librarian of Saint Genevieve, and given a small observatory for his use by the Order that he had earlier offended. For the transit of Venus, Pingré would sail to Rodrigues Island, in the Mascarene Islands in the western Indian Ocean.

Though the site was under French control at the time, Pingré requested a letter of safe conduct from the British Admiralty. "Since . . . the said Monsieur Pingré should not meet with any Interruption either in his passage to or from that Island, you are hereby most strictly required and directed not to molest his person or Effects upon any Account, but to suffer him to proceed without delay or Interruption." As we shall see, the letter would not do him much good.

Another Frenchman, Guillaume Joseph Hyacinthe Jean Baptiste Le Gentil de la Galasiere (1725–92), planned to observe the transit from Pondicherry in India.

In 1761 Siberia held the Halleyan Pole of maximum duration. Accordingly, both Pingré and the abbé Jean Chappe d'Auteroche (1728–69; see fig. 38) had volunteered to make the trip from St. Petersburg to Siberia with Russian astronomers. After some discussion, Chappe won. He would go to Tobolsk, Siberia, which left

Figure 38. Portrait of Jean Chappe d'Auteroche (1727–69). (From Chappe d'Auteroche, *Voyage en Californie pour l'observation du passage de Vénus . . .*, 1772.)

Pingré free to undertake his Indian Ocean adventure. Chappe, born in 1728 of a noble and wealthy family of Mauriac (Auvergene), was educated by the Jesuits and received his mathematical and astronomical training at the Paris Observatory. He became one of the astronomers most dedicated to chasing the shadow of Venus over the earth; in the end, pursuit of this goal would cost him his life.

Delisle proposed that Messier travel on board a Dutch East India Company ship to Batavia (now Jakarta), on the island of Java, but nothing came of the proposal. Instead of traveling to the East Indies, Messier remained at the Hôtel de Cluny, where he could continue his comet sweeping.

At almost the last minute, the Royal Society added another British expedition, to Bencoolen (now Bengkulu), on the island of Sumatra, a place under the control of the Dutch East India Company. Charles Mason, previously assigned to assist Nevil Maskelyne at St. Helena, was offered the position of leading it. Enticed with an increase in pay and a more generous allowance of liquor (!), he accepted. Meanwhile, a hurried search for an assistant turned up Jeremiah Dixon (1733–79), a land surveyor from County Durham, Ireland. Thus Mason and Dixon—names later indissolubly linked in the famous Mason-Dixon Line, between Pennsylvania and Maryland, which they surveyed together in 1763—were first brought together.

These were the leaders of the main expeditions. Others planned to watch the transit closer to home. Cassini de Thury, for instance, proposed to travel only as far as the Vienna Observatory, then under the direction of Jesuit astronomer Father Maximilian Hell (1720–

92). Pierre-Charles Lemonnier (1715–99), a close associate of Pingré, planned to stay put in Paris, but because of Louis XV's desire to see the transit had to lug his instruments to the Château de Saint-Hubert and carry out his measurements with the king staring over his shoulder.

Swedes and Russians were, of course, most favored by the geographical circumstances, since they could observe the transit in its entirety without traveling outside their borders. Though the French would have the largest contingent of observers, Sweden was second, surpassing even Britain. John Winthrop (1714–79), a professor at Harvard University, would lead an expedition from Boston to St. John's, Newfoundland. The late acquisition of Canada did not allow sufficient time to plan any British expeditions to any more remote parts of that large and favorably placed region.

Instrumental Interlude

In the 123 years between the 1639 and 1761 transits, telescopes had been significantly improved. Traditional refractors, having grown longer, were less suitable for distant overseas voyages. However, by the mid-eighteenth century, there were several alternatives. The first, the "Newtonian" reflector, developed by Isaac Newton, used a mirror instead of a lens to create an image. In order to deflect the image out the side of the tube where it can be examined with an eyepiece, a small flat mirror is used. A different form of reflector, using a concave mirror as a secondary, directing the image through a central hole in the primary mirror, had been designed by Scottish mathematician James Gregory even before Newton introduced his reflector. Hooke built the first working model in 1674. The advantage of using a curved-mirror secondary instead of a flat is that the mirror spacing is decreased—the light-path is "folded." This makes such a telescope more portable and convenient to use than a Newtonian. Unfortunately, Gregorian reflectors are difficult to construct, though a number of eighteenth-century transit observers did use them. A more popular type of "folded-optics" introduced a convex instead of a concave secondary. This design had been proposed in 1672 by one of Louis XIV's sculptors, Guillaume Cassegrain (1625–1712). The Cassegrain, especially in the modified Schmidt-Cassegrain form, developed in the twentieth century, is widely used by amateur and professional astronomers to this day. Transit observers using reflectors employed either Gregorians or Cassegrains, but not Newtonians.

Also by the mid-eighteenth century, a more portable form of refractor was coming into use, based on the invention of two-element "achromatic" objectives, using compound lenses of crown and flint glass to partially eliminate the chromatic aberration that had been so troublesome with the early single-element refractors. John Dollond (1706–61) was manufacturing high-quality achromats by the 1750s, but the innovation was only beginning to take hold at the time of the 1761 transit, when only three observers used them (Klingensterna in Sweden, Christian Mayer in Germany, and Lacaille in France). By 1769, the number had increased to twenty-six.

As better telescopes were being used, on more suitable mountings, the number of institutions worthy of the name of "observatory" also increased. By the 1760s, there were about eighty observatories in existence, almost all of them in Europe. Only the national observatories at Greenwich and Paris and several manned by Jesuit astronomers, such as that in Vienna, along with a few associated with universities, would be considered professional observatories in the modern sense of having paid staff; most others would be considered amateur establishments.

GLOBETROTTERS

Besides these more or less permanent observatories, the various temporary stations chosen for expeditions would, if all went well, allow for good coverage of the transit.

The first expedition to get under way was that of Le Gentil, who sailed from Brest in the fifty-gun *Le Berryer* on March 26, 1760. Mason and Dixon were next, departing Plymouth on the British warship HMS *Seahorse* in December. They were only hours out of Plymouth when the *Seahorse* was fired upon by the French frigate *Le Grand* and beat a retreat back to Plymouth with eleven dead and thirty-seven wounded. Repairs were undertaken immediately by the British Admiralty, which proposed sending the *Seahorse* out next time with an escort. Understandably, Mason and Dixon were beginning to reconsider the advisability of going anywhere at all. In a letter to the Royal Society, Mason mentioned "the Uneasiness I have my self, to see the Uncommon Misfortune that have attended our designs, and the sea sickness besides; have affected me in an Unusual Manner." He concluded, "We will not proceed thither, let the Consequence be what it will."[15]

The Royal Society took a different view of the subject and answered sharply, by return mail: "Their refusal to proceed upon this Voyage, after their having so publickly and notoriously ingaged in it . . . [would] be a reproach to the Nation in general, to the Royal Society in particular, and more Especially to themselves." Moreover, it could not "fail to bring an indelible Scandal upon their Character, and probably end in their utter Ruin."[16] The threat of prosecution led Mason and Dixon to reconsider. On February 3, 1761, signing their letter "dutiful servants," they announced they would set sail for Bencoolen in Sumatra without further delay; they embarked that very evening.

Meanwhile, Pingré had left Paris in November after a fond farewell dinner with friends, including Lacaille—"without appetite, he was terrified by the forthcoming voyage and fearful that he might be seeing his friends for the last time."[17] From Paris, he set out for the port of Lorient, on the coast of Brittany, and joined by his assistant Thuillier, boarded the *Comte d'Argenson* in early January. Only a day off the Brittany coast, as Pingré lay in his cabin complaining of lack of sleep from an attack of gout he had suffered the night before, the *Comte d'Argenson* came in sight of a British fleet. Fortunately, the captain managed to elude the hostile fleet by tacking with the wind and taking advantage of poor visibility during the long winter night.

Chappe had left Paris at about the same time. The encumbrance of his instruments made him hope to travel as much of the way to Tobolsk as possible by sea. However, war forced him to take an overland route through Vienna and Warsaw to St. Petersburg. In the end, his journey was to prove as perilous as any experienced by the overseas expeditions. He left Vienna in January, in the depth of winter. At first, ice made travel difficult, but he made up time by using sleds. He arrived at St. Petersburg in mid-February, only four months before the transit. Some of the members of the Russian Academy tried to talk him out of going to Tobolsk and to settle for a more accessible site; but Chappe was committed to his original plan. He and his entourage left St. Petersburg and hurried on to Moscow, making the journey in four days by sleds, which had to be replaced after they were damaged beyond repair by the pounding they took on the indifferent roads. Now the journey from Moscow to Tobolsk became a race against time. Chappe traveled along the ice pack of the Volga, still frozen and as "smooth as glass,"[18] between Nizhni Novgorod (Gorky) and Kozmodemiansk. He then crossed the Urals near Solikamsk and followed the narrow road, "bordered by dense and lonely forests of birch and pine . . . the isolation broken [only by the occasional] relief station for the horses and a hamlet of two or three houses."[19] In Siberia, he was abandoned by his entourage and found himself all alone in the wolf-infested forest. Setting out with pistols and nerves of steel, he recovered the deserters and continued his journey. He finally reached Tobolsk in April, just as the steppes were melting into mud. There had been scarcely any margin for error: a few days later the pack ice from the river broke up for the season.

Chappe set up an observatory on a nearby mountain, protected by armed Cossacks to keep the local peasants at bay—they were inclined to blame this foreigner, equipped with instruments pointed at the sun, for the unusually severe spring floods. From this station, he observed lunar and solar eclipses, from which he worked out his longitude.

Meanwhile, Pingré was making slow progress toward the East Indies. The boredom of his long sea journey was broken by attempts to measure longitude at sea and to help himself to liquor, which, he observed with the philosophy of the true sailor, "gives us the necessary strength for determining the distance . . . from the Sun."[20]

As Chappe arrived in Tobolsk, Pingré's ship, the *Comte d'Argenson,* had just rounded the Cape of Good Hope and was headed into the Indian Ocean. There it encountered a French ship, the *Lys*, damaged in a skirmish with the British. Since the rank of the *Lys*'s captain was superior to that of the *Comte d'Argenson*, the latter was pressed into serving as the *Lys*'s escort to the Isle de France (Mauritius). Pingré arrived there with only a month to spare until the transit. At once he changed ships and set out for Rodrigues Island, landing only four days before the transit was due to take place. He had time only to erect the crudest of shelters for his 18-foot telescope, and finding that some of the other instruments had become rusty due to the sea air on the long voyage, he lubricated them with local turtle oil just before the transit was set to begin.

At about the same time, Mason and Dixon had been wending their way to the cape; it had taken them three months to reach the tip of Africa. Once there, they learned that Bencoolen had fallen into French hands. Resolving to make the best of the circumstances and, frankly, never that keen on Bencoolen to begin with, they decided to stay put and observe the transit at Cape Town.

Le Gentil and *Le Berryer*, with a long headstart on Mason and Dixon as well as Pingré, had long since rounded the cape on the way to India. The first part of his voyage, writes Woolf, "was rather uneventful, save for the loss of a fellow passenger by suicide and the pursuit by an English fleet near the Cape."[21] In escaping the English fleet, *Le Berryer* headed into the Channel of Mozambique around Madagascar. It anchored at the Isle de France in July 1760; the transit was still almost a year away. Now Le Gentil began to have forebodings of trouble to come. Karikal, a French settlement on the Coromandel Coast south of Le Gentil's destination of Pondicherry, had been captured by the British, and Pondicherry itself was blockaded and under siege. A French fleet, forming at Isle de France and intended to relieve Pondicherry and to provide Le Gentil's means of reaching the island, was wrecked by an unexpected hurricane in January. Le Gentil was on the point of abandoning his plan of going to Pondicherry and went so far as to inform the Academy of Sciences that he was shipping out for Batavia. At that moment, though, he came down with dysentery. It was March by the time he was fit to travel and another French ship was available. The ship, *La Sylphide*, was due to set sail in order to attempt to raise the siege of Pondicherry, and Le Gentil decided to embark with it for his original destination. Again, however, his luck failed him: it was monsoon season, and the ship was blown off course. He wrote afterward, "We wandered . . . for five weeks in the seas of Africa, along the coast of Ajan [Aden], in the Arabian seas. We crossed the archipelego of Socotra, at the entrance to the gulf of Arabia. . . . We appeared . . . on the coast of Malabar, the 24th of May; we learnt from the ships of this country that this place was in the possession of the English, and that Pondicherry no longer existed for us. . . . I would not yet have despaired if we had followed our first objective to go to the coast of Coromandel; but they made, to my great regret, the decision to return to the Isle de France."[22]

DAY OF FAME—OR INFAMY

The transit occurred on June 6. The travails of long sea and land voyages and the hazards of war were over, and reports began to trickle back very slowly over the following weeks and months—or in a few cases, even years.

Le Gentil, alas, was defeated. Though the day was beautifully clear, he witnessed the transit uselessly from the bridge of his ship somewhere in the middle of the Indian Ocean. His precise longitude being unknown, he was unable to make any observations of scientific value.

Pingré and his assistant at Rodrigues Island awoke on the all-important date to find it raining. It must have been depressing for them to sit in their leaky shelter and watch the clouds pass by! Later in the day, the sky cleared partially. Though they had missed the first and second contacts, they were able to obtain micrometric measures of other stages of the transit and were satisfied enough to toast their success that evening at dinner.

Maskelyne and his assistant Waddington had even worse weather at St. Helena. They occasionally glimpsed the black spot of the planet on the solar disk through thinnings of the clouds but were unable to obtain any useful results.

Mason and Dixon were rewarded with perfect conditions at Cape Town. They observed the whole transit and drew up the only satisfactory report from the Southern Hemisphere.

Chappe, at Tobolsk, was unable to sleep the night before the transit, and sat watching every distant cloud or smoke curl from a fire with apprehension. On the day of the transit, however, he enjoyed perfectly clear, brilliant skies. He describes his excitement on beholding the long-awaited sight of the planet about to reach its internal contact: "The moment of the observation was now at hand; I was seized with an universal shivering, and was obliged to collect all my thoughts, in order not to miss it. . . . I . . . felt an inward persuasion of the accuracy of my process. Pleasures of the like nature may sometimes be experienced; but at this instant, I truly enjoyed that of my observation, and was delighted with the hopes of its being still useful to posterity, when I had quitted this life."[23]

An English observer, W. Hirst, together with the colonial governor, George Pigott (1719–77), made observations from Madras, on the Coromandel coast of India.

Hell and Cassini de Thury were successful in observing the transit at Vienna, as was Lemonnier at the Château de Saint-Hubert, and important observations were taken by the Swedes, including Pehr Wilhelm Wargentin (1717–83) at Stockholm and Torbern Olaf Bergmann (1735–84) at Uppsala. A number of Russians furnished valuable reports. Despite the infinite swarms of insects that had taken possession of the hill where they had set up their observatory, Winthrop and his assistants at St. John's, Newfoundland, made a good observation of the external contact and determined five positions of the planet on the disk of the sun.

Most of the observations employed the Delislean rather than the Halleyan method. On the whole, the results obtained fell short of the effort expended. The mishaps of war had interfered with some of the voyages; the weather had proved an insurmountable enemy with others. But the most serious problems were observational and were quite unexpected.

It proved in fact impossible to time the contacts to the high degree of precision Halley or Delisle had hoped. Mason and Dixon enjoyed the most favorable conditions that could possibly be, and yet even they—side by side—differed by four seconds in their estimates of the time of the internal contact and by two seconds for external contact. Their station, at least, had an accurately determined longitude. At many other stations the longitude, an underpinning of the Delislean method, was not so reliably known. (Many years later, the longitudes of these stations, whose positions were well marked or described, were remeasured more accurately, and the parallaxes recomputed, by Johann Franz Encke [1791–1865] and Simon Newcomb [1835–1909].)

UNEXPECTED PHENOMENA

The precision in the timings was also undermined by two unanticipated optical phenomena. The first was the appearance of a brilliant luminous ring around Venus when the planet was still well clear of the sun's disk. The ring brightened considerably as it overlapped the solar limb, and reached maximum brightness just as Venus was bisected by it. Chappe memorably called the ring a "little atmosphere." Similar effects were seen, independently, by the Swedish observers at Cajaneborg and Stockholm, by Lemonnier (and presumably Louis XV) at the Château de Saint-Hubert, and by Grandjean de Fouchy (1707–88) in La Muette at the Cabinet du Physique.

It is now well known that Venus does indeed have an atmosphere, a very dense one. Its existence was confirmed by German astronomer Johann Schroeter in the 1790s, who noted that when several degrees from the sun, Venus appears to be surrounded by an exceedingly thin luminous ring (see fig. 39). As is now known, this ring is produced by light scattered by haze at the planet's cloud-tops. Telescopes of the 1760s were not yet powerful enough to show that particular effect; however, when Venus is entering or leaving the sun at transit, the solar surface is refracted through the atmosphere of Venus and produces a glaringly conspicuous ring that is orders of magnitude brighter than the ordinary scattered-light halo seen near inferior conjunction.[24]

Russian Mikhail V. Lomonosov (1711–65) saw a related phenom-

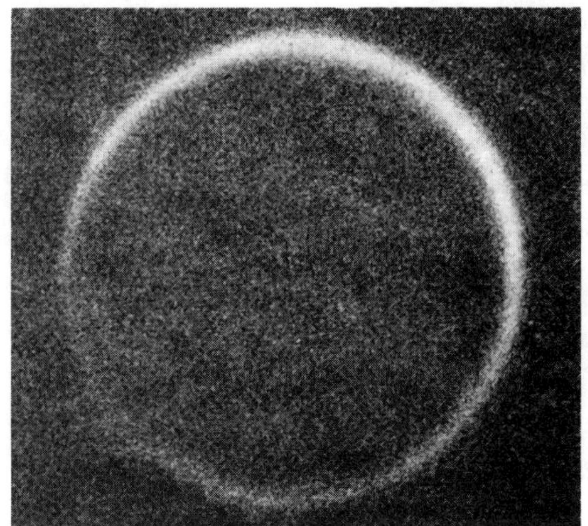

Figure 39. Photograph of Venus 1°.83 from the sun on June 19, 1964, showing the "cusp extension" as a complete ring. Taken by B. A. Smith with a 12-inch (30 cm) reflector with a red filter. North is at top. (Courtesy of the Association of Lunar and Planetary Observers and the Tortugas Station of the New Mexico State University Observatory.)

enon. A chemist and poet, he was the most distinguished nonforeign member of the Academy of Sciences at St. Petersburg. He observed the transit with a 4½-foot refractor of poor optical quality. At sunrise, Lomonosov enjoyed an excellent view, but it began to deteriorate about the time the planet entered the solar disk. As the seeing worsened, Lomonosov found that the edge of the sun became hazy, "whereas before that the edge had been even and clear all over; however, not having discerned any blackness, and thinking that a weary eye was the cause of the cloudiness, I left the telescope." Later, as Venus prepared to exit the sun, "there appeared on the edge of the Sun a blister, which became more clearly defined the nearer Venus came to passing off. Soon this blister disappeared and Venus suddenly appeared without an edge. The complete exit, or the last contact of Venus's rear edge with the Sun at the moment of exit, also took place with some tearing and with lack of clarity on the edge of the Sun."[25] This blister on the solar limb at the third contact was undoubtedly an effect of the refraction of light through Venus's atmosphere, and Lomonosov himself deduced that Venus had an atmosphere that might be "equal to, if not greater than, that of the Earth."[26]

The duration of the transit is defined as the time between the so-called internal contacts (second and third), when the limb of Venus is internally tangent with the limb of the sun. Thus the observations depend critically on ascertaining (1) when the light first goes all around Venus or (2) when the best-fit circle of Venus is tangent to the best-fit curve of the solar limb.[27] The appearance of the ring of light due to refraction of sunlight by Venus's atmosphere introduced one disturbing element: until the planet moved more than halfway off the solar disk, the entire arc of planetary atmosphere sticking out beyond the solar limb contained an image of the sun, producing a bright line completely encircling the part of the planet beyond the limb of the sun. In addition, there was another deleterious effect: the infamous "black drop," which consisted of an indistinctness, blurring, or ligament of darkness that formed when Venus's limb was in near-contact with that of the sun. Though it is often claimed otherwise, refraction by the atmosphere of Venus has nothing to do with it.[28] It can be—and often has been—observed in transits of airless Mercury (see fig. 40).

The black drop has been well named. For the astronomer obsessed with precision, it seemed, indeed, sinister. It was an affront to the methods of Halley and Delisle, which had depended on the expectation that when Venus trespassed the limb of the sun, its edge would be crisply defined and would remain so until only the finest thread of light remained between the solar limb and the planet. Instead, when sufficiently near the solar limb, Venus lost its circular shape and assumed a surprisingly distorted form; sometimes it joined the limb of the sun by means of a glutinous blob or rope-like extension. A particularly vivid description was given by Hirst at Madras: "At the total immersion of the planet, instead of appearing truly circular, [it] resembled more the form of a bergamot pear, or, as Governor Pigott than expressed it, *looked like a ninepin*; yet the preceding limb of Venus was extremely well defined." At the end of the transit, "the planet was as black as ink, and the body truly circular, just before the beginning of egress, yet it was no sooner in contact with the Sun's preceding limb, than it assumed the same figure as before."[29]

The causes of the black drop were not well understood at the time. Later, they were elucidated, at least in part, by the French astronomer Joseph-Jérôme de Lalande (1732–1807) in 1770 and seemingly definitively by University of Texas astronomer Bradley E. Schaefer in 2001:

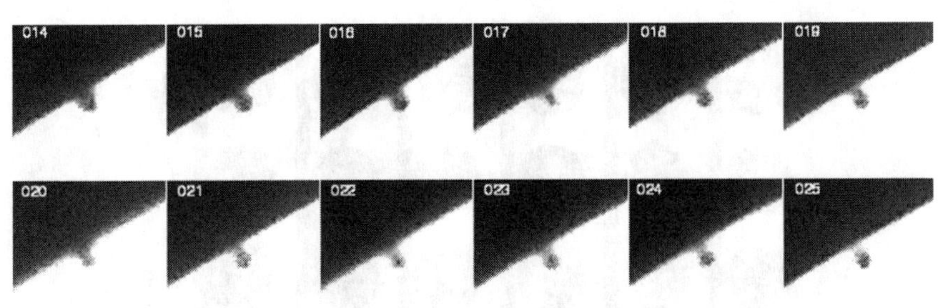

Figure 40. Black drop after second contact of the Mercury transit of November 15, 1999. The effect has been enhanced by contrast stretching of the original CCD images taken by John Westfall with a 36 cm (14 in) telescope from San Francisco, California.

> The complete and correct explanation . . . is that ordinary smearing (due primarily to atmospheric seeing and diffraction within the telescope) of the ideal image (a dark circle silhouetted against a bright circle) will naturally produce a detected image whose isophotal contours have a Black Drop shape around the times of interior contact. . . . Another way of saying this is that the narrow piece of the sun near the point of contact will have its light smeared over a wide region and so it will appear substantially darker, such that it is below the observer's chosen brightness level for defining the edges of Venus and the sun [see fig. 41 for a re-creation of this effect].[30]

The black drop effect cannot be eliminated altogether, but it is much more severe in observations made with telescopes of imperfect optical quality (as many of those used at the 1761 transit were) and in boiling or unsteady air. Confusion about the times of the internal contacts—whether defined as the moment when light goes around Venus or as the best-fit circle of Venus tangent to the best-fit curve of the solar limb—yielded contact times that differed among observers, because of the black drop, by as much as 52 seconds.[31] These observational problems seriously undermined all attempts to calculate exactly the solar parallax. In the end, there was a wide range of published values, from 8.28 arc-seconds (Planman at Cajaneborg, Sweden) to 10.60 arc-seconds (Pingré). The British astronomer James Short (1710–68), who himself observed the

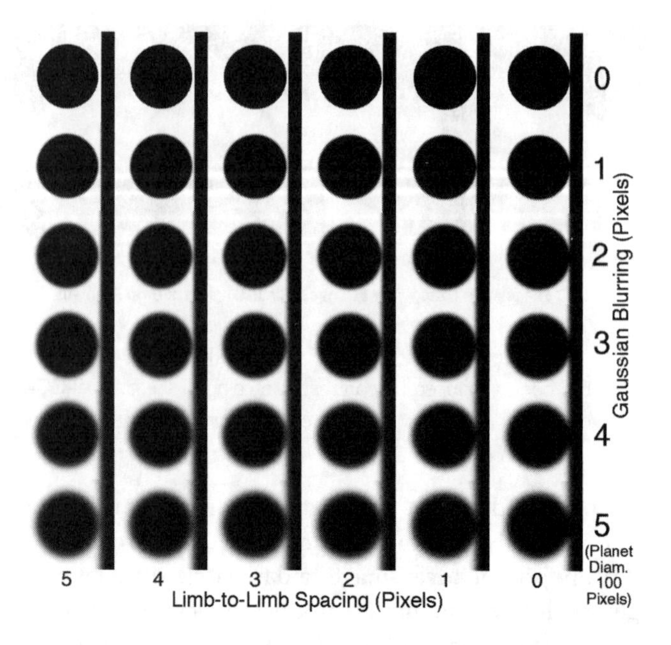

Figure 41. Digital re-creation of the black drop by John Westfall. Smearing of ideal images produces black drops. The top row consists of perfectly sharp images, which become increasingly blurred moving downward. From left to right, the planet moves closer to the limb of the sun. The images on the right show the planet in contact with the sun's limb.

transit from London, was alone responsible for six different parallax values, which, weighted equally, led to a result of 8.566 arc-seconds. Short considered that "probably the error [of this result] does not exceed 1/500 . . . as Dr. Halley had many years since confidently presaged." But few shared his confidence.

Though the 1761 transit observations were disappointing, they were not a total failure, as is sometimes suggested. The solar parallax was now known much more accurately than it had been before—after all, before the transit, Delisle from a review of earlier work had only been able to claim that the solar parallax was between 10 and 14 arc-seconds. Now it was known to be 9 arc-seconds or so, give or take perhaps half an arc-second. It was not likely to be more than 10 percent in error one way or the other.

The sense of failure was due only to the fact that aspiration had reached so far beyond what was achievable. This alone gave the impression, only recently corrected by historians, that the efforts of the men who undertook the great transit expeditions, for all their daring and skill, went for naught.[32]

Back to the Drawing Board

Everyone who participated in the 1761 transit observations knew that there would be another chance: Venus was due to transit the sun again, on June 4, 1769. By then the imprecision in the longitudes of the observing stations might be corrected; also, observers, by taking greater care with their techniques and by learning to expect the black drop's appearance, might pare down the vagaries introduced by it, even if they could not eliminate it altogether. Thus, by removing the element of surprise, they might compensate for it, possibly considerably reducing the brackets of error that hovered around the value of the solar parallax.

It might be a pipe dream. But hope—like the motions of the planets and the sun itself—was eternal. In any case, astronomers had eight long years to prepare themselves and to hone the sharp edge of that hope.[33]

8

Cook's Tour

"Why did James Cook set sail? . . ." He was teaching them a lesson. . . .

Tice said: "The calculations were hopelessly out." Siding with the girl. "Calculations about Venus often are." . . .

The girl marveled. "The years of preparation. And then, from one hour to the next, all over."

—Shirley Hazzard, *The Transit of Venus* (1980)

Paradise Found

Even today, when approached from the air, Tahiti is an unforgettable sight. It looms as a sudden presence in the midst of the sea, a fringe of reefs and verdant coasts rising into a jumble of rain-forest-covered mountains outlining the rim of an extinct volcano.

The island is no longer an unspoiled tropical paradise—if it ever was. Still, the green-clad mountains, often wreathed in clouds, are as impressive as ever, allowing one to picture, in the mind's eye, the beauty of the island as it must have existed before the Europeans came.

About ten kilometers west of the Faaa airport, on Matavai Bay, is Point Venus, one of the most popular beaches of Papeete, the capital, and once the site of the observatory from which Captain James Cook and his assistants observed the transit of Venus of June 3, 1769.

Cook is one of the great figures in the history of world exploration. He is remembered for circumnavigating New Zealand and reaching Australia on his first and second voyages and, on the third voyage, for discovering the Sandwich Islands (Hawaii), where he was finally killed by the native Hawaiians at Kealakekua Bay after having been first mistaken for a god. But all these adventures might never have taken place were it not for his first priority—the transit of Venus.

PLANNING FOR THE GREAT EVENT

Despite the scale of the efforts in 1761, the whole exercise had proved only a partial success. Some of the observing stations were plagued by bad weather. A few observers were disrupted by the Seven Years' War; many more were compromised by poor coordination and uncertainties in their exact positions (the exasperating "longitude problem"). But almost all suffered from the unexpected appearances of the luminous ring and black drop at the critical times of internal contacts.

Thus the value of the solar parallax remained uncertain. "In this uncertainty," wrote the Oxford astronomer Thomas Hornsby, "the astronomers of the present age are peculiarly fortunate in being able so soon to have recourse to another transit of Venus . . . when, on account of that planet's north latitude, a difference in the total duration may conveniently be observed, greater than could possibly be obtained, or was even expected by Dr. Halley, from the last transit."[1]

The French again unfurled ambitious designs, while the British had no intention of being caught off-guard again. Overall, the international efforts would dwarf those of 1761, and indeed represented an eighteenth-century equivalent of the twentieth-century "race to the moon."

In England, the nerve center of planning was the Royal Society in London. A Transit Committee was formed, consisting of Nevil Maskelyne, now Astronomer Royal; James Short; Henry Cavendish; James Ferguson; and John Bevis. Anticipating nineteenth-century transit preparations, Bevis went so far as to suggest using a "simulator," a mechanical device consisting of a miniature artificial planet crossing an artificial sun, in order to train observers for the actual event. It is not known whether one was ever actually built.[2]

The Committee's charter was "to consider and report on the places it would be advisable to take observations, the methods to be pursued, and the persons best fitted to carry out the work." The Committee first met just nineteen months before the transit, and its recommendations were set forth in a memorandum to George III:

> That the passage of the planet Venus over the disc of the Sun, which will happen on 3rd of June in the year 1769, is a phenomenon that must, if the same be accurately observed in proper places, contribute greatly to the improvement of Astronomy, on which Navigation so much depends.
>
> That several of the Great Powers in Europe, particularly the French, Spaniards, Danes and Swedes are making the proper dispositions for the Observation thereof: and the Empress of Russia has given directions for having the same observed in many different places of her extensive Dominions. . . .
>
> That the British nation has been justly celebrated in the learned world, for their knowledge of astronomy, in which they are inferior to no nation upon earth, ancient or modern; and it would cast dishonour upon them should they neglect to have correct observations made of this important phenomenon.[3]

British efforts were aided by George's direct support—he commanded the Royal Navy to furnish a ship to the Royal Society and provided a grant of £4,000 from the royal treasury to defray expenses.

The transit would be visible in its entirety over the Pacific, western America, and the Arctic. Only the beginning would be seen in eastern America and western Europe, and only the end in eastern Asia (see fig. 42).

The Transit Committee planned to send William Bayley to

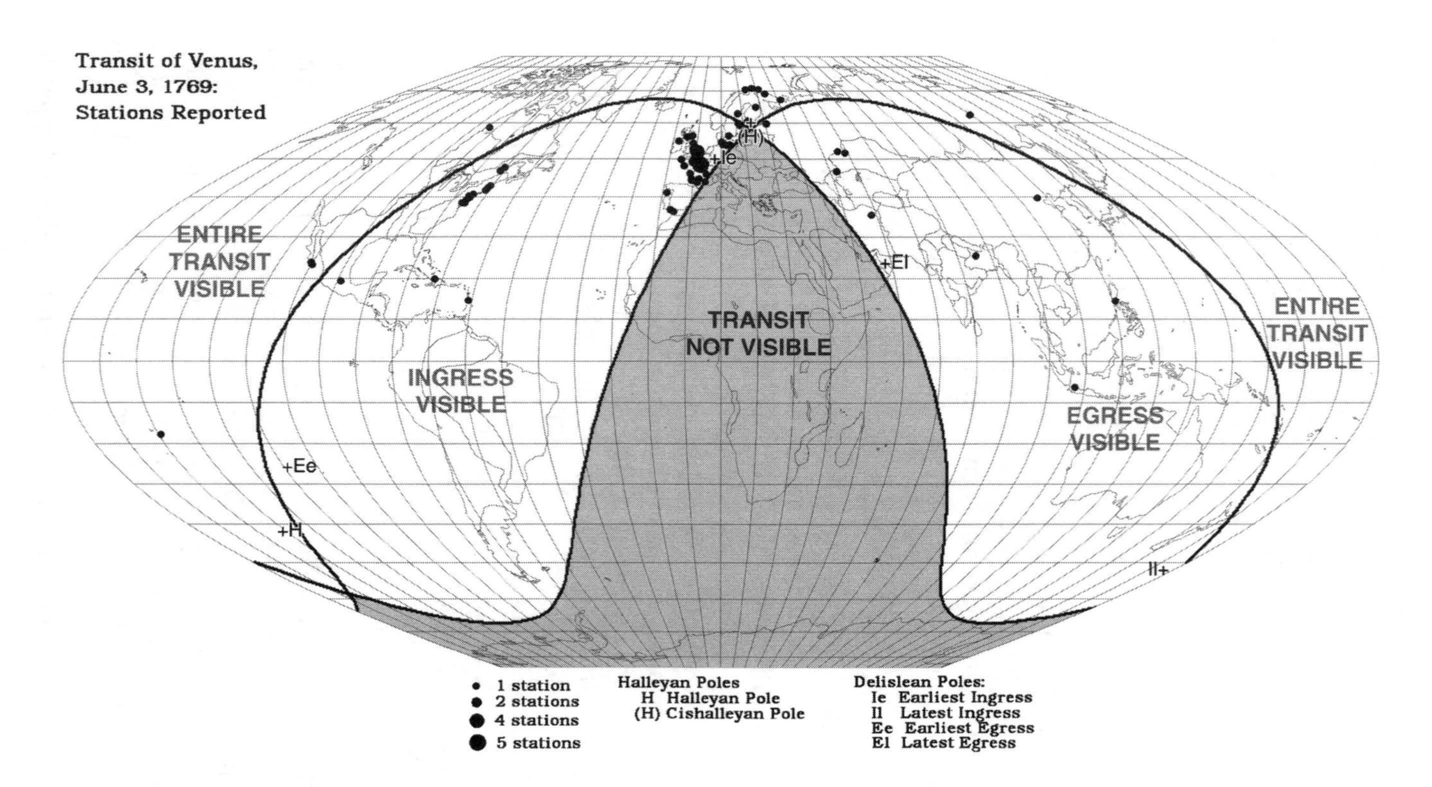

Figure 42. Visibility zones, Halleyan and Delislean Poles, and observing stations with published results for the transit of Venus of June 3, 1769. (Map by John Westfall.)

North Cape, on the Arctic tip of Scandinavia, where the duration of the transit would be 5 hours, 55 minutes. James Dixon was to be directed to another site in Norway, while William Wales and James Dymond received a commission to observe the transit from Fort Prince of Wales (now Churchill), originally founded by the Hudson Bay Company, on the shore of Hudson Bay.

With these observers assigned to the transit's northernmost zone of visibility, the Transit Committee set out to identify a favorable site somewhere in the South Pacific, near the transit's southern Halleyan Pole.

With a surface area of 70 million square miles, the Pacific was still largely an *aqua incognita* to Europeans. First glimpsed by Balboa from his peak in Darien in 1513, the watery expanse had been broached for the first time from the east on November 28, 1520, by Ferdinand Magellan, the intrepid Portuguese-born navigator sailing in the service of Spain. He celebrated the occasion by firing the guns of his flagship *Trinidad,* holding the silk banner of his command aloft, and naming the sea, which was at the moment so remarkably calm and placid, *Mar Pacifico.* A number of other Spaniards followed in Magellan's wake, establishing a regular trade route for their caravels from Acapulco, the coastal port of the Spanish in Mexico, to Manila in the Spanish-held Philippines. Sometimes veering or being blown off course, they chanced upon a few of the numerous islands scattered across the Pacific: New Guinea, the Caroline Islands, the Solomons, and the Marquesas. The gradual recognition of these islands dotting the Pacific, some of which consisted only of low-lying coral atolls and almost all of which, once found, were impossible to recover, fueled the myth of an El Dorado–like land as rich in gold as Mexico or Peru or as rich in spices as the Moloccas, thus giving rise to the legend of the "Great Southern Continent."

The grail of a Great Southern Continent caught the imaginations of navigators for two centuries; one of them was Englishman Alexander Dalrymple. His voyages had begun at the tender age of fifteen, when he had shipped to Madras in the employ of the East India Company. He later became an authority on the East Indies who attempted to revive British trade there at a time when it was increasingly falling into the hands of the Dutch. Dalrymple, whose passion was the Great Southern Continent, became the Transit Committee's first choice to head an expedition to the South Seas for the purpose of observing the transit of Venus. Maskelyne recommended him as "a proper person to send . . . having a particular turn for discoveries, and being an able navigator and well skilled in navigation." But he overplayed his hand by insisting that, though only a civilian, he be given command of one of His Majesty's ships. The last time that had happened was in 1698, when Edmond Halley had been assigned the *Paramour*, a pink (not, as often referred to, the *Paramour Pink*),[4] for his scientific voyage around the Atlantic. Though it was not Halley's fault, he had come close to facing down a mutiny in the Barbados. Recalling that experience, the First Lord of the Admiralty, Sir Edward Hawke, adamantly opposed the idea of Dalrymple being given command of a ship.

An impasse had been reached. To break it, the Transit Committee turned to Captain James Cook (see fig. 43). At forty, Cook was at the height of his powers. Born in Marston, Yorkshire, in 1728, he had received only a rudimentary education and left his family for the fishing village of Staithes, where he was apprenticed to a grocer. But no sooner was "the tang of sea-salt fairly in his nostrils" than he knew that somehow he must get himself to sea.[5] Persuading the grocer to cancel his indentures (the contract binding Cook to the grocer's service), he became apprenticed to ship-owning brothers John and Henry Walker in the nearby port of Whitby. For two years he learned the ways of the sea on a coal ship named the *Freelove*, regularly plying between Newcastle and London. Between voyages, in his spare time, he studied mathematics and navigation. As soon as his indentures with the Walkers were over, he signed on with other ships for voyages to Ireland and the Baltic, only to return to their service as mate of another collier, the *Friendship*.

He was tall and lean, sober, quiet, thoughtful, honest and thoroughly reliable. The Walkers were ready to offer him the command of the *Friendship* when England and France went to war, and Cook, seizing the chance, joined the Royal Navy. He then distinguished himself for cool service as a navigator and marine surveyor on the St. Lawrence River during the siege of Québec.

Figure 43. Portrait of Captain James Cook. (From James Cook, *A Voyage towards the South Pole and round the World* [London, 1777].)

After Dalrymple, Cook—staid, loyal, and trustworthy—became the Royal Society's choice to pull its transit-of-Venus chestnuts out of the fire. But exactly where in the South Pacific should they send him? The Marquesas Islands were one possibility, discovered by chance by Spanish navigator Alvaro de Mendaña y Neira in 1595. Another was Tonga, stumbled upon by the Dutch navigator Abel Tasman in 1643. But neither the Marquesas nor Tonga had been seen since their discoveries; in that era of uncertain longitude-determinations, no one knew quite where they were.

In May 1768, as the Transit Committee was debating Marquesas versus Tonga, an alternative presented itself—Tahiti. The British navigator Samuel Wallis, captain of the frigate HMS *Dolphin*, had just returned to England after a round-the-world voyage of nearly two years. When he had left England, his instructions had been to look for "Land or Islands of Great extent, hitherto unvisited by any European Power . . . between Cape Horn and New Zeeland."[6] He was only the latest dreamer in search of the Great Southern Continent. Sailing in the Pacific south of the route taken by an earlier British explorer, "Foulweather" Jack Byron (who would become grandfather of the poet Lord Byron), he caught tantalizing glimpses of distant cloud-capped peaks—or were they only clouds? Doubtful as it was, the vision added more fuel to the Great Southern Continent myth.

In addition, Wallis claimed a more substantial find, a hitherto-unknown but inhabited island in the Pacific. Wallis named it King George's Land, after George III; its natives knew it as Otaheite, or Tahiti.

At first sight, Tahiti dawned upon Wallis as a beatific vision. It was a place "such as dreams and enchantments are made of, real land though it was: an island of long beaches and lofty mountains, romantic in the pure ocean air, of noble trees and deep valleys, of bright falling waters. Man in his cool dwellings there was not vile."[7] Anchoring in the island's lagoon, Matavai Bay (see fig. 44), Wallis and his crew were promptly greeted by hordes of natives in Tahitian canoes. Some of the canoes were paddled by nubile young women, brown, scantily clad, and bare-breasted. However, the idyll quickly turned disagreeable; the Tahitians who formed the welcoming committee soon turned into a war party and began barraging the crew with stones in an attempt to board the ship. At length they had to be driven off with grapeshot from the *Dolphin*'s cannons. Another unprovoked attack two days later led to British reprisals in the form of a raiding party that went ashore and destroyed eighty canoes. Thereafter the two sides negotiated a truce and at last began to engage in friendly barter; the Tahitians furnished the British with much needed supplies—by then many of the men had become sick with scurvy, and Wallis himself remained incapacitated below deck during most of the six weeks of the *Dolphin*'s visit. In exchange for these supplies, the British supplied the Tahitians with metal objects; they had never before seen knives, hatchets, or nails. Nails in particular were valued for fish hooks, and the Tahitian men were willing to trade a hog for a nail—or, evidently all the same to them, their wives' and daughters' virtue. Before long, the *Dolphin* was in danger of falling apart for want of nails to hold it together!

By this time, some of the *Dolphin*'s crew had managed to measure the moon's distance from fixed stars using a sextant. With the aid of ephemerides calculated for the purpose, they worked out the longitude of Tahiti. Thus its position was known to a high degree of accuracy, not the least of its recommendations to the Transit Committee.

From Tahiti, Venus in transit on June 3, 1769, would follow a much different chord than at northern stations such as Norway or Hudson Bay. Its duration would be 5 hours, 28 minutes—a full 27 minutes less than the duration in Norway. (The difference was close

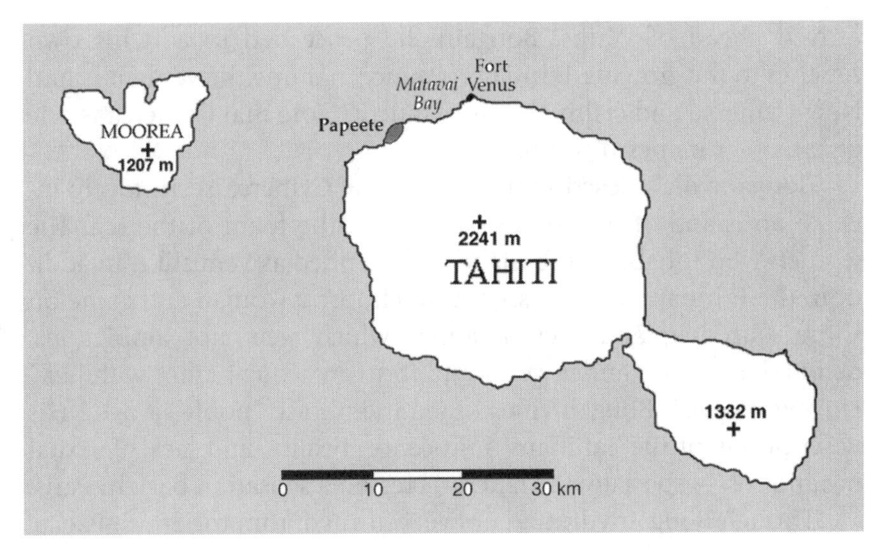

Figure 44. Map of the islands of Tahiti and Moorea, showing the location of Fort Venus in relation to Matavai Bay and the present capital of Papeete. (Map by John Westfall.)

to the maximum possible for the 1769 transit: 35 minutes.) Thus the transit of 1769 was extremely favorable for the application of Halley's method.

In addition to his charter to observe the transit, Cook received secret instructions from the Admiralty, inspired by Wallis's tantalizing glimpse of distant peaks. After the transit, he would continue his course southward from Tahiti in search of the Great Southern Continent, explore as much of its coast as practicable, land, and take possession of "convenient situations," "with the consent of the natives." Failing that, Cook was commanded "to fall in with the Eastern side of the Land discovered by Tasman and now called New Zealand."

While Wallis was still on his way back to England and the Transit Committee remained undecided between the Marquesas and Tonga, another European visitor to Tahiti, Montcalm's former aide-de-camp Louis Antione de Bougainville (1729–1811), arrived. Had he not been waylaid in the Falklands, evacuating a group of French colonists at the behest of Spain, he would almost certainly have arrived ahead of Wallis. Bougainville proceeded to add his own touches to the growing Tahitian romance, naming the verdant island New Cytherea and claiming it for France. (Note that Cytherea is one of the other names of Venus.)

Bougainville's descriptions made New Cytherea seem lovely indeed, an island of Aphrodite rising from the foam of the sea. The women who inhabited the island he described as Venuslike. In addition, the Tahitian men "pressed us to choose a woman and come on shore with her; and their gestures, which were not ambiguous, denoted in what manner we should form an acquaintance with her." During this visit Bougainville coined the phrase "noble savage." His descriptions of the Tahitians' innocence, beauty, and lack of sexual inhibitions—especially the latter!—created a sensation back in Paris.

Though Bougainville had not yet returned from the South Pacific before the Royal Society made its decision, even Wallis's more restrained discussions brought a greater than usual number of volunteers for Captain Cook's expedition. Several of the *Dolphin*'s crew, eager for a second taste of Tahitian paradise, signed up, a fact notable in those days when the hardship of a long sea voyage was not to be undertaken lightly. As Captain Cook's contemporary, Samuel Johnson, famously wrote, "No man will be a sailor who has contrivance enough to get himself into jail; for being in a ship is being in a jail, with the chance of being drowned."[8]

ENDEAVORING TO SEE THE TRANSIT FROM TAHITI

For the voyage to Tahiti, Cook was given command of a ship of the type that he knew well from his long experience in the coal trade, a round-bowed, deep-waisted, reliable tub of a bark of 360 tons, the *Earl of Pembroke*. The Royal Navy had purchased it for the express purpose of sending Cook to the South Pacific with it. Refitted and armed for the long voyage at the navy shipyards at Deptford, it was renamed the *Endeavour*.

Cook was assigned one trained astronomer, Charles Green. He had been one of Maskelyne's assistants at the Royal Observatory, and Maskelyne regarded him as a skillful observer. Cook's other companions included Joseph Banks (1743–1820), a man of great fortune and thoroughly grounded in natural history. A gentleman born and bred—unlike Cook, who had risen from a hardscrabble existence—Banks had enjoyed every possible advantage as a member of the British aristocracy and went to the best schools. He proved unscholarly at Harrow and Eton; however, in the summer of 1759, an event occurred to change the direction of his life. After a swim with friends, he strolled back alone to the college by a country road. For some reason he began to take notice of the country flowers. As one of his companions later recalled, "He stopped and looked round, involuntarily exclaimed, How beautiful! After some reflection, he said to himself, it is surely more natural that I should be taught to know all these productions of Nature, in preference to Greek and Latin."[9] Thus he embarked on his lifelong passion for botany. Still, he had little if any interest in astronomy. His sole motivation in joining Cook was the opportunity to make botanical discoveries and to enlarge his collections.

In addition to a suitably spacious accommodation for Banks himself, Cook had to find a way of fitting the rest of Banks's large entourage on the ship. The "regime," as they were unofficially called, included a Swedish botanist and erewhile student of the great Linnaeus, Dr. Daniel Carl Solander; a gifted twenty-two-year-old natural history painter, Sydney Parkinson; a Scottish landscape painter, Alexander Buchan; a Swedish secretary, Herman Spöring; and several servants. Altogether there were seventy-one on board the *Endeavour*: eleven supernumeraries, twelve marines, and Cook's crew, who included a number of Wallis's veterans.

Loaded with provisions sufficient for eighteen months, *Endeavour* set out from Plymouth on August 26, 1768. The provisions included 4 tons of beer, 185 pounds of great Devonshire cheeses, fresh meat by the hundredweight, salt beef by the ton, biscuits and vinegar, sauerkraut and fruit for the prevention of scurvy, and last but not least, 604 gallons of Jamaican rum.

Endeavour reached her first stop, Madeira, in the Atlantic, on September 13. Cook stopped long enough to reprovision the ship with fresh beef, fruit, and vegetables. As always he was preoccupied with the scurvy threat and issued every man twenty pounds of onions, with strict orders they be eaten on pain of flogging.

The next port of call was Rio de Janeiro, on November 13. Relations between Great Britain and Portugal were generally excellent at the time, but they found an unmistakably cool reception from the Portuguese viceroy. This man refused to believe that a refurbished collier could belong to the Royal Navy or that anyone in their right minds would travel so far only to make an observation of a planet. They must be smugglers! An armed guard was stationed on the ship, and no one was allowed on shore without escort, but the irrepressible Banks did manage to slip out at night and went ashore with Solander, Parkinson, and the servants to gather botanical specimens.

As soon as she had been reprovisioned, *Endeavour* beat a hasty exit from Rio. She did lose one able-bodied seaman, who slipped in the rigging and fell overboard, and suffered another misadventure when, being towed out from the harbor, she was fired on from the fort at the entrance. The commander of the fort later claimed that he had received no instructions to let the ship pass. After an angry protest from Cook to the viceroy, and aided by an easterly breeze, *Endeavour* made her way back into the open sea on December 5. Banks sighed with relief that they were finally rid of these "troublesome people."

They continued around South America—heading farther and farther south, past the Valdes peninsula of Patagonia and the Falkland Islands. They paused long enough to celebrate their first Christmas at sea, three hundred miles from land. "All good Christians," noted Banks, "got abominably drunk."

By January 12, 1769, *Endeavour* was approaching the Straits of Le Maire, which opened out into the Pacific. They were standing off the east coast of Tierra del Fuego—the "Land of Fire," as Magellan had named this desolate windswept tongue of land.

The south side of the straits, known as Staten Land, had been described by an earlier British navigator, Lord Anson, as making "the most horrid appearance of anything I ever saw. It is very high, broken, rocky land, seems to be the vast ruins of some prodigious edifice, and is a proper nursery for desparation." The same writer had described Tierra del Fuego as "really no more than a huge chain of monstrous mountains whose tops are continually covered with snow or hid in the clouds." Cook did not find it so inhospitable. Through his glass he saw trees, consisting of the flat bushy-topped Antarctic beech, interspersed with the smoke of the fires of inhabitants, the fires that had inspired Magellan to give the place its name. After obliging Banks, who had been badgering him incessantly ever since leaving Rio to anchor off any small island or sandbank in order to allow him to pursue his botanizing, the weather turned suddenly cold, and two of Banks's servants froze to death on shore. (On a later occasion, when he had grown increasingly tired of Banks's specimen-gathering, Cook famously exclaimed, "Damn and blast all scientists!")[10] Now Cook skillfully threaded his way through the Straits of le Maire and doubled around the often treacherous Cape Horn, "with as much ease as if it had been the North Foreland on the Kentish Coast." At last he set his course northwest for Tahiti.

Tahiti Looms

For many days afterward the seascape remained virtually featureless, boasting nothing more significant than the low-lying coral atolls of the Tuamotos. At last, on April 10, there was a sure sign of approaching landfall: seabirds swooping down over *Endeavour*, playing in the masts and the rigging. Over seven months after Cook and his crew had left Plymouth, they finally laid their eyes on the peaks of Tahiti, whose shadows emerged from the slanting, thin gray lines of an April squall.

As they prepared to enter the perfect semicircle of Matavai, Tahiti loomed ahead of them as an impressive volcanic mass reaching up to a full mile above the sea. There were sandy beaches in the north and more scattered beaches, littered with rough coral, on the south. The whole island was surrounded by a barrier reef, in most places extending from a half-mile to two miles from the shore. In the northern part on Matavai Bay there was a break that allowed the passage of canoes such as those that had come out to greet the *Dolphin*. *Endeavour* had not even dropped anchor when she was "surrounded by a large number of canoes who traded very quietly and civilly, for beads chiefly, in exchange for which they gave coconuts, breadfruit, both roasted and raw, small fish and apples."[11]

Matavai Bay provided an excellent break and cover from the howling sea winds. It was lined with black volcanic sand backed against a wall of coconut trees, their shaggy tops flouting the blue and cream-curds of an unspoiled tropical sky. On entering the bay's safe confines, Cook and his crew drank in the spectacular beauty of the island. The island's terrain was "uneven as a piece of crumpled paper," as Parkinson, *Endeavour*'s artist, wrote: "being divided irregularly into hills and valleys; but a beautiful verdure covered both, even to the tops of the highest peaks." The lower ridges gave way, far off, to majestic green-clad Orofena. Everywhere the lower slopes luxuriated in palm and chestnut, breadfruit, bananas, and yams. The rugged sides of the ridges were dissected with valleys through which water flowed in gurgling streams or tumbled in delightful waterfalls as it made its way shorewards. In Cook's day, a thirty-foot-wide river, the Vaipopo, flowed all the way to the shore. Its course has changed since, and now this part of the shoreline is a swamp. There were, and still are, flies. But "the warm air remains, in bright day or soft night; the green of spontaneous growth, the smell of earth and blossom."[12]

Cook and his crew were in far better shape to enjoy the island's pleasures than the scurvy-ridden men who had served under Wallis or Bougainville. The eighteenth-century sailor's life was still wretched, and nothing made it more so than the afflictions of scurvy. In the days of Vasco da Gama, scurvy had claimed the life of a hundred men of a crew of 160. The French explorer Jacques Cartier had vividly described the ravages it inflicted on his men during their expedition down the St. Lawrence River in 1536: "Some did lose all their strength, and could not stand on their feet. . . . Others also had all their skins spotted with spots of blood of a purple color: then did it ascend up to their ankles, knees, thighs, shoulders, arms, and necks. Their mouths became stinking, their gums so rotten, that all the flesh did fall off, even to the roots of their teeth, which did also almost all fall out."[13] During their stopover at Tahiti only two years before, Bougainville's men were suffering badly from the disease, and Bougainville shortened his stay and beat a quick retreat across the Pacific back to France. By the time he returned to Saint-Malo, in Brittany, in March 1769, his scurvy-ridden ship had lost seven of its crew.

Cook was avant-garde in terms of scurvy prevention. Long before the British Navy adopted the practice of rationing citrus to their crews, making them "limeys," he followed Scottish physician James Lind's advice and furnished his crew with greens and fresh vegetables, including onions in Madeira and "scurvy grass" in Tierra del Fuego. He also pushed sauerkraut on his crew, an acquired taste for most of the sailors. At first the threat of scourging was necessary to induce them to eat it. Eventually Cook found a better means of ensuring their compliance:

> The Sour Krout the Men at first would not eate untill I put in practice a Method I never once knew to fail with seamen, and this was to have some of it dress'd every Day for the Cabbin Table, and permitted all the Officers without exception to make use of it and left it to the option of the Men either to take as much as they please or none atall; but this practice was not continued above a week before I found it necessary to put every one on board to an Allowance, for such are the Tempers and dispossions of Seamen in general that whatever you give them out of the Common way, altho it be ever so much for their good yet it will not go down with them and you will hear nothing but murmurings gainest the man that first invented it; but the Moment they see their Superiors set a Value upon it, it becomes the finest stuff in the World and the inventor a damn'd honest fellow.[14]

Point Venus

Preventing scurvy was only one of Cook's worries; now that *Endeavour* was anchored in Matavai Bay, he had another: commerce with the natives. As Banks and a large party rushed ashore, the natives treated the visitors cautiously at first. However, an exchange of plantain fronds satisfied them of peaceful intent. Using one of the men who had shipped with the *Dolphin* as a guide, Banks and several others forged a track four or five miles long, through splay-limbed padanus and groves of coconut and breadfruit trees, into the interior of the island. Banks recorded that it seemed "the truest picture of an arcadia the imagination can form."

The crewmen who had shipped with Wallis and the *Dolphin* sought in vain the Queen of Tahiti, Oborea or Purea, whom they had met on their earlier visit. Most of the population had moved west since Wallis's visit, taking their hogs and fowl; Cook decided, nevertheless, that no better harbor was to be found on the island. It was best to remain where they were. He situated his observatory near the northeast point of Matavai Bay, a lovely spot where the Vaipopo ran parallel to the beach and furnished a source of fresh water. On a sandy spit where the trees thinned out, leaving the sky open for telescopes, he erected tents and began to oversee the building of a fort—Fort Venus—made up of shallow ditches and breastworks surmounted by palisades except on the Vaipopo-facing side, where two four-pounders mounted a double row of casks (see fig. 45).

The curious natives began showing up in increasing numbers in their canoes, some even offering their assistance in building the fort. A few demonstrated their penchant for the thievery in which Wallis had found them "prodigious[ly] expert." A spyglass was stolen from the ship's Swedish naturalist, Daniel Solander, and William Monkhouse, the ship surgeon, had his snuffbox filched. Through the intercession of the chiefs, spyglass and snuffbox were returned. The next day was marred by a more serious incident: Cook with a small band and a large group of Tahitians had crossed the river and watched Banks give a demonstration of his shooting prowess by taking down three ducks with one shot. Then other shots rang out. At the tent, one of the natives had made off with a musket and was shot dead on the spot by a sentry. The rest of the natives scattered in terror. Eventually Cook managed to call them back and restore the calm. At the same time, as a precaution, he brought *Endeavour* closer to shore, so that if necessary, he could bring her guns to bear on the part of the bay near the fort.

Figure 45. Woodcut showing Fort Venus, Matavai Bay, Tahiti. (From Sydney Parkinson, *A Journal of a Voyage to the South Seas in His Majesty's Ship, the* Endeavour [London, 1784].)

Cook and the astronomer, Green, were the first of the British party to spend a night on shore. They were eager to obtain a precise determination of the observatory's longitude by the method of lunars, but clouds frustrated their first attempt. Banks and Solander joined them. "Soon after my arrival at the tent," Banks wrote in his journal, "3 hansome girls came off in a canoe to see us. . . . [T]hey chatted with us very freely and with very little perswasion agreed to send away their carriage and sleep in [the] tent, a proof of the confidence which I have not met with upon so short an acquaintance." Young Banks was especially taken with one of the native women, "a very pretty girl with a fire in her eyes," and later he and Monkhouse became rivals and faced a duel over the young women. Fortunately, however, the pistols were not accurate, and Cook could only note that it would have been most unfortunate if either had actually shot the other!

The Tahitians were, on the whole, a friendly and attractive people. Cook was personally impressed with their fine white teeth, their graceful gait, and their hygiene (they bathed three times a day).

Forewarned by the experience of the *Dolphin,* he insisted that "no sort of iron or cloth or other useful or necessary articles are to be given for anything but provisions."

A single nail no longer sufficed to purchase a hog or a woman's virtue. The price for even a small pig had inflated to at least a hatchet. The young and laughing girls—so ready to break into impromptu dances on the beaches, moving in ways that seemed sensuous, even lascivious, to European eyes—proved inordinately fond of cut-glass beads. Before long they were developing close friendships with the British sailors, friendships not likely to remain "Platonick" for long: "The women of this Island," one of the *Dolphin*'s former crewmen noted, were "very Kind in all Respects as Usal when we were here in the Dolphin."

It was not long before the ship's artist, Sydney Parkinson, with Cook himself one of the more puritanical of the crew, noted disapprovingly, "Most of our ship's company procured temporary wives amongst the natives, with whom they occasionally cohabited; an indulgence which even many reputed virtuous Europeans allow themselves, in uncivilized parts of the world, with impunity; as if a change of place altered the moral turpitude of fornication."[15]

By early May, venereal disease—almost certainly gonorrhea—was burning its way among the sailors of Point Venus. Cook was puzzled by the information. Before letting his crew set foot on land in 1767, Wallis had insisted they be inspected by the ship's surgeon and declared free of the disease; Cook had done the same. Cook was a humane man. Avant garde in this respect no less than in the prevention of scurvy, the question of the evils transmitted by Europeans on the peoples of the South Sea weighed on him for the rest of his life. At first he feared that "we had brought it along with us which gave me no small uneasiness." He added, "I did all in my power to prevent its progress, but all I could do was to little purpose for I may safely say that I was not assisted by any one person in ye Ship. . . . [T]his distemper very soon spread it self over the greatest part of the Ships Compney." Later he learned from the Tahitians themselves that two ships—Cook believed them to have been Spanish—had arrived in the island since Wallis's visit and had "brought the Venerial distemper to this Island where it is now as common as in any part of the world and which the people bear with as little concern as if they had been accustomed to it for ages past." In fact the rumor had been false; *Endeavour* was the first ship to arrive in Tahiti since Wallis and Bougainville. All that can be said is that the Tahitians originally contracted the disease from Europeans, then returned the favor to them.

Fort Venus was finished a month before the transit. Now the precious scientific instruments could be escorted onto the shore. An astronomical clock, furnished with a gridiron pendulum adjusted exactly to its length at Greenwich, was set up in the middle of a large tent. There were two reflecting telescopes made by noted instrument maker and member of the Royal Society's Transit Committee James Short (see fig. 46 for a similar model). There was also a large quadrant for measuring the altitude of celestial bodies above the horizon, which provided the other measure, latitude, needed to work out the observatory's exact location.

As soon as the quadrant was set up, it disappeared! In spite of walls and posted sentries, one of the Tahitians had managed to get hold of it. Poor Green was beside himself; he immediately set off running down the beach, pistol in hand, in search of the thief. A man was captured. It was the wrong man; worse, a chief. Once again, Cook managed to smooth things over with the Tahitians. Eventually the quadrant was recovered—its essential parts, at least; the thief had dismantled it.

In the weeks leading up to the transit there were a number of minor heists—nails, iron implements, casks (even, on one occasion, Cook's stockings, stolen in broad daylight from right under his nose!). But at least there were no major incidents, and Green made the observations necessary to fix the location of Point Venus on the globe: latitude 17° 29' 15" south, longitude 149° 32' 30" west of Greenwich.

Now that relations with the unpredictable Tahitians seemed to have been satisfactorily settled, the greatest concern became the even more unpredictable weather. There would be, after all, only one chance during the observers' lifetimes, to witness a transit—June 3,

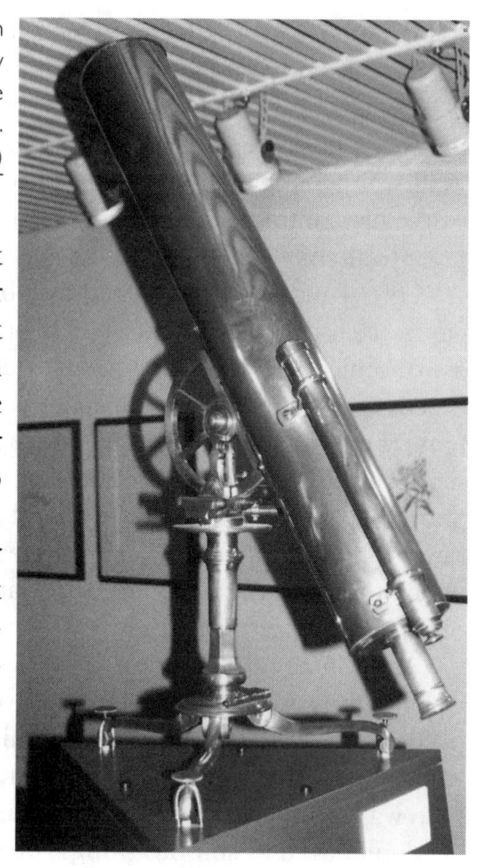

Figure 46. Heath and Wing Gregorian telescope in the Museum of New Zealand in Wellington, similar to those used by Cook for the transit of Venus. (Photograph by William Sheehan, 2000.)

1769. The next one would not take place until December 1874. Every precaution that could be taken was taken: On the day of the transit, the natives were all kept well clear of the observatory in order to keep them from interfering. Cook, Green, and Solander prepared to observe the transit at Fort Venus. Banks, Monkhouse, and Spöring were dispatched by longboat to Moorea, the island in the west, while another contingent was sent east to Taaupiri. All were disposed in order to maximize the chances of success.

WHAT IS SO FAIR AS A TRANSIT DAY IN JUNE?

On the eve of the transit—Friday, June 2—conditions at Point Venus could hardly have looked less promising. There was a ripping wind, and the sky was completely overcast. The tension for all the crew, but especially for Green, the expedition's one astronomer, must have been agonizing. Fortunately, transit day dawned superbly clear, without a puff of cloud in the sky. The tropical sun beat down mercilessly on the island—later, by midday, the temperature climbed to 119° F, the highest that had been recorded. By nine in the morning, the sweltering observers stood on the alert, awaiting the critical first bite of Venus from the limb of the sun.

The moment that had been so long anticipated—the moment of destiny, for which Cook and his crew had endured so many privations during the previous long months—had arrived. It was the moment for which, if they were worthy of themselves, science and posterity would remain forever in their debt. In Cook's *Journal,* the following account is given:

> The day proved as favourable to our purpose as we could wish; not a cloud was to be seen the whole day, and the air was perfectly clear; so that we have every advantage in observing the whole of the passage of the planet Venus over the Sun's disc. We very distinctly saw an atmosphere, or dusky shade, which very much disturbed the times of the contact, particularly the two internal ones. Dr. Solander observed as well as Mr Green and my self, and we differ'd from one another in observeing the times of the Contacts much more than could be expected. Mr Greens Telescope and mine were of the same Mag[n]ifying power but that of the Dr was greater than ours.[16]

The times fixed on by Green and Cook for the first contact was 9 hours, 25 minutes, 45 seconds AM, with total immersion nineteen minutes later. The planet's exit from the sun was recorded at 3 hours, 14 minutes, 8 seconds PM. At Moorea, where the observers had established themselves upon a convenient rock, Banks returned from shore with provisions, bringing with him the king of Moorea and his sister, who observed the transit with them. Banks informed them that he and his companions had come all the way from their own country solely to view it in that situation, a statement that can hardly have made any sense to the natives.

Apart from the distressing appearance of the black drop (see fig. 47), the only major incident occurred when some members of the crew, taking advantage of the preoccupation with the transit observations, managed to make away with 120 pounds of nails. Cook worried that if nails were circulated among the natives, it would do harm to the Europeans' bargaining power by reducing the value of

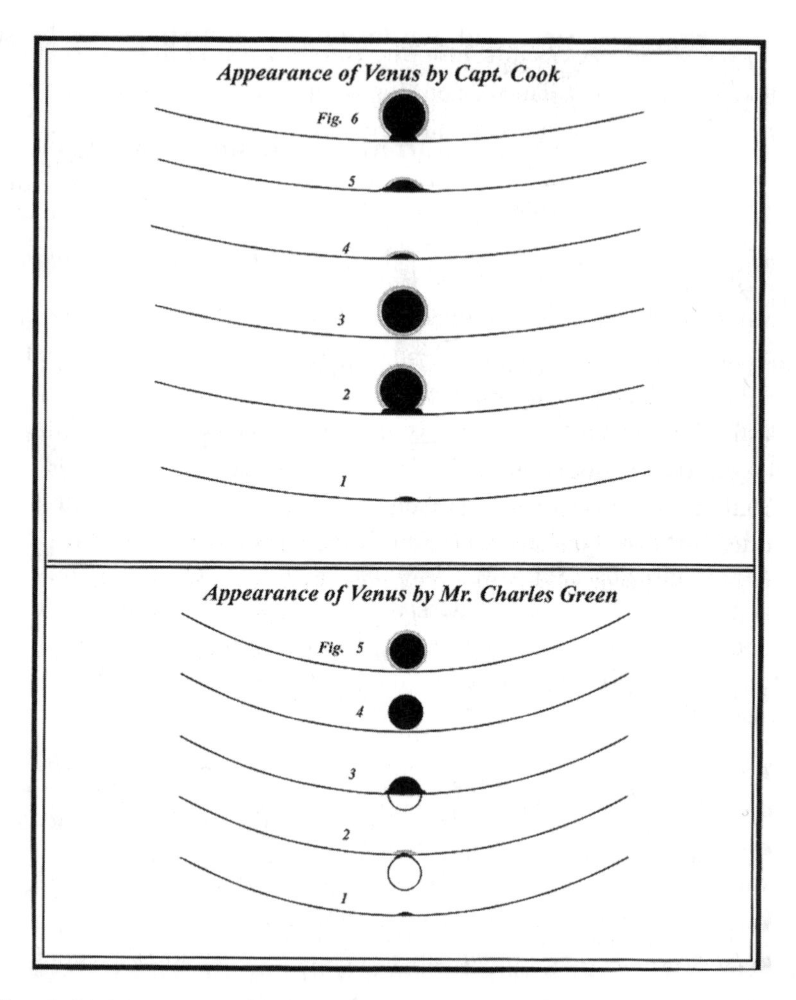

Figure 47. Appearances of Venus near the limb of the sun, ingress and egress, as seen by Capt. James Cook (above) and Charles Green (below), astronomer of the *Endeavour*, recorded at Point Venus, Tahiti, June 3, 1769. (From *Philosophical Transactions of the Royal Society* 61 [1771].)

their most precious commodity, iron. Later, one of the culprits was caught, his pockets bulging with nails; he received two dozen lashes.

With the transit of Venus successfully observed, Cook was eager to leave Tahiti and to continue on to the next stage of the instructions he had received from the British Admiralty, proceeding southward "in order to make discovery of the [Great Southern Continent]," assuming that it existed (and Cook himself had always been skeptical). At the end of June, after circumnavigating Tahiti, they set their course south, as far as latitude 40°. Failing to find the continent, they continued on to New Zealand, sighting the coast of the North Island in early October. In November, Cook observed the transit of Mercury and established his position from the shore of what is now known as Mercury Bay.

On this first great voyage of discovery, which had started with the transit of Venus, Cook and his crew went on to circumnavigate and map the coastline of New Zealand. They then carried on to Australia, landing at Botany Bay, and survived a wreck on the hitherto undiscovered Great Barrier Reef. On their return to England, they stopped at Batavia, the Dutch trading colony in Java, and suffered an attack of dysentery. Cook called the place a "stinking hellhole." The epidemic claimed the lives of the astronomer Green and ship's artist Parkinson, among others. Now *Endeavour* was a "hospital ship" and limped back to England via Cape Town. When it finally anchored off Deal, Kent, on July 13, 1771, nearly three eventful years had passed since it had departed Plymouth. By then the transit of Venus must have seemed a distant memory.[17]

9

Pursuing a Planet

Time is short and space is infinite—how infinite only those who study astronomy know—and perhaps I shall be worn out before I make my mark.

—Thomas Hardy, *Two on a Tower* (1882)

Chasing Results

The transit of Venus in 1769 was distinguished by the extreme northerliness of the planet's track across the sun (see fig. B1 in appendix B, a chart of transit tracks across the sun). In the lead-up to the transit, Oxford astronomer Thomas Hornsby noted, "The astronomers of the present age are peculiarly fortunate in being able . . . to have recourse to another transit of Venus . . . when, on account of that planet's north latitude, a difference in the total duration may conveniently be observed, greater than could possibly be obtained, or was even expected by Dr. Halley, from the last transit."[1]

The expeditions' destinations at the 1769 transit had been mapped out by Joseph-Jérôme de Lalande (see fig. 48). To complement Cook's observations near the south Halleyan Pole, a number of observers went to the far north. William Bayley and James Dixon went to Norway. Bayley observed at North Cape; Dixon did so at Hammerfest, an island off the Norwegian coast; his 1761 observing colleague and collaborator on the Mason-Dixon Line, Charles Mason, observed the transit from County Donegal, northwest Ireland.

WHAT THE HELL?

Jesuit Father Maximilian Hell (1720–92) had observed the 1761 transit from Vienna. In 1769 he located above the Arctic Circle in 1769, at Vardö in Norway (see fig. 49). Hell had left Austria in the company of his assistant, Father Johann Sajnovics, more than a year before the transit; when they arrived at Vardö in October, the sight of a foot of snow on the ground greeted him. On transit day, ingress was observed through a break in clouds just after 9:30 PM. Despite the late hour, the sun was still above the horizon; it did not set at all

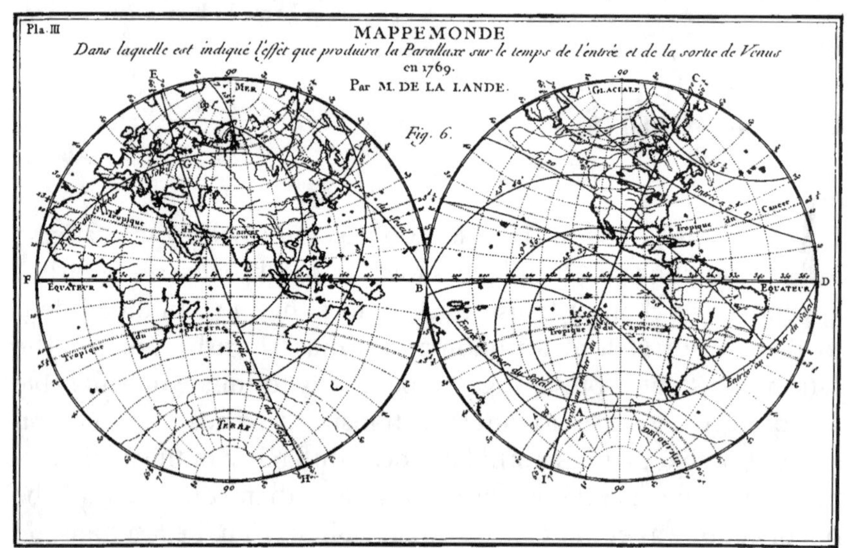

Figure 48. *Mappemonde* for the 1769 transit of Venus, by Joseph-Jérôme le François de Lalande (1732–1807). (From *Mémoires de l'Académie Royale des Sciences* [1757], p. 250, pl. 25.)

that night, at that far-northerly latitude at that time of year. At that very moment, the captain of the astronomers' ship fired off a nine-gun salute and ran his flag up the mast—but to no avail. The sky remained hopelessly overcast until just before egress, when, providentially, another break in the clouds occurred.

A controversy developed later when Hell declined to allow the French astronomer Lalande to see his observations in advance of publication. Lalande began to grow suspicious; was Hell waiting to see other observations, in order to correct his own? Had he perhaps not

Figure 49. Portrait of Father Maximilian Hell, S. J. (1720–92). (Courtesy of the Mary Lea Shane Archives of the Lick Observatory, University of California–Santa Cruz.)

observed the transit at all? In due course, the Danish Academy of Sciences published the Vardö observations, but suspicions lingered. The issue was taken up in 1835 by C. L. Littrow, director of the Vienna Observatory, who after examining Hell's journals declared that the observations had in fact been forged. Littrow asserted that the journal was full of erasures and that corrections had been added in ink of a different color. Hell's reputation remained ruined for a half-century. At long last the mystery was cleared up by nineteenth-century transit expert Simon Newcomb. In searching for original observations in order to recalculate the value of the solar parallax from them, Newcomb visited Vienna and looked up Hell's notebooks. He discovered that Hell had indeed made the observations he had claimed; the problem was with Littrow, who was color-blind. Many of the corrections that had made Littrow suspicious had been made before the ink had dried. The contact times, in particular, had not been erased, as he supposed; where the ink had not flowed freely from Hell's pen, he had indeed rewritten the figures—but in the same ink—and Newcomb concluded: "No further research was necessary. For half a century the astronomical world had based an impression on the innocent but mistaken evidence of a color-blind man respecting the tints of ink in a manuscript."[2]

THE "BARBAROUS REGION" COMES INTO ITS OWN

North America was the destination favored by a number of observing parties. Recall that Horrocks's transit, in December 1639, had been visible in its entirety from North America, but the planet had squandered its charms—as Horrocks himself had so memorably put it—on those barbarous regions. By the eighteenth century, civilization was further advanced; the regions were no longer barbarous, but in June 1761, the transit was already over before the sun rose on the eastern seaboard. However, an expedition led by John Winthrop of Harvard, who had journeyed to Newfoundland, secured observations from American soil, but then only of the end of the transit. In 1769 conditions were more favorable.

Whereas the French had been in the vanguard of the 1761 transit preparations, by 1769 the British—by virtue of the Treaty of Paris signed six years earlier—had seized control of all of French North America east of the Mississippi, already possessing the colonies along the eastern seaboard. The Astronomer Royal, Nevil Maskelyne, realized that North America lay directly in the path of the transit. The beginning and a good part of the rest would be visible from the eastern seaboard, while the entire event would be visible farther west. As might be expected, Benjamin Franklin, who had already carried out his famous experiments with electricity, became linked to the transit. As early as 1753, he had received a letter from James Alexander of New York, a correspondent of Delisle, about the prospect of achieving at the 1761 transit the result that had "baffled all the Art of Man hitherto To discover (Viz: the Sun's Distance from the Earth. . . . It would be great honour to our young Colleges in America if they forthwith prepared themselves with a proper apparatus for that Observation and made it."[3] Before the 1769 transit, Maskelyne had consulted Franklin about the feasibility of sending an American party to Lake Superior.[4]

Nothing came of it, but a major British expedition—the most ambitious, apart from Captain Cook's to the South Seas—was sent as far north as was feasible on the great North American landmass. The leaders of the expedition were William Wales and James Dymont, and their destination was the Prince of Wales Fort (now Churchill) on Hudson Bay (see map, fig. 50). They arrived well in advance of the transit, in August 1768. Their observatory consisted of a prefabricated wood-and-canvas hut, within which they housed an astronomical clock, an astronomical quadrant (for measuring latitude and making observations to check the going of the clock), and two Gregorian telescopes of 2-foot focus manufactured by James Short.

The conditions they endured could hardly be contrasted more sharply with the balmy South Sea conditions experienced by Captain Cook. At Hudson Bay, winter began early; on September 8 they already had snowfall. By November, Wales recorded in his journal that the rapids of the river had frozen over. A half-pint glass of British brandy was found frozen solid in the observatory. Even so, the worst was yet to come. In January, five months before the transit,

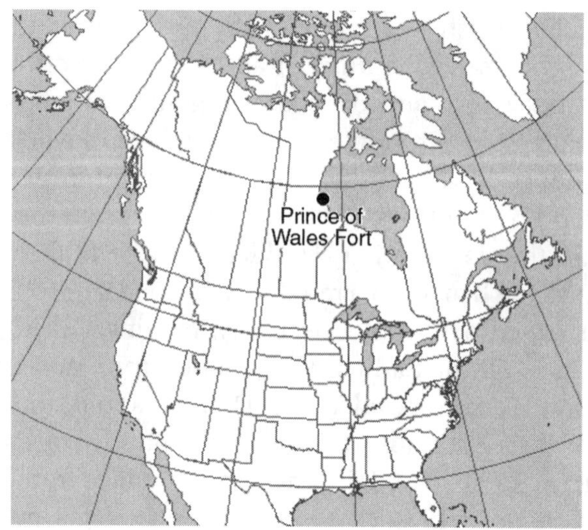

Figure 50. Location of Prince of Wales Fort, on the shore of Hudson Bay, adjoining modern-day Churchill, Manitoba, Canada. *Note:* modern boundaries are shown. (Map by John Westfall.)

the cold was so intense that, Wales wrote, "even in our little cabbin, which was scarcely three yards square, and in which we constantly kept a very large fire . . . my bedding was frozen to the boards every morning."[5] On January 23, the cold reached its bitterest and most intense, when the mercury of the thermometer recorded an air temperature of –43° F. Wales carried a half-pint of brandy into the open air and found that in less than two minutes it was "as thick as treacle": "It was now almost impossible to sleep . . . without being awakened by the cracking of the beams in the house, which were rent by the prodigious expansive power of the frost. It was very easy to mistake them for the guns on the top of the house, which are three pounders. But those are nothing to what we frequently hear from the rocks up the country, and along the coast; these often bursting with a report equal to that of many heavy artillery fired together."[6] Somehow they managed to survive and to enjoy an early spring thaw; by May they found the country "to be really agreeable; . . . the dandelion, having grown pretty luxuriant, made most excellent salad to our roast geese." Unfortunately, the idyll was interrupted by the return of the "muschettos" (mosquitoes), which made Wales wonder whether, after all, winter at the fort had actually been "the more agreeable part of the year."

At least the observers experienced cooperative weather when it was most needed, on the day of the transit itself: "Although there was some passing cloud, in the intervals between . . . the air was very clear, and the Sun's limb extremely well defined."[7] They recorded the first contact at 12h 56m 49s, local time. So far so good—but then, alas, the troublesome luminous ring and black drop effects interposed:

> Soon after Venus was half immersed, a bright crescent, or rim of light encompassed all that part of her circumference which was off the Sun; thereby rendering her whole periphery visible. This continued very bright until a few minutes of the contact and then vanished away gradually.
>
> We took for the instant of the first internal contact the time when the least visible thread of light appeared behind the subsequent limb of Venus; but before that time, Venus's limb seemed within that of the Sun, and his limb appeared behind hers in two very obtuse points, seeming as if they would run together in a broad stream, like two drops of oil; but which nevertheless did not happen, but joined in a very fine thread at some distance from the exterior limb of Venus. This appearance was much more considerable at the egress than at the ingress. . . . We took for the instant of internal contact, at the egress, the time when the thread of light disappeared before the preceding limb of the planet.[8]

Astronomer Faints during Venus Transit!

The British colonies in North America supplied numerous amateur observers of the 1769 transit, including Benjamin West and his colleagues, of Providence, Rhode Island, who observed from what is now called Transit Street. Two other noteworthy observers in North America were David Rittenhouse in Philadelphia and Abbé Chappe at San José de Cabo, Baja California.

Plans for "effective provision" in observing the transit were announced at a meeting of the American Philosophical Society in Philadelphia more than a year before the transit. The most notable observations were made, however, in the small rural village of Norristown (now a suburb of Philadelphia) by Rittenhouse, a clockmaker and member of the American Philosophical Society, one of the most remarkable scientific figures of the American colonial period. He had been born in 1732, the same year as George Washington. His great-grandfather, an immigrant from Germany, had built a paper mill on Monoshoe Creek, near the rural village of Norristown, as early as 1690, and Rittenhouse was intrigued early on by mechanical inventions. He also studied mathematics; his brother recalled the chalk figures and calculations with which he covered the ploughs and fences. Later he was drawn to clockmaking, which combined the requirement for a high level of manual skill with more than rudimentary knowledge of arithmetic and geometry.

Rittenhouse was a perfectionist. Though reputed to be a man of considerable feeling, he always held his feelings in check—he was like one of his own finely crafted and regular clocks. By his mid-thirties he was married, living comfortably in the house his great-grandfather had built, and suffering from an ulcer. At this time of life he felt a restlessness and sense of dissatisfaction; he aspired to do something grander than make precision clocks. Just then, he received a commission to construct an orrery (so called after the well-known prototype built for Charles Boyle, the fourth Earl of Orrery). It would be a model representing Newton's system of the world, accurately depicting the clockwork motions of the planets and their satellites. Though such had been produced before, Rittenhouse aimed at something uniquely ambitious: his would show the elliptical motions of the planets, as calculated by Kepler.

Nonetheless, Rittenhouse set aside this project, never completing it, in order to prepare for the transit of Venus. He set up a wooden observatory on an elevated piece of land near his farm house; the housing for a 2-foot Gregorian reflector, equipped with a Dollond micrometer; and, naturally, an accurate clock. In February of 1769, he began a series of observations of the immersion into Jupiter's shadow of the planet's first satellite, in order to establish his longitude. As usual, he overworked himself; the vigils in the cold night air took their toll, and by late May, he was laid low with a cold and a cough. As the transit approached, the tension for this nervous and high-strung man became unbearable.

The first week of June, Rittenhouse was joined by two observers from Philadelphia, William Smith and the surveyor general of Pennsylvania, John Lukens. The sky was depressingly overcast, and heavy rains prevailed. For several days, only once, and briefly, did the sun shine long enough for Rittenhouse to take its altitude. The transit was due on Saturday. All spirits were low. And yet on Thursday of transit week, the sky cleared and remained "in such a state of serenity, splendor of sunshine, and purity of atmosphere that [there was] not the least appearance of a cloud," as Smith recorded in his report.[9]

Because of uncertainties in the exact time when it was predicted that the great and original of all orreries, the solar system itself, would bring Venus onto the limb of the sun, the observers were already scanning the disk well in advance of the anticipated time in order to register the first hint of the contact. A half-minute beforehand, Smith cried out the alert to the other observers. The most significant moment in the life of an astronomer—success or failure, the ability to reflect with satisfaction or to cling to regret for a lifetime—hung in the balance.

No one was more tightly wound than the clockmaker of Norristown. He was a man of the greatest self-demand, a perfectionist who aspired to the highest standard of workmanship in his clocks. For more than a year he had lived and breathed the transit of Venus; he had put up his observatory, adjusted telescopes and clocks, timed and retimed the satellites of Jupiter, in the quest for absolute accuracy. At the very moment of the greatest anticipation, something unthinkable happened to David Rittenhouse. He collapsed. The moment of first contact "excited . . . an emotion of delight so exquisite and powerful, as to induce fainting"—so writes Benjamin Rush, the founder of the American Psychiatric Association.[10] Rittenhouse's biographer, Brooke Hindle, interprets Rittenhouse's collapse psychoanalytically, and ties it to the ambiguous appearances of the circlet of light and the black drop: "The crisis of tension was brought to an almost intolerable intensity in the terrifying appearance of Venus on the sun. Whatever he had read . . . Rittenhouse was unprepared for the yawning uncertainty of the exact moment at which he must declare that Venus had first touched the sun. There was no single, identifiable point of time at which he could pronounce the two in contact."[11] Faced with this irresolvable tension and ambiguity, Rittenhouse, according to Hindle, resolved the problem the only way he could: he collapsed into temporary oblivion. (Perhaps, less romantically, he simply hyperventilated.)

True, the moment of Venus's entry onto the sun was as elusive as it had been in 1761, and was so even for observers who remained conscious and alert at their posts. The first impression of Venus's arrival was almost ghostly. Instead of being clear-cut and well defined as expected, Venus, at first contact, belonged to a realm uncertain and indefinite. It appeared almost subliminally before it registered unmistakably to consciousness. It first featured as a tremulous wavering of the sun's limb, affecting only a small portion along a quavering and ragged line. By little and little, it grew into a shadowy pyramid. Only gradually this shadow cone assumed more definite form and rounded itself into a planetary silhouette. The transit, in other words, did not occur so much as unfold. It did not provide a clear, crisp, emphatically punctuated datum point; it was blurred and blear, hazy and indefinite. How did one define the exact moment of contact between the planet's edge and the limb of the sun? How did one put a ribbon around a cloud?

Astronomers of the Enlightenment—Rittenhouse among them—were confronted, unexpectedly, with the essential intractability of the ideal of the perfect datum point. They realized the elusiveness of the absolute. In the age of *L'Etat, c'est moi* ("I am the state," a famous quote by Louis XIV) an era of absolutist tendencies publicly asserted by monarchs and privately espoused by the human conscience, a doubtful boundary—elusive, irresolvable, hazardous—was mapped onto the human experience. One need not put undue emphasis on Rittenhouse's collapse. Yet it is intriguing as a symbol: faced with the possibility that the gears of the celestial orrery might not be any better oiled than those of man's contrivance, the "agony of uncertainty" was introduced into man's calculations. The eighteenth-century system itself would soon be under siege. Only a few years after the transit of Venus, Benjamin Franklin would add, in confident self-assertion of the power of reason, to a document written by Rittenhouse's friend Thomas Jefferson: "We hold these truths to be self-evident." The allure of a truth as exact and uncompromising as the motions of a mechanical clock—or a geometry text—remained strong in human consciousness. But the elusiveness of the outline of Venus's globe threw a wrench of vexation into the enterprise. Its implications were to be worked out and elaborated only over another century of human experience.

Rittenhouse himself regained consciousness and composure and returned to the telescope, only to be confronted with yet further quavering of the image (see fig. 51).

Chappe's Tragic Success

On the other side of the North American continent, Abbé Chappe, the veteran of the Siberian expedition to observe the transit of 1761, had reached the tip of the Baja California peninsula in a dogged quest to observe the last transit of the century.

Chappe's expedition was an unprecedented visit by foreign scientists into the heart of the Spanish colonial empire, a realm hitherto closed to all non-Spanish parties as if behind an early form of an iron curtain. The British Royal Society had requested similar permission but had been turned down. Initially the group included five Frenchmen—Chappe, M. Pauly, M. Dubois, the artist Alexander-Jean Noël, and one servant—but it was required by the Spanish to include two Spanish officers, Vicente de Doz and Salvador de Medina, both of whom made useful observations. Additional servants were recruited along their route, bringing the number of total personnel to twenty-eight.

The Chappe expedition's voyage from Paris to, finally, the Franciscan (recently Jesuit) mission San José del Cabo typifies the lengthy journeys and unforeseen delays endured by eighteenth-century overseas expeditions. Although they left Paris 8½ months before transit day, they arrived at San José del Cabo with barely enough time to set up their instruments.

Observing from within the mission grounds, they converted an abandoned barn to an observatory. The chief instruments were 3-foot and 18-inch quadrants, chiefly to determine their positions; a Berthoud

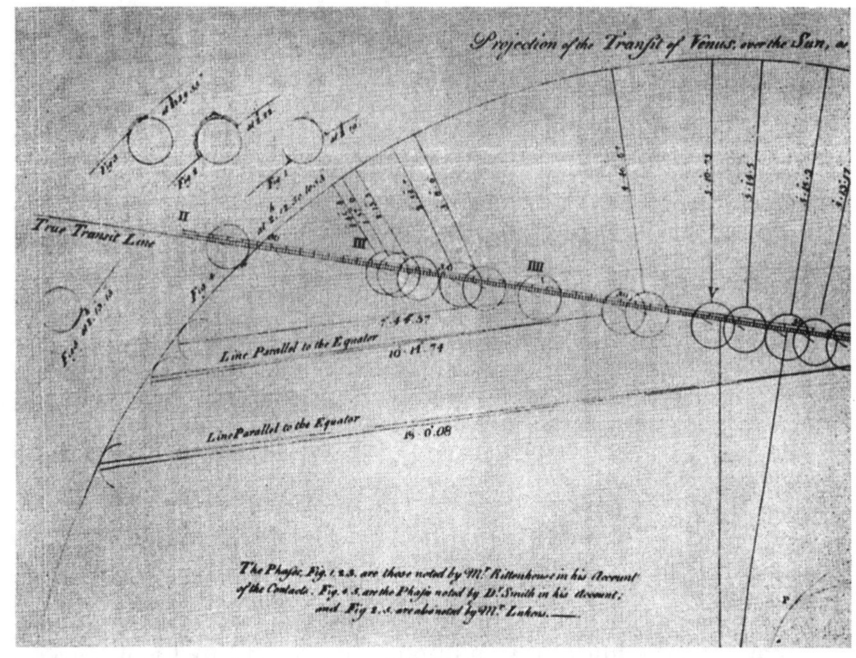

Figure 51. "Projection" of the transit of 1769 by David Rittenhouse (1732–96). (From *Transactions of the American Philosophical Society* I [1769].)

pendulum clock for the same purpose and for timing the transit contacts; and the latest in high technology, 3-foot and 10-foot Dollond achromat refractors. They also found that an "epidemical distemper" was raging at the mission, probably typhus, which had already killed one-third of the local population.[12] Nonetheless, the equipment was installed in enough time to determine their latitude prior to the transit.

As is typical for Baja California in June, skies were clear, so Chappe and Medina independently timed all the contacts of the June 3 transit. However, their longitude was still unknown, and Doz, Medina, and eleven servants fell prey to the epidemic on June 5. Chappe fell ill shortly thereafter but even so managed to time the lunar eclipse of June 19 in order to help determine the mission's longitude. Indeed, from June 21 to June 30, the remnants of the party also timed several eclipses of the Galilean satellites of Jupiter in order to refine their position. Having done his utmost to make the expedition a scientific success, Chappe died on August 1 and was buried with honor in a now-unknown grave (see fig. 52). Before the survivors left for mainland Mexico, they had lost twenty-one of their twenty-eight personnel—a mortality rate of 75 percent. Of the original party only Pauly and Noël survived to bring the observations back to France.

The Chappe expedition was a scientific success, although at a terrible human price. Besides the transit observations themselves, their latitude and longitude determinations helped finally to give the western coast of North America its modern appearance on maps; they shifted mission San José del Cabo's position a half-degree north and five degrees east.

More Fortunate in Love than in Astronomy

Chappe's death entitles him to the name of a martyr in the cause of science. His misfortune overshadows that of Guillaume Le Gentil, but at least Chappe secured the observations he set out to secure. For an astronomer—especially the globetrotting astronomer in quest of

Figure 52. The funeral procession of Abbé Chappe d'Auteroche at mission San José del Cabo, drawn by Alexander-Jean Noël. (From the Cabinet des Dessins, Musée du Louvre, Paris, INV 31478. Reproduced by permission of Réunion des Musées Nationaux/Art Resource, NY.)

observations—to be clouded out is a fate almost worse than death. By most measures, Le Gentil was the most unfortunate of the astronomers of the transit of 1769.

The reader will remember Le Gentil, the French nobleman who had set out, at his own expense, for Pondicherry, India, to observe the transit of 1761, only to learn en route that the city had changed hands. Since it was now British and Le Gentil was French, he was unable to proceed with his plans. Le Gentil beat a swift retreat, hoping to regroup in Mauritius in time for the transit, but ran out of time and ended up onboard a pitching ship in the middle of the ocean when the transit occurred. In order to optimize his chances for the next transit, he decided to remain, more or less, where he was for the next eight years. He whiled away the time in Mauritius and Madagascar, taking voluminous notes on the botany, zoology, geology, and anthropology of these islands, which he later published. He decided, three years before the transit of 1769, that a good place to observe it would be Manila, in the Philippines, which he had previously found to be one of the most beautiful of all the island groups of the seas off Asia, and therefore boarded a Spanish warship bound for Manila. Moreover, the city's position had been accurately determined. Le Gentil seemed to be in complete readiness for the transit when, eleven months after his arrival in Manila, he received a letter from Lalande, a pupil of Delisle. Lalande informed him that Pingré, who had produced arguments in favor of returning to Halley's method over Delisle's, thought Le Gentil "was going too far; he wanted [Le Gentil] to return to Pondicherry."[13] He obeyed, and this time, greatly aided by his former enemy—the English—secured passage from Manila to Macao. There he caught up with an Indian ship, manned by a mixed crew of Portuguese and local sailors, that was bound for Pondicherry. He was nearly wrecked in the Strait of Malacca but reached his destination safely a full year before the transit. He was received enthusiastically by the (now French) governor and was granted all the help he needed in remodeling a ruined pavilion into his observatory. The skies were beautifully clear, as only tropical skies can be, for a full month, right up until the day before the transit. But Le Gentil's famous bad luck continued to hold; the night before the transit, the wind began to change, and Le Gentil—excited by the transit to come, and sleeping lightly—was wide awake by two o'clock in the morning. He rose from his bed and watched as a front moved in: "I heard the sand-bar moaning in the south-east; which made me believe that the breeze was still from this direction. . . . I regarded this as a good omen . . . but curiosity having led me to get up a moment afterwards, I saw with the greatest astonishment that the sky was covered everywhere, especially in the north and north-east, where it was brightening; besides there was a profound calm. From that moment on I felt myself doomed. I threw myself on my bed, without being able to close my eyes."[14]

Instead of improving during the night, the weather grew steadily worse. The winds were rising, and a squall was buffeting the sea, making it white with foam, until dawn. Then the wind began to die down, but the clouds remained. At three or four minutes before seven o'clock, just about the time Venus was calculated to be leaving the sun from Pondicherry, "a light whiteness was seen in the sky which gave a suspicion of the position of the Sun," but nothing could be seen in the telescope. Only after the transit was over did the clouds disperse—the sun remained visible the rest of the day!

Poor Le Gentil had been defeated "by the fate which often attends astronomers. I had traveled about ten thousand leagues; I seemed to have traversed such a great expanse of seas, exiling myself from my homeland, only to be the spectator of a fatal cloud, which arrived in front of the Sun at the precise moment of my observation, snatching from me the fruit of my efforts and exertions. I was unable to recover from my astonishment. I had difficulty realizing that the transit of Venus was finally over."[15] He could hardly bring himself to record this depressing fact in his journal: "I was more than two weeks in a singular dejection and almost did not have the courage to take up my pen to continue my journal; and several times it fell from my hands, when the moment came to report to France the fate of my operations."[16]

Even now, Le Gentil's adventures were not over. At Pondicherry he contracted dysentery, which nearly killed him. After weeks in bed, he regained his strength sufficiently to leave the place where he had suffered such disappointments and booked passage for Mauritius, where a French commissioner attempted to recruit him for a trip to Tahiti. Not surprisingly, he declined. Though it was late in the season, he was eager to return to France. On the first try, his ship lost its mast in a hurricane off the Cape of Good Hope and was forced to return back to Mauritius. The next ship out was a Spanish warship that managed to limp around the cape and, after being blown off course north of the Azores, finally succeeded in reaching Cadiz. Understandably sick to death of sea voyages, Le Gentil made the last leg of the journey, across the Pyrenees, over land. At last he arrived in Paris after an absence of "11 years, 6 months, 13 days," only to find that he had been declared dead, his estate had been robbed, and his relatives were planning to divide what was left of it.

Given that he had worse luck than almost any other astronomer, Le Gentil has become the standard bearer for the whole fiasco of the eighteenth-century transits of Venus. It is pleasant to record that at least after his return to Paris, he seems to have prospered. He wrote two volumes of memoirs describing his research in the Indies—including a study of the customs and religions of the Indian subcontinent and a good deal about the physical characteristics of the islands of Madagascar, Mascarene, and the Philippines. Woolf says that "a rather high opinion of much of this work is still maintained."[17] If he had had any romantic interest in Paris before setting out on his scientific adventures, the woman would have gotten tired of being "stuck in a fluffy gown waiting for a man who is always late," in the words of one of the love interests of a character based on Le Gentil in Canadian playwright Maureen Hunter's play *Transit of Venus* (1992). In Le Gentil's case, eleven years late. But the long-absent Le Gentil, who was only forty-six when he returned to France, even found domestic happiness: he met, courted, and married a wealthy heiress, Mademoiselle Potier. They had a daughter upon whom he doted; indeed, the whole family lived, in apparent happiness, at the Paris Observatory, which was still under control of the Cassini dynasty but which was finally unseated during the French Revolution. Le Gentil, who had missed the transit of Venus, also missed the Revolution and the Terror, in which so many French nobles and savants perished; he died peacefully at his home on October 22, 1792, at the age of sixty-seven. Even then, the results of the transit observations of 1761 and 1769 had not yet been fully analyzed, nor would they be for decades to come.[18]

10

A Noble Triumph —Surpassed

When we consider the ingenuity of the method employed in arriving at this determination, and the refined nature of the process by which it is carried into effect, we cannot refrain from acknowledging it to be one of the noblest triumphs which the human mind has ever achieved in the study of physical science.

—Robert Grant, *History of Physical Astronomy* (1852)

Astronomy by Numbers

On the eve of the 1874 transit of Venus, an educated person, looking back to the last such event in 1769, might well conclude that the intervening 105 years had seen greater changes to the world than in any previous similar interval. The changes were literally revolutionary: the American Revolution; the French Revolution, followed by the French Revolutionary and Napoleonic

Wars; the Latin American Revolutions; and the revolutions of 1830 and 1848.

At least as significant as these political upheavals was the Industrial Revolution—with its steam engines, spinning jennies, locomotives, and a host of other innovations—and the promise that the advance of technology would lead to improvement in all areas of life. Indeed, the whole nineteenth century would become associated more than any other with notions of progress, evolution, amelioration; but also skeptics like William Blake spoke prophetically of the dark underbelly of progress—the menacing aspect of the "dark satanic mills."[1]

It is not surprising that the aspirations of that restless century would also be associated with an attempt to refine those numbers—the solar parallax, the distance of the earth to the sun—which had been the particular preoccupation of the eighteenth century. The nineteenth century's pair of transits—those of 1874 and 1882—threw, as it were, a gauntlet down to the astronomers of the age, who did their best to rise to the occasion.

At a South Kensington exhibit of scientific instruments in 1876, a British engineer, William Siemens, speaking from the chair, noted, "Nearly all the grandest discoveries of science have been but the rewards of accurate measurement and patient long-continued labour in the minute sifting of numerical results."[2] The adoption of precise standards of measurement was seen as crucial for economic and military success, so that the adopted values of science would seem intimately tied in to Victorian values, in the larger sense. A few years after Siemens made his comments, the leading British physicist of the age, William Thomson, Lord Kelvin (1824–1907), looked back over the astonishing advances of physics. Physicists had refined Newtonian gravitational theory, subdued the provinces of electromagnetic phenomena, discovered radio waves, X rays, and the phenomenon of radioactivity. Kelvin emphasized the distinguishing feature of science and the basis of its astounding success as the preoccupation with quantification and accurate measurement: "When you can measure what you are speaking about, and express it in numbers, you know something about it; but when you cannot measure it, when you cannot express it in numbers your knowledge is of a meager and unsatisfactory kind: it may be the beginning of knowledge, but you have scarcely, in your thoughts, advanced to the stage of science."[3]

It was not Kelvin, as popularly believed, but Sir William Dampier who made the famous remark that has come to sum up the complacency—the confidence—of nineteenth-century physicists that they were measuring the right things, and measuring them more and more accurately. "In their various branches," said Dampier, "the explanations of new discoveries fitted together giving confidence in the whole, and it came to be believed that the main lines of scientific theory had been laid down once and for all, and that it only remained to carry measurements to the higher degree of accuracy represented by another decimal place."[4]

In astronomy, the quest for carrying measurements "to the higher degree of accuracy represented by another decimal place" was epitomized by the obsession with the solar parallax and the precise value of the earth-sun distance. Science was obsessed with its own perfectibility. The grail-like quest for a number—not unlike that which in our own times has characterized the modern cosmological quest for the Hubble Constant and the age of the universe—became almost an end in itself. In addition, science attempted to measure its own rate of advance and its own relation to the goal of perfectibility. We must keep clearly in mind the fixation with measurement and number of many scientists of the nineteenth century when considering the sometimes mind-numbing and often esoteric preoccupation with sources of error, instrumental refinements, optimization of observing stations and reduction of data. In the end—so it was hoped—all this concern with detail would lead to an exact value of this all-important number.

The best result of the eighteenth century's massive efforts to measure and time the transits of 1761 and 1769, then to reduce those measures and timings to a value of the solar parallax, had been, as we have seen, equivocal. An attempt to grapple with the mass of data and put the best possible face on it was made by German astronomer Johann Franz Encke (1791–1865) in 1824 (see fig. 53). At the time, Encke was director of the Seeburg Observatory at Gotha, Germany. His training had been impeccable; he had studied under the great German mathematician Carl Friedrich Gauss (1777– 1855) and had received a prize in 1817 for a careful study of the comet of 1680, the very one that had pushed the reluctant Newton in the direction of the theory of gravitation. He also recognized one of the many comets discovered by Jean-Louis Pons of the Marseilles Observatory as a short-period comet—it completed each orbit around the sun in only 3.3 years—and is still the comet with the shortest period known; the comet is still known as Encke's comet. While only in his early thirties, Encke had established a reputation as a first-rate astronomer and calculator. Thus when he undertook the painstaking task that eighteenth-century astronomers had largely shirked—of weighting the observations from the various stations, updating the longitudes in light of more recent determinations, then reducing this vast quantity of data to arrive at a more or less definitive result—the value he published was received as authoritative, adopted by the *British Nautical Almanac* from 1834 to 1869 and the *American Ephemeris* from 1855 (its first issue) to 1869. This value was memorized by a generation of schoolchildren: Encke concluded that the solar parallax was 8.5776 arc-seconds, with a stated uncertainty of only 0.4 percent.[5] Converting this quantity to miles (or kilometers) required an accurate determination of the earth's equatorial diameter. Encke, utilizing the value published by Walbeck in 1819, made the astronomical unit equal to 153,340,000 ± 660,000 kilometers (95,280,000 ± 410,000 miles).

Figure 53. Portrait of Johann Franz Encke (1791–1865). (Courtesy of Special Collections, San Diego State University Library.)

Another World Observed

The world that prepared to receive the shadow cast upon it by Venus in 1874 and 1882 had greatly changed since the last transits of 1761 and 1769. The magnitude of the difference is evident in a glance at the world map. Britain had lost most, and France and Spain almost all, of their American colonial empires (Russia had at least managed to sell Russian America to the United States in 1867). Excepting Canada (a dominion within the British Empire since 1867), British Honduras, and the Guianas (British, Dutch, and French), the American mainland was occupied entirely by independent states, and three of them—the United States, Mexico, and Brazil—would sponsor overseas expeditions to observe the nineteenth-century transits. These three states, as well as Canada, would also occupy transit stations within their own territories.

Including the remnants of their continental holdings, along with Caribbean and some other islands, Britain, France, Denmark, Holland, and Spain all retained footholds in the Western Hemisphere. Since an observing station for a transit of Venus needed little space, all five colonial empires were able to observe the events from within their territorial holdings.

In Europe itself, the three new states of Germany, Italy, and Belgium would stage their own transit expeditions. A fourth new state, Greece, would take advantage of the opportunities for observation presented within its own territory.

Elsewhere in the Old World, the "scramble for Africa" had just begun, with Portugal, Britain, and France the only major players as yet. The same three states had holdings on the southern periphery of Asia, of which British India was by far the most extensive. Farther east in Asia, or off its shores, lay the Spanish Philippines, the Dutch East Indies, British Hong Kong, and Portuguese Macao. To the southeast lay the British dominions of Australia and New Zealand.

Not very impressive on the map, but presenting potential sites for transit expeditions, were various scattered insular possessions or protectorates in the Atlantic, Indian, and Pacific oceans. Britain held more of these than anyone else, but France had several as well, while Spain, Portugal, and Germany also held islands in these waters.

In order to derive a value for the solar parallax from observations of a transit of Venus, it was necessary to compare the results of stations thousands of kilometers apart. There were several states that could manage to do so with stations solely within their own territories. Great Britain was the clearest example of such a power, having an "empire on which the sun never sets."[6] Actually, if one looked more closely at a world map of the time, France, Spain, and Holland also had such empires. Less extensive in longitude, but large enough, or scattered enough, to contain well-spaced transit stations within their own dominions were the Russian Empire, Portugal, and the United States.

In peaceful times, countries could and did mount astronomical expeditions to the territories of other, friendly powers. And the 1874–82 period was, for the most part, peaceful. The last general European war had ended two generations before, while the more recent Crimean and Franco-Prussian wars were limited in their extent. The American Civil War had ended; even the Taiping Rebellion in China—by some considered the most murderous war in history up to then—was over. Thus, unlike in 1761, when people like Le Gentil were thwarted in their plans, expeditions could pass safely over the world's seas and establish observing stations, without harassment, on most of its land areas. Admittedly, some areas were still unstable, such as the Balkans, where Russia, Serbia, and Montenegro were waging war with Turkey in 1877–78. Britain was involved in colonial wars, including the Ashanti War in western Africa in 1873–74, the Zulu and the Second Afghan wars in 1877–78, and a brief war with the Boers in the Transvaal Republic in 1881. They also occupied Egypt in 1882. The central Asian emirates, which were becoming Russian Turkestan, were best avoided, as were certain South American states. Overall, however, the last third of the nineteenth century offered safe observing sites throughout much of the world.

International tensions were in fact less divisive than they had been in 1761 and 1769. Still, the inconvenient circumstances of the transit itself rendered some of the best locations, from the strictly geometrical point of view of both the Halleyan and Delislean methods, all but inaccessible in practice to observers.

On the other hand, at least it was infinitely easier for expeditions to reach their destinations than it had been for the adventurers of the eighteenth century. In the days of Chappe, Le Gentil, and Cook, land travel was slow, and sea travel was unpredictable, involving the use of military vessels or the need to negotiate terms with the owners of private vessels. In 1761, it had taken Chappe five months to travel from Paris to Tobolsk (Siberia), and he had arrived with little time to spare. In 1768–69, it took him eight months to make the journey from Paris to San José del Cabo in Baja California, and that time he did not return alive.

But by 1874, steamships had largely replaced sailing ships. The former had both greater carrying capacity, three or even five times the carrying power per ton displacement than was the case for sailing ships, according to Michael G. Mulhall,[7] and much faster sailing speeds. Indeed, it was possible by 1869 to cross the Atlantic in less than 8 days; Chappe had taken 77! Following the opening of the Suez Canal in 1869, passage times from Europe to the Indian Ocean and the Pacific were greatly reduced. A steamer could travel from London to New Zealand (one of the favored destinations of transit expeditions in 1874) in about 50 days. Of course, when the destinations lay off the main shipping routes, passages took considerably longer. Thus in 1874, the U.S. Navy steamship *Swatara,* carrying the American expeditions to the Southern Hemisphere for the transit, took 91 days from New York via Cape Town to Kerguélen Island, and another 24 days to Hobart, Australia. In the same year, the Pacific Mail side-wheeler *Alaska,* on its regular route, took only 23 days to carry the American Far East transit party from San Francisco to Yokohama. In 1882, Georges Ernest Fleuriais (1840–95), traveling to his second transit of Venus, required 56 days for the voyage from Bordeaux to Santa Cruz, Patagonia.

As great as was the advantage of steamships over sailing ships, land travel—to the interiors of continents—had been revolutionized by the railroad. Chappe had slogged with sleds over the Siberian ice packs; but by 1870, 122,000 miles of railway had opened on six continents, and over the next decade track mileage grew by 82 percent, to 223,000.[8] There were dense networks in Europe and the eastern

United States, which also boasted two transcontinental lines. India and the more settled portions of Latin America, South Africa, Australia, and New Zealand also had railways connecting their major centers (see fig. 54). The location of observatories, not surprisingly, closely followed the distribution of the transportation and communication networks (see fig. 55).

Communication was now possible by means of telegraph lines, so that instead of requiring weeks to cross the Atlantic, or months, as had been the case between Europe and India or the Far East, messages could travel thousands of miles in a matter of hours or, in some cases, only minutes.[9] Besides land telegraph lines, which were usually found along railways, submarine telegraph lines linked vital coastal points among all six settled continents, as well as strategic islands in the Caribbean and the Mediterranean Seas, the Dutch East Indies, and the Philippines, Japan, and New Zealand.

The advantages of telegraph communications for transit planners are obvious. They allowed planning of the activities at distant field stations from such metropolises as London, Paris, Berlin, St. Petersburg, and Washington as well as the rapid communication of observations. Also—perhaps the most important advantage of all—one could transmit time signals that could be used to determine the longitudes of stations more accurately than by any previous means; this resulted in an improvement in position-finding comparable to that of the Global Positioning System (GPS) of our own day. Errors in the longitudes of the stations had been even more significant than observational quandaries posed by the black drop effect in the eighteenth century. Thus, better longitudes meant greater accuracy when analyzing the transit contact timings using the method of Delisle, which required the accurate comparison of times recorded at stations distant from each other.

Besides the greater ease with which they were able to reach these far-flung sites, astronomers—their ranks including many professionals, who hardly existed in the eighteenth century, as well as amateurs, clergymen, seamen, and surveyors—could muster a far more formidable array of scientific instruments and bring them to bear on the planet of their predilection. They had better telescopes—mostly

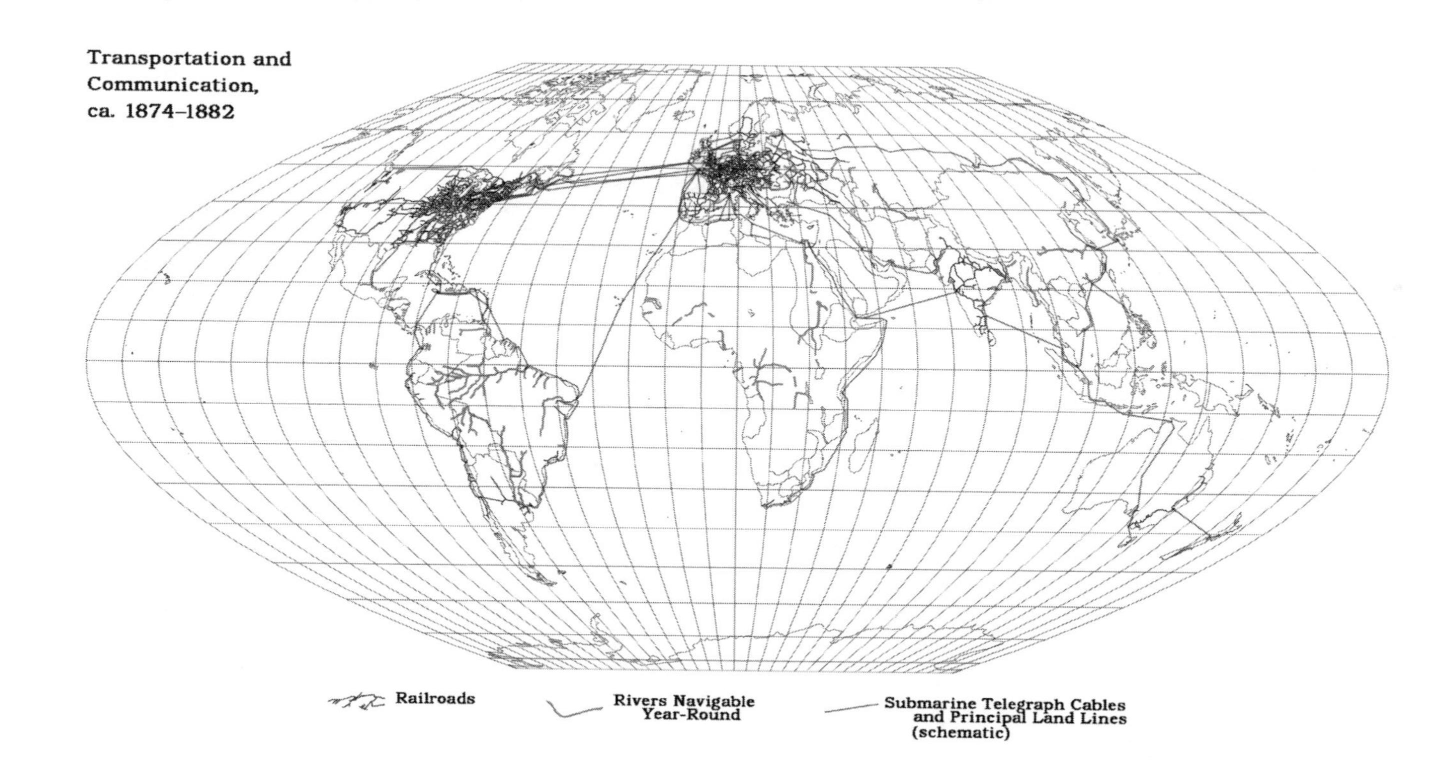

Figure 54. World transportation and communication networks at the time of the 1874 and 1882 transits. (Map by John Westfall.)

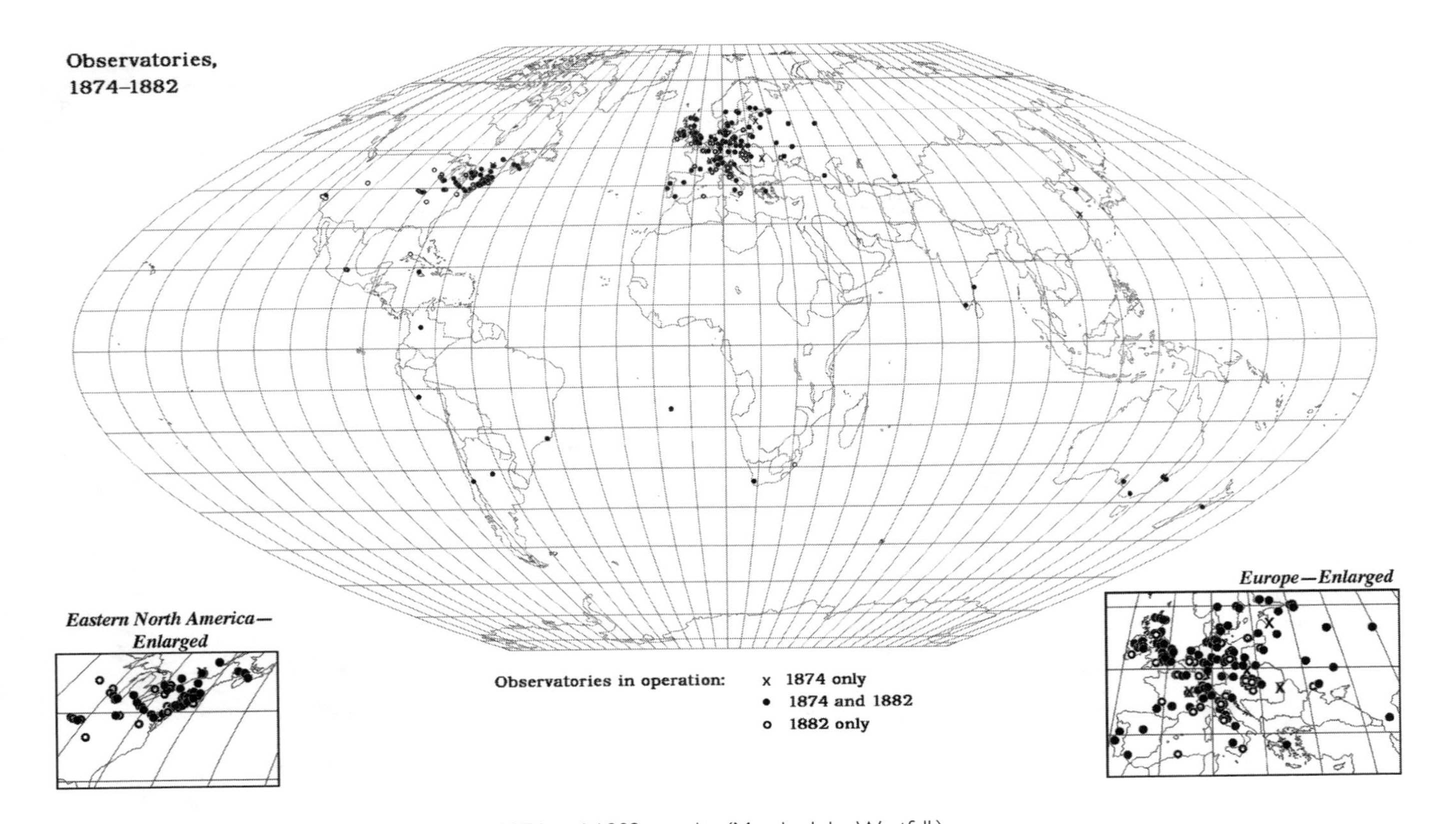

Figure 55. Observatories in operation for the 1874 and 1882 transits. (Map by John Westfall.)

closed-tube refractors, with two-element achromatic lenses much improved from the early models of John and Peter Dollond, by clever opticians like the German Joseph von Fraunhofer (1787–1826); in addition, reflectors with silver-on-glass mirrors had replaced the hard-to-maintain and easily tarnished speculum metal. Many of these instruments were equipped with drive mechanisms allowing them to track effortlessly the movements of the sun or stars without the observer having to make constant manual corrections. An important variant of this innovation was the heliostat, which used a clock-driven mirror to follow the sun across the sky to deliver its image to a long-focus fixed objective lens. Horizontal telescopes of this design, having focal lengths of up to 12 meters (40 feet), were specially developed for observations of the transits of Venus. The images of the sun they produced were large—about 110 mm (4.4 inches) across, or about four times larger than the image of the sun produced by an 8-inch f/15 achromatic refractor. A large image scale made for greater ease and accuracy in timing contacts or measuring the position of Venus on the sun with a micrometer. Though William Gascoigne had invented the eyepiece micrometer around the time of Horrocks, it was still suitable only for measuring distances of a few arc-minutes at most. It was not able to measure distances as large as those between Venus and the limb of the sun in mid-transit. Precise measurements of these larger angles had to await the invention of the heliometer, whose concept was first proposed by Servington Savery and Pierre Bouguer in the eighteenth century but was perfected—again—by Fraunhofer. In the heliometer design, a telescope objective lens is cut in equal halves—one can only imagine the anxiety of the optician required to do so! One segment is mounted so that it can be moved precisely parallel to the other, and the double image produced by this split-objective can be moved to allow the measurement of angles of up to over 2 degrees, perfect for the measurement of Venus on the sun.

Besides better telescopes and measuring devices, two new and entirely unanticipated avenues of research had opened up during the 105 years between the last of the eighteenth-century and the first of the nineteenth-century transits. These fields—spectroscopy and photography—would come to dominate astronomy in the twentieth century and form the basis of the emerging field of astrophysics.

With the spectroscope, one could use a prism or fine-ruled glass grating to analyze "white" light as from the sun into a full spectrum of color. Passing the light through a slit provided a sharper spectrum, which led to the identification of a multitudinous series of dark lines in the solar spectrum—the Fraunhofer lines. Though the French philosopher Auguste Comte (1798–1857) had gone on record in the 1830s as saying that the chemical composition of celestial bodies was something that could never, by any means, be known,[10] by the 1850s a German chemist, Robert Bunsen (1811–99) teamed with physicist Gustav Kirchoff (1824–87) to establish the fundamental laws of spectroscopy. Some of the dark lines in the solar spectrum occupied the same positions as the bright lines produced when substances such as sodium were heated to incandescence in the laboratory. Kirchoff realized that the bright lines were produced by the emission of light in particular wavelengths, while the dark lines were produced by the absorption of light. In the case of the Fraunhofer lines, these were produced when light from the brilliant surface of the sun, the photosphere, passed through the cooler solar atmosphere. Clearly, Comte had been wrong: henceforth it was possible in principle to identify chemical elements in the sun and the stars, and even in the atmosphere of a planet like Venus.

Indeed, quite apart from Venus's use as an auxiliary in determining the distance from the earth to the sun, astronomers were eager to study its spectrum as a means of finding out more about the planet itself. And an intriguing world it seemed to be. The atmosphere, which had first been deduced from the presence of the ring of light seen by Chappe and others at the transit of 1761, had been confirmed by telescopic observers such as William Herschel (1738–1822) in England and Johann Hieronymus Schroeter (1745–1816) in Germany. In the last decades of the eighteenth century, they had often remarked upon the planet's dazzling and featureless appearance and the falling off of light at the terminator as the sure signs of a dense, enveloping, and cloudy atmosphere. Beyond this, they parted ways; for Herschel, there was only an atmosphere to be observed, but Schroeter, a magistrate at Lilienthal (near Bremen), using a series of large but unwieldy reflectors to observe the planet, saw irregularities in its outline and sometimes a tiny detached point of light flickering in the darkness off the planet's brilliant cusp. He surmised, in 1792, that there were lofty mountains near the south pole of Venus (see fig. 56). Indeed, he calculated that his "enlightened mountain" reached a height of 47 kilometers (29 miles)! He also "improved" the accepted value of the rotation period of the planet, suggesting that his observations implied a period of 23 hours, 21 minutes, 19 seconds, which he later revised to 23 hours, 21 minutes, 7.9 seconds.

Though Herschel vehemently disputed Schroeter's findings, and though Schroeter's observatory was wrecked in 1813 during the Napoleonic Wars, a number of nineteenth-century observers seemed to confirm his reports of irregularities of the terminator. Hence, Schroeter's southern mountains of enormous height—the so-called Himalayas of Venus—entered into textbooks and the astronomical lore of the period, invoked, for instance, in the romantic drawings (see fig. 57) that illustrated *Les terres du ciel* (The Worlds in the Sky), a best-selling book by French popular-science writer Camille Flammarion (1842–1927).

Flammarion also suggested that Venus, being earthlike in so many respects—in diameter and mass, in rotation period, in the presence of a mountainous surface and a cloudy atmosphere—was probably inhabited:

> Of what nature are the inhabitants of Venus? . . . All that we can say is, that organized life on Venus must be little different from terrestrial life, and that this world is one of those which resembles our own most. We will not ask, then, with the good

Figure 56. View of Venus, drawn by Johann Hieronymus Schroeter (1745–1816). (From Schroeter, *Selenotopographische Fragmente* [Helmstedt: Johann Georg Rosenbusch, 1791].)

Figure 57. Fanciful view of the "Himalayas" of Venus, towering mountains in the southern hemisphere of the planet inferred from visual telescopic observations made during the nineteenth century. (From Camille Flammarion, *Les terres du ciel* [Paris, 1877].)

> Father Kircher, whether the water of that world be good for baptizing, or whether the wine would be fit for the sacrifice of the Mass; nor, with Huygens, whether the musical instruments of Venus resemble the harp or the flute; nor, with Swedenborg, whether the young girls walk around without clothing. . . . The only scientific conclusion which we can draw from astronomical observation is, that this world differs little from ours in volume, in weight, in density, and in the duration of its days and nights. . . . It should, then, be inhabited by vegetable, human, and animal races but little different from those of our planet.[11]

Given the existence of an appreciable atmosphere, astronomers were keen to find out whether it was, indeed, a breathable one. Did it contain the same gases as the atmosphere of the earth? Here direct visual observation was insufficient; analysis could only be made with the spectroscope. In the years leading up to the 1874 transit of Venus, the German spectroscopist Hermann Vogel (1841–1907) of the Potsdam Observatory reported that there seemed to be weak enhancements of some of the spectral lines that implied the existence of water vapor and oxygen in the Venusian atmosphere. Astronomers hoped for more definitive results from observations that could only be made during the transit itself. At the 1868 eclipse in India, during the brief period of totality when the sun was hidden behind the moon, a number of astronomers including James Francis Tennant (1829–1915) and Jules Janssen (1824–1907) at Guntoor, Norman Robert Pogson (1829–91) at Masulipatam, and Georges A. P. Rayet (1839–1906) at Wha-Tonne on the coast of the Malay Peninsula, trained the slits of their spectroscopes upon the pinkish tongues of gas, or prominences, projecting into the dark space above the limb of the sun. They all observed a series of bright emission lines, of which the most prominent were red, orange, and blue. It was obvious that during a transit of Venus similar observations might be attempted by training the slit of the spectroscope on the brilliant ring formed by the atmosphere of the planet, during the brief interval when the planet was straddling the sun's limb.

Photography, the other technological innovation that was introduced between the eighteenth-century transits and those of the nineteenth century, was if anything even more far-reaching and was deemed to have the greatest potential for transit-based determinations of the solar parallax. Photography's value was obvious: it produced a permanent and apparently objective record of the event, a record that could moreover be carefully measured, at leisure, repeatedly if necessary, in order to determine contact times or positions of the planet during its progress across the sun. These advantages appeared so overwhelming that most of the transit expeditions in 1874 gave higher priority to photography than to the traditional visual contact timings.

The first practical photographic process was invented in 1839 by Louis Jacques Mandé Daguerre (1789–1851). Its product, the "daguerreotype," was an image etched on a copper plate. The new technique spread rapidly; by the early 1840s the American amateur astronomer and photographer John William Draper (1811–82) had recorded an image of the moon and a spectrum of the sun.[12] In 1845, Armand H. L. Fizeau (1819–96) and Léon Foucault (1819–68) obtained an image of the sun showing sunspots and limb darkening. Though daguerreotypes have the advantage of being perhaps the most permanent form of photography, they have at least two disadvantages: they form on an opaque medium, which makes them difficult to reproduce, and they are very "slow." Their low sensitivity to light was not a serious problem for solar work, however, and the daguerreotype process was still used by some observers for the transits of Venus in 1874 and 1882.

By then, however, there were alternatives. A much "faster" medium than the daguerreotype became available in 1851, when Frederick Scott Archer (1813–57) demonstrated his "wet plate" process.[13] Here a glass plate is coated with a wet, viscous collodion solution, which can be printed or enlarged after being dried. The process remained awkward because the plate had to be coated with its "emulsion" shortly before being exposed and could not be allowed to dry until it was finally developed. A wealthy English amateur, Warren de la Rue (1815–89), began to use wet plates for solar photography in 1861, demonstrating their potential for transit photography. Moreover, an even more convenient form of photography, the dry plate process, became available right on the eve of the nineteenth-century transit season. In 1874—the year of the first transit—the process became available commercially and in "do-it-yourself" form as published by William de Wiveleslie Abney (1842–1920).[14] But the appearance of the dry plate was not quite soon enough to influence nineteenth-century transit observers greatly, and the wet-plate process remained the most widely used form of photography for both the transits of 1874 and 1882.

The Grand March of Technology

Astronomers preparing for the nineteenth-century transits had many other advantages over their eighteenth-century predecessors. They had fully absorbed the eighteenth-century experiences and believed they knew what to expect. They were not going to be taken off-guard by the black drop effect, which had been so entirely unexpected by observers in 1761 and even by those in 1769, and which had interfered so decisively with the accuracy of the timings. The longitudes of observing stations, which were essential to the success of the Delislean method in particular, could be much more precisely determined than had been possible in the eighteenth century, when the methods of "lunars" or the observation of eclipses of the satellites of Jupiter had been the only methods that were feasible. Since then there had been several improvements: Marine chronometers, able to keep time accurately enough to provide a practical method of determining longitude at sea, had been devised by John Harrison (1693–1776) and independently by Pierre Le Roy (1717–85) and Ferdinand Berthoud (1727–1807). The basis of the method was to set a chronometer in port to, say, Greenwich or Paris time and carry that time, in the form of the chronometer, to a distant destination. Previous clocks lost or gained too much time and were completely useless. The first satisfactory marine chronometers were just appearing during the decade of the eighteenth-century transits. In 1761–62, Harrison's famous "Number 4" lost just 1 minute, 54.5 seconds on a 147-day voyage from England to the West Indies and back. This error corresponded to a difference of slightly less than half a degree in longitude—small enough to be acceptable for purposes of navigation but of course much too gross an error to be of any use astronomically. Thus they did not figure significantly in any of the transit expeditions. But by the first half of the nineteenth century, ships would sometimes "transport" large numbers of chronometers from port to port, taking their mean for a more accurate determination of time differences. In this way times accurate to within a few seconds could be achieved. Lunar culminations—determining the time the moon crossed the celestial meridian at a given location—were also widely used to determine longitude by the nineteenth-century transit expeditions. But when available, the most accurate positions of all were those determined by method of telegraphic "exchange of signals" between the observing station and a place of known longitude. The process was complex and required the cooperation of the telegraph company—for instance, it was necessary to measure and correct for the transmission time of the signal—and could not be used for isolated places such as Hawaii or Kerguélen Island, which had not yet been reached by telegraph lines.

The solar parallax is itself a number of arc-seconds measured upon the sky; it does not by itself provide the earth-sun distance, which is determined only by adoption of a suitably accurate value for the equatorial diameter of the earth. It follows that the distance to the sun cannot be known to a higher degree of accuracy than the size of the earth itself. For the 1761 and 1769 transits, the most recent "spheroid" (a description of the ellipsoid that best fitted the actual earth) had been calculated by French astronomer Nicolas-Louis de Lacaille (1713–62) in 1753; it was about 0.08 percent smaller than the modern value. By the nineteenth-century transits, the standard spheroid was the "Clarke 1866," measured by British geodesist Alexander Ross Clarke (1824–1902), only 0.001 percent larger than the current "WGS84."

Since the 1760s, there had also been improvements in understanding of the process of observation and measurement itself. The observers of the eighteenth century had trusted Halley's authority and naively hoped, with him, to be able to visualize sharply the contact times and trust to their chronometers. But the contacts proved to be incredibly "fuzzy" events. There were the black drop and related phenomena resulting from the fact that the limbs of the sun and Venus were poorly defined, owing to a combination of factors: atmospheric seeing, solar limb darkening, and the limited resolving power of the instruments. The planners for the nineteenth century could not limit the effects of seeing by locating at stations where the sun would be as far as possible above the horizon, because then the distance from the Halleyan and Delislean Poles would have rendered the timings useless for parallax determination and defeated the whole purpose. The drop-off of light near the limb of the sun, known as limb darkening, could have been reduced by using warm-hued (yellow, orange, or red) filters. However, this effect seems to have been unrecognized and was not deliberately employed by any observers at the transits (such filters were used by a few, apparently quite fortuitously). Standardization of telescopes, largely to 4- to 6-inch refractors, helped minimize differences in the impressions between various reporters. But even when all of these factors were taken into account, there remained obnoxious individual differences among the observers. Unrecognized by astronomers in the eighteenth century, these biases or personal errors had been noted and systematically studied by astronomers in the early nineteenth century and were described by the general term of "personal equation."[15]

Thus one observer, measuring the time a star crossed the wire of a transit instrument, might differ by a significant fraction of a second from another. A systematic difference of only eight-tenths of a second between David Kinnebrook, an assistant at the Greenwich Observatory, and his director, Nevil Maskelyne, led to the assistant's dismissal. Maskelyne had concluded that the difference was the result of Kinnebrook's having "commenced a vicious way of observing the times of [these stellar] transits too late. . . . As he had unfortunately continued a considerable time in this error before I noticed it, and did not seem to me likely ever to get over it, and return to a right method of observing, therefore, though with great reluctance . . . I parted with him."[16] It was later recognized that the difference could be explained in part by whether an observer attended first to the star's crossing the wire then to the clock beat or the reverse. However, even under the best circumstances there were discrepancies, and astronomers at major observatories where highly accurate positional work was carried out eventually began to publish "personal equations" along with their measurements. Personal equations would be obviously relevant, in spades, to the observation of the contacts at the forthcoming transits, so a number of steps were made to determine the personal equations and to properly "calibrate" the observers.

In Paris, Charles Wolf (1827–1918) employed a transit simulator consisting of a series of lamps and screens in the window of the library in the senate building of the Jardin du Luxembourg that were visible to an observer using telescopes at the Paris Observatory. Apparent contacts could be compared with actual contacts by means of telegraph signals transmitted back and forth between the locations. A transit simulator was also set up on a building a little less than a mile away from the U.S. Naval Observatory and viewed by astronomers there.[17] These efforts—though in retrospect seeming to have limited practical benefits—were well-intentioned attempts to apply the well-worn principle of "practice makes perfect." They were attempts to familiarize observers in advance with the conditions likely to be experienced during the incredibly rare actual events of the transits themselves.

ELIMINATING THE "PERSONALITY OF THE EYE"

In the end, many of the transit planners decided that the personal equations affecting different observers making visual observations could never be determined with sufficient accuracy and that a more impartial method than the human eye was to be preferred: they saw this more objective method as photography. Photography was seen as a way of getting around what the American astronomer Ormsby Macknight Mitchel (1809–62) had called the "personality of the eye." It seemed that it might provide an absolute and completely unvarnished record of the truth.

Even better might be a moving picture, a series of images of the planet in motion across the solar disk. After all, as was realized by at least a few of the more subtle philosophers of the science of observation, reality does not consist of a choppy succession of static shapes; instead reality is fluid and continuous. The still photograph was, at best, a rude approximation to reality, unable to contend with or express phenomena whose essential aspect was movement.

The pursuit of an objective record of the transit of Venus would lead, in the end, to the most innovative and exciting of all the devices employed by astronomers at the transits of 1874 and 1882.

That device—the revolver-camera—would be the precursor of cinematography itself. Thus the transits would be memorable not only for the measurements of astronomers but also for the uniquely moving record of that transient event.

The beautiful Planet of Love, or her shadow, passing between the earth and the sun, would become the first star of a motion picture. Her dignified progression would be etched, plate by plate and frame by frame, with the massiveness and solidity of stone sculpture. But if some of these frames were presented in serial form and in rapid succession, the transit began to move, and it was transformed into a different kind of occurrence, as evanescent and as fleeting as a wisp of cloud or the white whirling garment of a goddess.[18]

11

FROM ENLIGHTENMENT TO PRECISION

When the last transit occurred the intellectual world was awakening from the slumber of ages, and that wondrous scientific activity which has led to our present advanced knowledge was just beginning.

—William Harkness, U.S. Naval Observatory (1882)

A PARALLAX CHALLENGED

The (British) *Nautical Almanac* and the *American Ephemeris* adopted Johann Franz Encke's values of the solar parallax and the astronomical unit, based on the contact timings of the transits of Venus in 1761 and 1769. This might give the impression the matter had been settled at least until the next transit, in 1874. However, in addition to the triangulation-based transit-of-Venus method, there were several more sophisticated—and increasingly competitive—means for deriving the same results. The moon's motion could be analyzed on the basis of gravitational theory to derive values of the solar parallax. A study based on the *parallactic inequality*, the displacement of the moon in its orbit by the tidal effect of the sun, led to parallaxes published by Peter Andreas Hansen.[1] Investigations of the *lunar equation*, the shift of the sun's position caused by the monthly wobble of the earth about the earth-moon center of gravity, formed the basis of results published by Hansen, Urbain Jean Joseph Leverrier (1811–77), Simon Newcomb, and Edward J. Stone (1831–97).[2] The solar parallax could be derived from the effect of the gravitational pull of the earth on both Venus and Mars; from the *annual aberration* of starlight—a small displacement in the positions of stars caused by the earth's annual motion around the sun; and from attempts to measure the parallax of Mars against the stars as Flamsteed and Cassini and Richer had attempted in the eighteenth century. It could even be derived from attempts to determine directly, by experiment, the exact value of the speed of light, as attempted by Léon Foucault in France. A table of solar parallaxes containing the most important results before 1874 is presented in Table 3. Most of them clustered around a value of 8.85 arc-seconds, which was notably larger than Encke's.

As the 1874 transit approached, it was obvious that the value of the solar parallax still had not been pinned down. Even the staid *Nautical Almanac* yielded to the pressure to revise the solar parallax, although only in 1870, when they changed the "official" value from 8.5776 to 8.95 arc-seconds, a value that they carried until 1882. This implied an earth-sun distance of 147 million kilometers (91 million miles), based on the "Clarke 1866 ellipsoid" for the figure of the earth, moving the earth over 6 million kilometers (4 million miles) closer to the sun than Encke's distance.

Adopting the "correct" value was more than a matter of academic interest on the part of pedantic astronomers or an attempt to give a truthful answer to the child's question, "How far is the sun from the earth?" The adoption of an incorrect value rippled—because of Kepler's third law—through the whole fabric of celestial mechanics and led to erroneous values of the distances and hence the masses of all the other planets. This produced inconsistencies: one value was contingent on another, and an error in fundamentals would throw off everything else. The correction of the solar parallax and the astronomical unit would presumably assure the smoother running of the intricate system. It would provide, as it were, a new balance wheel setting it back in order—or at least removing the need for something truly drastic. As French astronomer Hervé Faye (1814–1902) pointed out in 1862, "If we persist in our false evaluation of the parallax, the hypothesis that there exists some other planet which has gone unperceived until now . . . will need to be considered. And, since we cannot see such a probable planet, science finds itself forced into an impasse. . . . All the discordances, all the contradictions that menace the future . . . of astronomy will disappear if the direct determination of solar parallax . . . gives us 8".9 instead of 8".57."[3]

Laying Plans

Although "refining" the solar parallax was considered so important, given the proliferation of other methods for obtaining its value, the question remains why such effort was to be expended to observe the transits of Venus. Partly it was a matter of sheer inertia; in a sense, astronomers had begun eagerly awaiting the 1874 transit as soon as the fourth contact of the 1769 transit had passed. Serious planning began with a paper read to the Royal Astronomical Society in 1857 by British Astronomer Royal Sir George Biddell Airy (1801–92; see fig. 58).[4] This influential astronomer threw his influence behind the transit of Venus method. His prestige—and his accomplishments—were already impressive. Judged snobbish and conceited by his classmates, without friends and not regarded as particularly gifted by his professors, he was an obsessive worker—as he was to remain for the rest of his life—and graduated from Trinity College as Senior Wrangler in 1823. By that time, at the age of twenty-two, he already thought well enough of himself to write an autobiography! In 1828, he became director of the Cambridge Observatory. A prodigious scientific writer, the young Airy fought his way upward in the British astronomical community and achieved the recognition he dreamed of by being appointed Astronomer Royal in 1835. He reorganized the Royal Observatory,

Table 3. Selected estimates of the solar parallax, 1771–1874

Method	Value (arc-seconds)	Notes (see bibliography)
Transit of Venus contact times (1761/69) [mean = 8.743 ± 0.039]		
Lalande, 1771	8.59	Midrange of 8.55–8.63 (Dick et al. 1998)
Pingré, 1772	8.80	(Dick et al. 1998)
Encke, 1822	8.5309	(Encke 1822, p. 159)
Encke, 1824	8.5776	(Encke 1824)
Ferrer, 1833	8.58	(Weinberg 1903)
Powalky, 1864	8.86	(Weinberg 1903)
Hall, 1865	8.8415	(Weinberg 1903)
Stone, 1868	8.91	(Weinberg 1903)
Faye, 1869	8.8	Midrange of 8.7–8.9 (Weinberg 1903)
Newcomb, 1869	8.87	(Weinberg 1903)
Powalky, 1870	8.7869	(Powalky 1870)
Powalky, 1872	8.77	PE = ±0.03 (Weinberg 1903)
Parallax of Venus		
Gould, 1856	8.754	PE = ±0.311 (Weinberg 1903)
Parallax of Mars [mean = 8.968 ± 0.177]		
Henderson, 1834	9.028	(Weinberg 1903)
Taylor, 1835	10.841	(Weinberg 1903)
Taylor, 1836	8.595	(Weinberg 1903)
Gould, 1856	8.495	(Weinberg 1903)
Bond, 1857	8.605	(Weinberg 1903)
Stone, 1863	8.943	(Weinberg 1903)
Winnecke, 1863	8.964	(Weinberg 1903)
Ferguson, 1865	8.7225	PE = ±0.0752 (Weinberg 1903)
Liais, 1865	8.760	(Weinberg 1903)
Stone, 1865	8.943	(Weinberg 1903)
Newcomb, 1867	8.855	(Weinberg 1903)
Schultz, 1867	8.87	PE = ±0.02 (Weinberg 1903)
Parallax of minor planet		
Galle, 1874	8.858	MP Flora (Galle 1874)
Parallactic inequality [mean = 8.784 ± 0.028]		
Laplace, 1802	8.633	(*Méchanique Céleste* 3, p. 282)
Burckhardt, 1812	8.605	(Weinberg 1903)
Laplace, 1823	8.647	(Weinberg 1903)
Bürg, 1825	8.595	(Weinberg 1903)
Airy, 1848	8.7977	(Campbell & Neison 1880)
Airy, 1849	8.624	(Weinberg 1903)
Hansen, 1854	8.92	(Campbell & Neison 1880)
Hansen, 1855	8.859	(Weinberg 1903)
Airy, 1859	8.7977	(Campbell & Neison 1880)
Airy, 1861	8.788	(Weinberg 1903)

Hansen, 1867	8.916	(Stone 1878)
Newcomb, 1867	8.920	Combined results (Weinberg 1903)
Newcomb, 1867	8.842	(Weinberg 1903)
Newcomb, 1867	8.844	Combined results (Weinberg 1903)
Newcomb, 1867	8.848	Mean result (Weinberg 1903)
Stone, 1867	8.850	PE = ±0.056 (Weinberg 1903)
Newcomb, 1868	8.8449	(Campbell & Neison 1880)
Lunar equation [mean = 8.938 ± 0.010]		
Hansen, 1856	8.97	(Hansen, "On the Value . . . ," 1863)
Leverrier, 1858	8.95	(Weinberg 1903)
Hansen, 1863	8.9159	(Hansen, "Calculation of . . . ," 1863a)
Newcomb, 1867	8.9405	(Weinberg 1903)
Stone, 1867	8.916	(Stone 1878)
Planetary equation [mean = 8.704 ± 0.073]		
Leverrier, 1861	8.581	(Weinberg 1903)
Leverrier, 1872	8.8335	PE = ±0.0024 (Weinberg 1903)
Powalky, 1872	8.698	(Weinberg 1903)
Stellar aberration [mean = 8.808 ± 0.003]		
Richardson, 1840	8.8064	(Weinberg 1904)
W. Struve, 1841	8.79506	(Weinberg 1904)
W. Struve, 1842	8.80464	(Weinberg 1904)
W. Struve, 1843	8.81584	(Weinberg 1904)
Lindhagen, 1853	8.8060	(Weinberg 1904)
W. Struve, 1853	8.8081	(Weinberg 1904)
O. & W. Struve, 1872	8.81584	(Weinberg 1904)
Nyrén, 1873	8.8004	(Weinberg 1904)
Nyrén, Struve et al., 1873	8.81929	(Weinberg 1904)
Times of eclipses of Jupiter's satellites		
Delambre, 1792	8.898	(Harkness 1891, p. 28)
Glassenapp, 1874	8.7619	PE = 0.0178 (Harkness 1891)
Combination of methods		
Powalky, 1872	8.77	(Weinberg 1903)

Note: When not given by the authority, parallax values from indirect methods (parallactic inequality, stellar aberration, lunar equation, planetary equation), as well as probable errors (PE) have been calculated by John Westfall. When there have been three or more estimates using the same method, the method's unweighted mean and probable error have been calculated by John Westfall.

Figure 58. Portrait of Sir George Biddell Airy (1801–92). (Courtesy of the Mary Lea Shane Archives of the Lick Observatory, University of California–Santa Cruz.)

making it the world's leading institution in positional astronomy, then the most prestigious branch of astronomy. His research interests were many. In addition to his astronomical labors, he often served as a consultant to the British government on technical matters (we owe the 4-foot, 8½-inch standard railway gauge to him!).

Airy had successfully weathered a withering storm of criticism in the years after 1846, when he was widely criticized for his failure to pursue actively the search for the "eighth planet" (Neptune) after the brilliant young Cambridge astronomer and mathematician John Couch Adams (1819–92) had submitted to him his predictions of its position. The initial credit for the "virtual" discovery thus went to Leverrier of France, whereas the telescopic identification was made by Johann Gottfried Galle (1812–1910) in Berlin. Both were blows to British pride. Airy, however, was convinced that he had done his duty. With his massive sense of his own integrity intact he remained in the position of Astronomer Royal and, far from becoming more subdued, only redoubled his prodigious efforts on behalf of British science.

Inevitably, he became the chief figure in planning the British efforts for the 1874 and 1882 transits of Venus. After stating in his 1857 paper that the effort "will require all our cares and all our ingenuity," and "upon this measure depends every measure beyond the Moon,"[5] he appeared to favor the method of Delisle (timing the transit contacts rather than the duration of the entire transit).

With men like Airy organizing them, it was assured that the transit programs of the nineteenth century were to be much more organized and better equipped than those of the eighteenth century.

In the two decades that followed Airy's call to arms, the Great Powers (and some near-greats, too) established transit commissions to direct their efforts. Great Britain was the exception; it needed no commission, since it had Airy, who was an institution in himself, and could count on the cooperation and activity of the Royal Astronomical Society. A discussion of the viability of various methods of determining the value of the solar parallax was especially lively in France. Leverrier, perhaps the leading astronomer of France, was a partisan of celestial-mechanics approaches and on the whole intensely skeptical of the transit method; Foucault was a pioneer of the determination of the speed of light. Moreover, the country was politically unsettled. At last, a French *Commission du passage de Vénus de l'Académie des sciences* was authorized in 1869, and though it first met in January 1870, its work was interrupted by the Franco-Prussian War of 1870–71. The French Commission was strongly influenced by the amateur astronomer Armond Hippolyte Louis Fizeau (1819–96).[6] A German commission, established in 1869, was chaired by Hansen. In the United States, Congress appropriated the sum of $2,000 to establish the U.S. Transit of Venus Commission at a relatively late date—March 1871. It was poorly funded but well endowed with talent. Headed by Rear Adm. Benjamin F. Sands (1812–83), superintendent of the Naval Observatory, its other members included Joseph Henry (1797–1878), president of the National Academy of Sciences; Benjamin Peirce (1809–80), superintendent of the Coast Survey; and Simon Newcomb and William Harkness (1837–1903) of the U.S. Naval Observatory. Denmark and Italy planned observations only in their own territories, although subsequently Italy decided to send an expedition to India in 1874. Incidentally, at the time of Airy's speech Germany and Italy had not yet become nations; Bismarck founded the Prussia-dominated German Empire in 1871, while the unification of Italy was completed only in 1872 when the kingdom occupied the Papal States.

Decisions about the coming transit expeditions were made in the metropolises of Berlin, London, Paris, St. Petersburg, and Washington, and on a smaller scale in Copenhagen, The Hague, Mexico City, and Rome. The chief constraint faced by the planners was the timing of the transit itself. The event, on December 9, would occur near the winter solstice in the Northern Hemisphere with the sun well south of the equator. The map of zones of visibility for the 1874 transit of Venus shows that none of the event would be visible in the Americas, and it would end before sunrise for much of Europe (see fig. 59). The Middle East and Africa would have good views of egress, which would also be visible from southeastern Europe and much of Russia, but still with the sun low in the sky and poor weather prospects. On the other hand, the central Pacific basin would enjoy good views of the most dramatic phase of the transit—ingress—but only south and southeast Asia, Australia, and New Zealand, as well as most of the Indian Ocean would enjoy a view of the transit in its entirety.

Halley's method required the sun to be suitably placed above the horizon for the 3½ hours between second and third contact (out of a total of 4½ hours for the entire transit). This meant an ideal northern station should be in Siberia, while the ideal southern station would have ended up on the then largely unexplored Antarctic coast. The Delislean Poles were also awkwardly placed. Ingress timings would have been best made in the Indian Ocean south of the Cape of Good Hope or in the landless northern Pacific; egress timings in northern European Russia and in the Drake Passage southwest of Cape Horn. Had the earth consisted entirely of dry land, with all stations equally accessible, and weather nothing to worry about, these locations would no doubt have been occupied. But since none of these conditions were met with in reality, the transit parties had to compromise and settle on sites located between the ideals dictated by geometry and the realities imposed by geography.

One important decision was whether the stations under a commission's control should be placed in preference to the method of Halley or that of Delisle. Airy himself initially favored the method of Delisle, concluding that the Halleyan approach was unfavorable for the transit of 1874.[7] He gradually modified this extreme view, coming to admit later that stations should be placed with regard to both methods.[8] Nonetheless, the influential British popularizer of astronomy, Richard Anthony Proctor (1837–88), took exception to Airy's preference for the Delislean method and instead argued

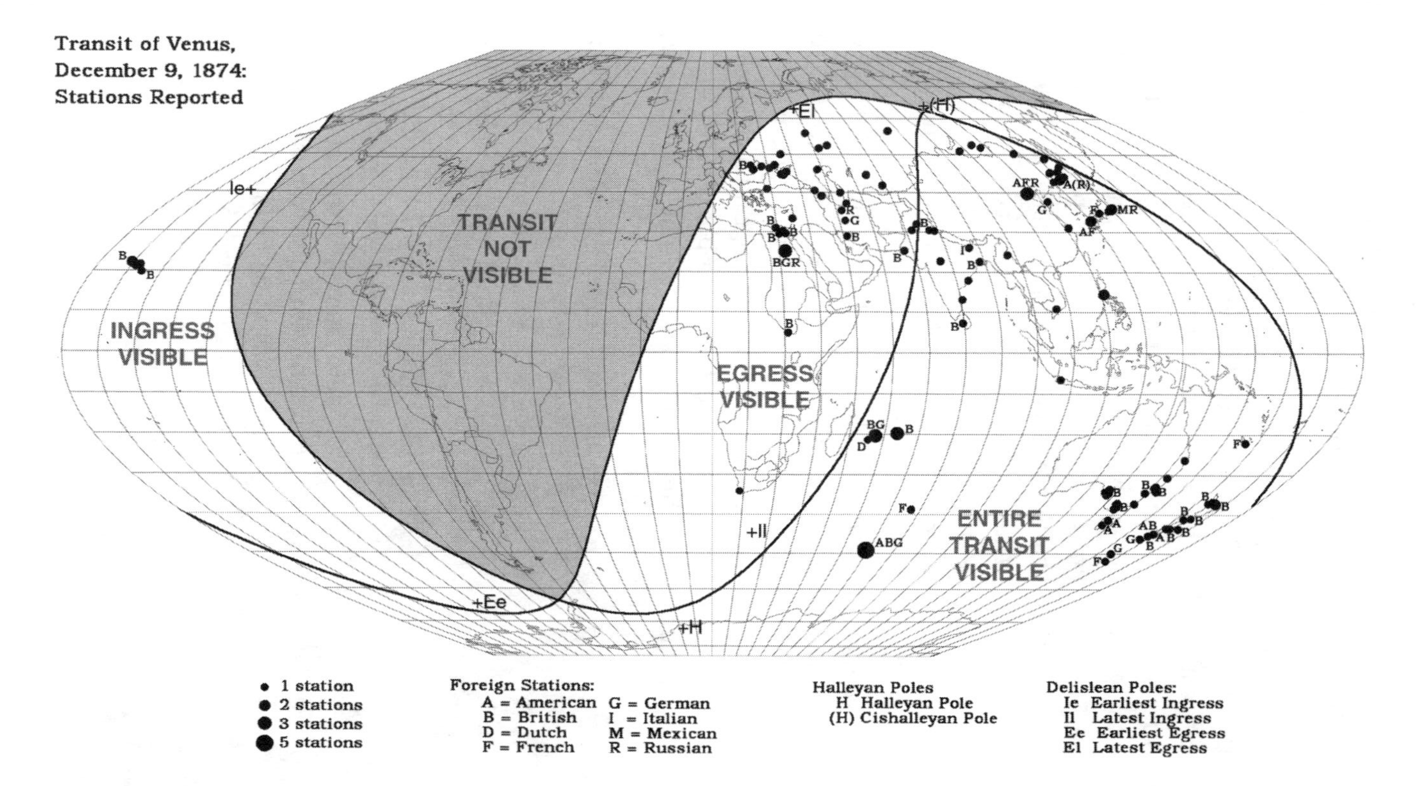

Figure 59. Visibility zones, Halleyan and Delislean Poles, and observing stations with published results for the transit of Venus of December 9, 1874. (Map by John Westfall.)

strongly—perhaps partly out of a sense of British patriotism—for the establishment of Halleyan observing stations for the 1874 transit (see fig. 60). In his books *Transits of Venus: A Popular Account* (1874) and *The Universe and the Coming Transits* (1874), he attacked Airy's opinions with fury.[9] His view of Airy was that "such persons are apt, especially when high in their respective departments of service, to forget that they are servants of and paid by the public, acting as though in authority over others than their proper subordinates."[10] In 1868–73, Proctor published a series of papers on the coming transit in the *Monthly Notices of the Royal Astronomical Society* and eventually in the *Times*, criticizing Airy and advocating the establishment of stations in the far south.

Proctor had been educated at King's College, London, and at St. John's College, Cambridge. Although Proctor had finished only as twenty-third wrangler, he had been elected a fellow of the Royal Astronomical Society in 1866 and had quickly risen to become secretary and editor of its *Monthly Notices*, where he published his articles without independent review. In that same year he suffered huge financial losses owing to a bank failure and was forced to turn to popular writing in order to support himself and his family. These personal circumstances may account in part for the vehemence of his attacks on the government-subsidized Astronomer Royal. However, as with the Neptune controversy over a quarter-century before, Airy's head was bloodied but unbowed. When he came to write his autobiography, he hardly mentioned the Neptune controversy at all, and remembered about the transit of Venus discussions only that "there was some hostile criticism."[11] Proctor's invective also extended to

Figure 60. Photograph of Richard Anthony Proctor (1837–88), taken in 1875. (Courtesy of the Mary Lea Shane Archives of the Lick Observatory, University of California–Santa Cruz.)

Rear Adm. Sir George Henry Richards (1819–96), the hydrographer for the Royal Navy. Richards defended himself in dignified Admiralty style: "Enthusiasts no doubt there are who, however accomplished they may be as astronomers, are wanting and cannot but be deficient on many subjects which it is as necessary to take into account as astronomy in a question of this kind; and hence we are told to send to the Antarctic Continent and to visit a variety of small rocks interspersed over the Southern Ocean at distances from each other varying from 1,000 to nearly 4,000 miles, many of which are actual myths, while on those which do exist, it is certain that there is no anchorage for a ship, and that even landing would be generally impossible."[12]

Proctor replied with his usual vigor, commenting on a chart of proposed southern observing stations (see fig. 61): "one naturally feels doubtful about Admiralty statements which would appear to be variable according to official requirements. It did not seem well to insert any island, or group of islands, in the chart with some such note as 'Here, if convenient to those in authority, there is an island,' or 'this group of islands can be regarded as a reality or a myth as may be required,' and so on."[13] After Richards registered a complaint to the council of the Royal Astronomical Society, Proctor resigned as secretary and departed on a lecture tour to the United States. A year after the transit he emigrated to Florida.

It is just as well that Proctor's suggestions were not followed in detail. Two of the islands and island groups he mentioned, "Royal Company Islands" and "Emerald Island," were apparently taken from the notoriously inaccurate 1864 edition of Admiralty Chart 2683 of the Pacific Ocean and were two of the 123 nonexistent islands deleted for the 1875 edition.[14] Likewise, the authors have been unable to find Proctor's "Kemp Island" on modern maps; perhaps it was one of the islands off the Kemp Coast (or Kemp Land), although placed about 7 degrees of latitude too far north and 15 degrees of longitude too far east. At any rate, the Kemp Coast and four other actual Antarctic places suggested by Proctor (Adélie Land, Enderby Land, Possession Island, and Sabrina Land) lie south of the average front of pack ice, even at its minimum season. However, while creating considerable ill feeling, Proctor apparently influenced

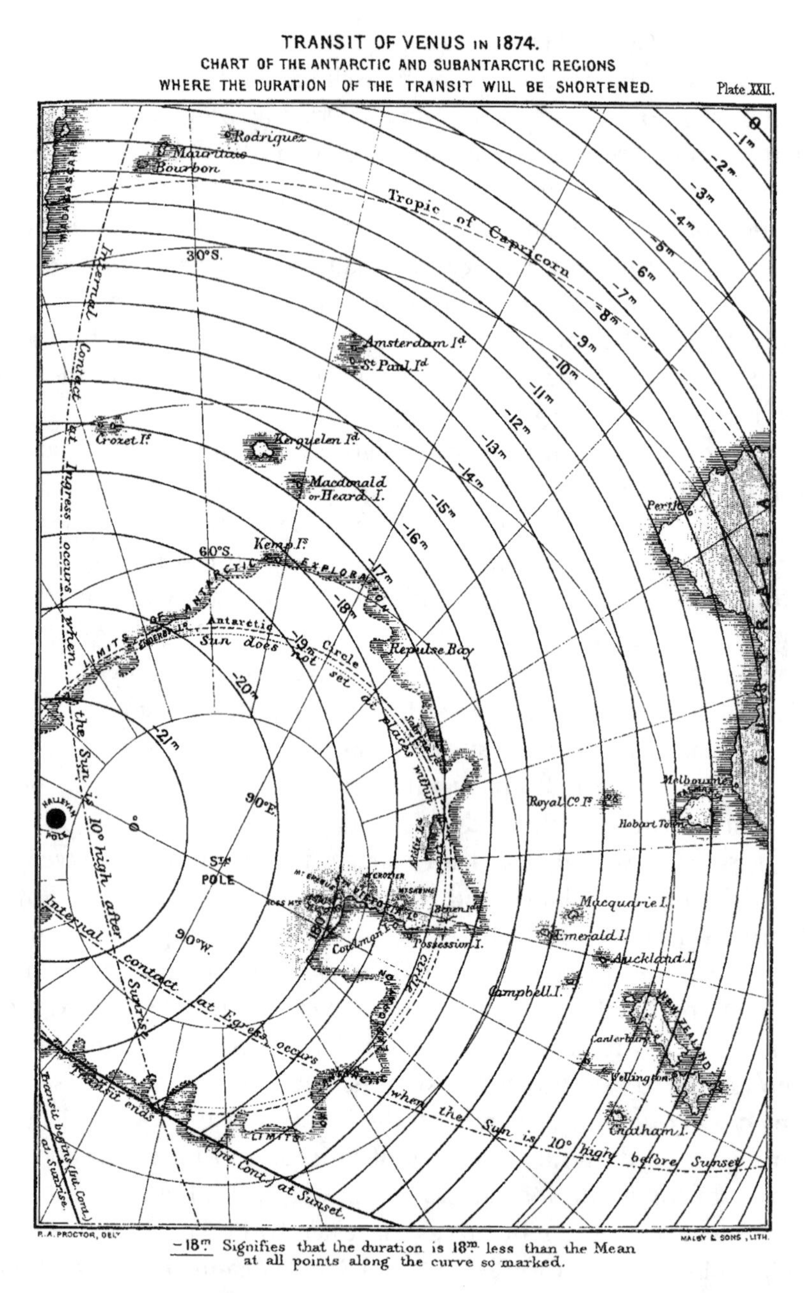

Figure 61. Map of the antarctic and subantarctic by Richard A. Proctor (1874), showing the limited knowledge of the region at that time, with the southern stations suggested by him, at least two of which (Emerald and Royal Company Islands) did not exist.

Airy and others to some extent. At least they dispersed the British observing stations more widely than they had first planned, and their decision to do so probably influenced other nations as well.

The Transit Parties

If one compares the 1874 transit visibility map with the distribution of observatories at the time, it is obvious that only the observatories at Cape Town, Chennai, Adelaide, Melbourne, and Sydney would be able to watch the transit. This meant that expeditions were definitely needed.

The "official" expeditions that actually observed the 1874 transit —or tried to but were clouded out—are listed in Table 4. The overseas expeditions were almost always sponsored by national governments. In addition, some governments organized parties within their own borders, while a few observatories organized dispersed groups of observers in their vicinities. The table includes some individuals who observed on their own but who forwarded their observations to the national commissions. Only limited information is available for some observing stations, and in some cases nothing more than the fact that they were used by at least one observer for the 1874 transit.

For all nations, expedition planning had to be done hurriedly, mostly between 1870 and 1874. Thanks to Airy, Britain was ahead of the other countries. The total funding for all the "official" British expeditions amounted to an impressive £15,500; in addition their parties included military personnel and often used Royal Navy ships and supplies. As one might expect, Airy prepared detailed instructions and observing forms. He and his superiors (the Admiralty and the "Board of Visitors" of the Royal Greenwich Observatory) settled on expeditions to five "districts": A, Egypt; B, the Sandwich (Hawaiian) Islands; C, Rodrigues Island (in the western Indian Ocean); D, New Zealand; and E, Kerguélen (in the far southern Indian Ocean, named the "Isle of Desolation" by Captain Cook). Each district actually included several observing stations, separated from each other so as to maximize the prospect of clear skies for at least one station in each district.

Table 4. Organized parties for the 1874 transit of Venus

Sponsoring country	Leader (with lifespan if known)	Results (notes)
	Australia (states given in parentheses)	
Adelaide (SA)	J. H. E. Brown (1856–1903)	VIS (1)
Adelaide (SA)	A. W. Dobbie (1843-?)	VIS (1)
Adelaide (SA)	J. D. Smeaton	VIS (1)
View Hill (VIC; near Bendigo)	C. Moerin	VIS (2)
Melbourne (VIC)	R. L. J. Ellery (1827–1908)	VIS (2)
Mornington (VIC)	W. P. Wilson (?–1874)	VIS (2)
Glenrowan (Glen Rowan, VIC)	Morris (Cap., Royal Artillery)	VIS (2)
Goulburn (NSW)	A. Liverside (1847–1927)	VIS (3)
Eden (NSW)	W. Scott	VIS (3)
Sydney (NSW)	H. G. A. Wright (?–1901)	VIS (3)
Sydney (NSW)	F. Allerding	VIS (3)
Armidale (NSW)	Belfield	VIS (3)
Woodford (QLD)	L. A. Vessey	VIS (3)
	Egypt	
Abbassiyya (Abbasseyeh, Egypt)	M. Bey al Falaki (1815–85)	VIS (4)
Rejef (Rageef, Sudan)	C. G. Gordon (1833–85)	VIS (5)
	France	
Île St.-Paul (St. Paul's I., FSAL)	E. A. B. Mouchez (1821–92)	VIS
Ho Chi Minh City (Saigon, Vietnam)	G. Héraud	VIS
Beijing (Pekin, China)	G. E. Fleuriais (1840–95)	VIS
Nagasaki (Japan)	P. J. C. Janssen (1824–97)	VIS (6)
Kobe (Japan)	Delacroix (Ensign)	VIS (6)
Nouméa (New Caledonia, FOD)	C. L. F. André (1842–1912)	VIS
Campbell I. (New Zealand)	J. J. A. Bouquet de la Grye (1827–1909)	NV
	Germany	
Thebes (Egypt)	von Auwers (1838–1915)	VIS (7)
Esfahan (Ispahan, Iran)	E. E. H. Becker (1843–1912)	VIS
Mauritius	M. Löw (1841–1900)	VIS
Kerguélen I. (FSAL, Betsy Cove)	Von Schleinitz (?–1912?)	VIS
Yantai (Chefoo, China)	W. Valentiner (1845–1931)	VIS
Auckland I. (New Zealand)	H. Krone (1827–1916)	VIS
Bluff Harbor (New Zealand)	H. Seeliger (1849–1924)	VIS

	Great Britain	
Waimea (Atooi, Hawaii, USA)	E. J. W. Noble (1828–1904)	VIS (8)
Honolulu (Hawaii, USA)	G. L. Tupman (1838–1922)	VIS (8, 9)
Kailua Kona (Kailua, Hawaii, USA)	G. S. Forbes (1849–1936)	PV (8)
Cape Town (Capetown, South Africa)	E. J. Stone (1831–97)	VIS (10)
Alexandria (Egypt)	Barkew	VIS (11)
Muqattam (Mokattam Hills, Egypt)	C. O. Browne (1838–1906)	VIS (11)
Suez (Egypt)	Samuel Hunter	VIS (11)
Thebes (Egypt)	W. de W. Abney (1843–1920)	VIS (11)
El Uqsr (Luxor, Egypt)	E. Ommanney (1814–1904)	VIS (11)
Bushehr (Bushire, Iran)	—	VIS
Mauritius (Île de France, Mauritius)	C. Meldrum (1821–1901)	VIS
Belmont (Île de France, Mauritius)	Lord J. L. Lindsay (1847–1913)	VIS (12)
Rodrigues I. (Mauritius, Pt. Venus)	W. J. L. Wharton (1843–1905)	VIS (13)
Rodrigues I. (Mauritius, Hermitage I.)	R. Hoggan	VIS (13)
Karachi (Kurrachee, Pakistan)	T. Addison	VIS
Kerguélen I. (FSAL, Supply Bay)	C. Corbet (1850–76)	VIS (14)
Kerguélen I. (FSAL, Observatory Bay)	S. J. Perry, S.J. (1833–89)	PV (14)
Kerguélen I. (FSAL, Thumb Peak)	W. S. Goodridge (1849–1929)	VIS (14)
Roorkee (Rorke, India)	J. F. Tennant (1829–1915)	VIS
Chennai (Madras, India)	N. R. Pogson (1829–91)	PV
Colombo (Sri Lanka)	—	VIS
Calcutta (India)	—	VIS
Queenstown (New Zealand)	C. H. F. Peters (1813–90)	VIS (15)
Naseby (New Zealand)	H. Crawford	NV (16)
Rockyside (in Dunedin, N. Z.)	Thomson (Lt., R.A.)	NV (16)
Burnham (New Zealand)	L. Darwin (1850–1943)	VIS (16)
Christchurch (New Zealand)	H. S. Palmer (1838–93)	VIS
Wellington (New Zealand)	A. Stock	NV (16)
Auckland (New Zealand)	T. Heale (1816–85)	NV (16)
Thames (Grahamstown, N.Z.)	H. A. Severn	NV (16)
	Italy	
Madhepur (Mudhapur, India)	P. A. Secchi (1818–78)	VIS
	Mexico	
Yokohama (Japan)	Diáz Covarrubias (1833–99)	VIS (17)
"Bluff Hill" (nr. Yokohama, Japan)	Don F. Jiminez	VIS (17)
	The Netherlands	
Réunion	J. A. C. Oudemans (1827–1906)	PV

Russia (under overall leadership of O. W. von Struve [1819–1905])

Odesa (Odessa, Ukraine)	—	NR
Mykolayiv (Nikolayev, Ukraine)	—	NR
Thebes (Egypt)	W. Döllen (1820–97)	VIS (7)
Oreanda (Orianda, Ukraine; in Yalta)	J. Kortazzi (1837–1903)	VIS
Yalta (Jalta, Ukraine)	—	VIS
Kharkiv (Kharkov, Ukraine)	—	VIS
Kerch (Kertch, Russia)	—	NV
Yerevan (Erivan, Armenia)	—	NV
T'bilisi (Tiflis, Georgia)	—	NV
Naxçivan (Nakritchevan, Azer.)	—	NV
Astrakhan (Russia)	—	NV
Kazan (Kasan, Russia)	—	NV
Ural'sk (Fort Uralsk, Kazakhstan)	—	NV
Tehran (Teheran, Iran)	J. Stebnitzki	VIS
Turkmenbashi (Krasnovodsk, Turkm.)	—	NR
Bandar-e Torkeman (?) (Aschuradeh, Iran)	—	NR (18)
Orenberg (Russia)	—	NV
Kyzylorda (Fort Perowsky, Kaz.)	—	NR
Tashkent (Kazakhstan)	—	NR
Tomsk (Omsk, Russia)	—	NV
Kyakhta (Kiachto, Russia)	—	VIS
Chita (Tchita, Russia)	Winogradski	VIS
Beijing (Pekin, China)	H. Fritsche (1839–1913)	VIS
Nerchinsk (Nertschinsk, Russia)	F. Schwarz (1847–1903)	VIS
Blagoveshchensk (Russia)	—	NV
Pos'yet (Possiet, Russia)	B. Hasselberg (1848–1922)	VIS
Vladivostok (Wladiwostok, Russia)	Onazevitch	VIS
Khanka (Hanka, Russia)	—	NR (19)
Nakhodka (Russia)	Schubin	NR
Khabarowka (Habarovsk, Russia)	P. Kuhlberg	VIS
Busse (Russia)	—	NR (19)
Yokohama (Japan)	O. W. von Struve	VIS

United States

Chatham I. (New Zealand)	E. Smith (1851–1912)	PV
Molloy Point (Kerguélen, FSAL)	G. P. Ryan (1842–77)	VIS
Beijing (Pekin, China)	J. C. Watson (1838–90)	VIS
Nagasaki (Japan)	G. O. Davidson (1825–1911)	VIS
Vladivostok (Wladiwostok, Russia)	A. Hall (1829–1907)	VIS
Hobart (TAS, Australia)	W. Harkness (1837–1903)	VIS

Campbell Town (TAS, Australia)	C. W. Raymond (1842–1913)	VIS
Queenstown (New Zealand)	C. H. F. Peters (1813–90)	VIS (15)

Note: Order is alphabetical by sponsoring country, then by longitude. Modern place names are given first, followed by name in 1874 or alternate name, if significantly different. Modern names are as in *The Times Atlas of the World*, 10th ed. (New York: Times Books, 1999) if given therein. FOD = French Overseas Department, FSAL = French Southern and Antarctic Lands. Under "Leader," a dash indicates not known. Under "Results (notes)," VIS = transit visible, PV = transit partly visible with cloud interference, NV = transit not visible due to clouds, NR = no report.

(1) Adelaide Observatory group
(2) Melbourne Observatory group
(3) Sydney Observatory group
(4) Astronomer to the Khedive of Egypt
(5) Under orders of the Khedive
(6) Nagasaki group
(7) With British Abney group
(8) Hawaii group
(9) Chief of all British parties
(10) Royal Observatory
(11) Egypt group
(12) Private expedition
(13) Rodrigues group
(14) Kerguélen group
(15) Joint British-American station
(16) Burnham group
(17) Member of Yokohama group
(18) Possibly Bandar-e Torkeman, Iran
(19) Location uncertain

Kerguélen's three stations were placed in the charge of a non-military person, Jesuit astronomer Stephen Joseph Perry. In Victorian times it was unusual for the British government to employ a Jesuit, but Reverend Perry was singularly qualified. He had been director of the observatory at Stonyhurst, the Jesuit college near Manchester, where poet Gerard Manley Hopkins—a subsequent professor of classics there and friend of Perry—later taught classics. Hopkins described the college as having a "garden with a bowling-green, walled in by massive yew heges," and "a bowered yew-walk," gloomy and geometrical, the trees severely chopped into symmetrical blocks.[15] One of Stonyhurst's students, Arthur Conan Doyle, the future creator of the Sherlock Holmes stories, who spent seven years there before leaving in 1875, found the discipline severe even by English public-school standards. "They try to rule too much by fear—too little by love and reason," he wrote; he admitted that he would never have sent a son of his there because of the psychological pressures, the constant scrutiny by spying prefects and masters "whose presence secures . . . training in orderliness, self-control, and obedience to law."[16] At least in Perry's case, his Jesuit superiors seemed to have relaxed their discipline in allowing him permission "to follow the natural bent of his mind, and to devote himself wholly to science." He had served as chief of the British expedition to Cadiz to observe the total eclipse of the sun in 1870 and had also made a deep impression on contemporary scientists with his sunspot studies, his work on geodesy and terrestrial magnetism, and his spectroscopic studies of the solar corona.

Another British expedition went to Mauritius; formerly known as the Île de France, it had been a British colony since the end of the Napoleonic Wars. This expedition was privately funded by Lord Lindsay (James Ludovic Lindsay, 1847–1913, the soon-to-be fourth Earl of Crawford and Balcarres). Lindsay was one of the "Grand Amateurs" of the Victorian era. He had attended Eton College and Trinity College, Cambridge, where he had focused on astronomy. After observing the solar eclipse at Cadiz with Father Perry, he had begun the construction of a magnificent private observatory at Dun Echt in Aberdeenshire, Scotland, equipping it with a 15-inch refractor by Grubb and a 40-foot-focus solar telescope. His intention was nothing less than to rival the Imperial Russian Observatory at Pulkova.[17] For the express purpose of viewing the transit, he acquired a three-masted, 398-ton yacht—the *Venus*—with a crew of twenty-two. The expedition to Mauritius would be described as "perhaps the most completely equipped one ever undertaken by a private individual in the interests of astronomy."[18] Lindsay's expedition was also notable in serving to introduce his hired astronomer, David (later Sir David) Gill (1843–1914), to the solar parallax problem. Gill, Lindsay's first director at Dun Echt, would leave his employ in 1876 and go on to a brilliant career that included the post of Astronomer Royal at the Cape of Good Hope (see fig. 62).

An official British party was directed to Thebes, Egypt, joined by German astronomer Georg Friedrich Julius Arthur von Auwers (1838–1915). An informal party was charged with observing the transit from the Sudan where, in 1874, the famed Col. Charles George Gordon (also known as "Chinese Gordon" and "Gordon Pasha" and later, after his

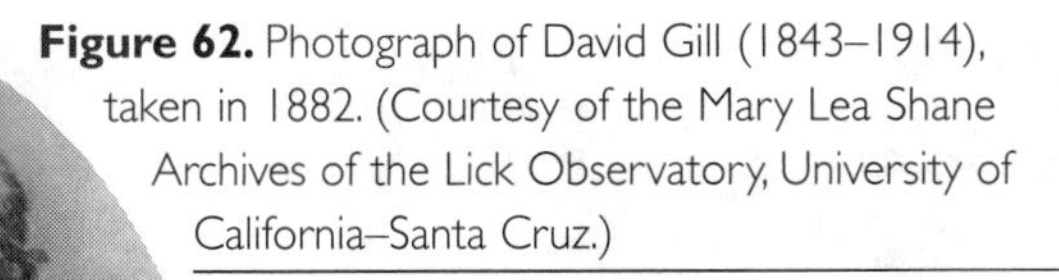

Figure 62. Photograph of David Gill (1843–1914), taken in 1882. (Courtesy of the Mary Lea Shane Archives of the Lick Observatory, University of California–Santa Cruz.)

heroic but unsuccessful defense of Khartoum, as "Gordon of Khartoum") served as a governor under the command of the Khedive (Isma'il Pasha) of Egypt. Gordon would sponsor the Venus-transit observations of two officers at Rejef, five degrees north of the equator.

The United States Transit Commission, though initially endowed with a paltry sum, did well in its lobbying efforts, receiving from Congress a total of $177,000 in appropriations by 1874, in addition to salaries and the use of navy ships and facilities (see fig. 63).[19] The Americans decided on eight overseas expeditions: three Northern Hemisphere parties would be sent to Vladivostok in Russia, Pekin (Beijing) in China, and Nagasaki in Japan; five Southern Hemisphere parties were assigned to Crozet Island (1,500 miles south of Madagascar; it was never actually occupied), Kerguélen, Hobart in Tasmania, Bluff Harbor on the South Island of New Zealand, and Chatham Island southeast of New Zealand.

The French Commission, under the Ministry of Instruction, acting on recommendations of Hervé Faye, planned to establish northern stations at sites in Palestine on the Red Sea coast, Pekin (Beijing), and Tokyo.[20] (Their actual northern expeditions went to Beijing, Yokohama, and Saigon [Ho Chi Minh City].) Faye's 1869 recommendations for the French southern stations were Réunion, Île St.-Paul, Campbell Island, Nouméa (New Caledonia), and Honolulu. However, as Réunion was to be used by the Dutch and Honolulu by the British, the French commission decided to concentrate their efforts on Île St.-Paul, Campbell Island, and Nouméa.

The Germans sent an expedition to Betsy Bay, Kerguélen Island,

Figure 63. Photograph of the American Transit of Venus Commission, field-party members, and some of their instruments, at the U.S. Naval Observatory in the spring of 1874. Adm. Charles F. Davis is at the left edge; Henry Draper is next to him, while Simon Newcomb sits to their left. (USNO photograph; originally published by J. F. Jarvis, Washington, DC.)

while another group settled upon Lindsay's destination, Mauritius. Still other German expeditions or individuals traveled to Terror Cove in the Auckland Islands; Bluff Harbor, New Zealand; Esfahan, Iran; and Yantai, China. Russia's efforts were coordinated by the Pulkova Observatory's director, Otto Wilhelm von Struve, who planned no fewer than thirty-two stations, including twenty-seven within the Russian Empire itself, stretching from European Russia across Asia to the Pacific coast; some positions were reached only with difficulty because the railway then stopped at the Urals.[21] More modest international participants in the transit included a Mexican expedition to Yokohama, a Dutch expedition to Réunion (Île Bourbon, east of Madagascar), and an Italian expedition to Madhepur, India. Besides the government-sponsored expeditions, there were many individual observers in various positions in the track of the transit, waiting to do their duty during the most eagerly awaited few hours of nineteenth-century astronomy—prepared and armed to the hilt to secure the observations needed for the grand project.

The expeditions were equipped with the most sophisticated equipment available. The American expeditions adopted a 5-inch Alvan Clark refractor as their standard, one of which is on exhibit at the U.S. Naval Observatory (see fig. 64). Airy ordered twenty-three telescopes for his parties in 1874. The Honolulu station had enclosures for a 6-inch refractor (the standard British instrument, used with a micrometer by J. W. Nichol and George Tupman) and a 4½-inch refractor (used also by Tupman, with a spectroscope), while 3⅝-inch and 3-inch refractors were mounted on tripods in the open.

Observing a planet in front of the sun's disk posed additional difficulties associated with the overpowering brilliance—indeed, dangerous intensity—of the sun's light. Eyepiece projection onto a white screen was the safest method available in 1874. It had some disadvantages, chiefly the need to exclude almost all other external light to avoid degrading the image contrast. Though favored by many amateurs because the method was convenient and safe, and because it allowed a group to watch simultaneously with the same telescope, eyepiece projection was rarely used by the formal expeditions. An alternative was the only practical full-aperture filter available in 1874—smoked glass—however, nobody wanted to smoke an optical flat, let alone their telescope objective, yet ordinary glass would degrade the image. Thus the most common solution for direct solar viewing was a "solar prism," an unsilvered prism placed in front of the telescope's focus that would pass about 90 percent of the heat and light out the back of the telescope and reflect about 10 percent toward the eyepiece. This still left perhaps 10,000 times too much radiation for safe and comfortable viewing, so tinted-glass filters were also used in front of, or more often behind, the eyepiece to make viewing apparently safe. In addition, the aperture of the objective was often "stopped down," sometimes to a half or less of the normal diameter. Nevertheless, there were still considerable risks involved, and a comment by Samuel Hunter, an observer at the British Suez station, is suggestive: "I examined the colored glasses, and found that the center of each of them had been partially melted with the sun's rays at some previous period."[22] Captain Hixon, observing at Goulborn, Australia, had all his dark glasses split, "causing him to

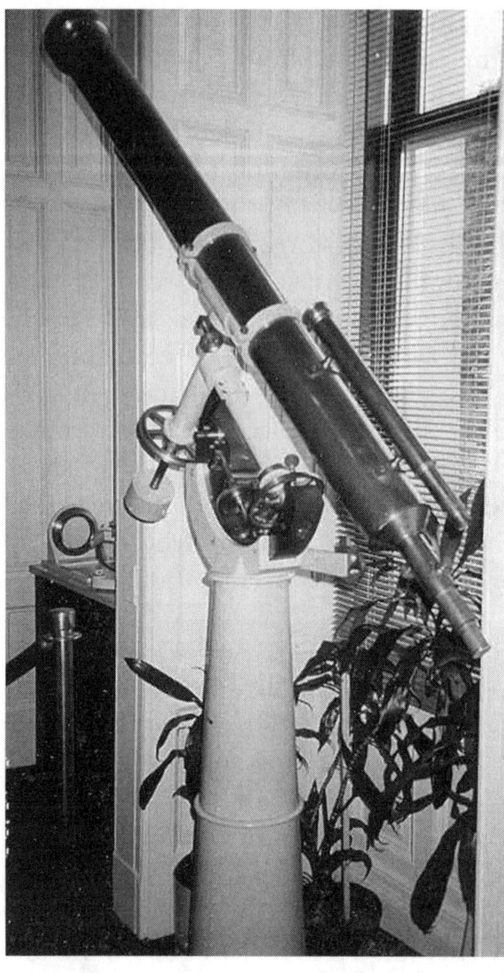

Figure 64. An example, on display at the U.S. Naval Observatory, of the standard 5-inch Alvan Clark refractor used by the American transit of Venus expeditions in 1874 and 1882. (Photograph by William Sheehan, 1997.)

lose the Ingress entirely."[23] The wonder is that no observers lost their eyesight!

A number of teams used heliometers to measure the spacing between the limb of the sun and the planet, though these were rare and expensive. The Germans had the most heliometers, though several Russians and the well-heeled British expedition of Lord Lindsay also employed them. Visual observers who were not equipped with heliometers used micrometers—including special double-image micrometers.

The American, British, and French commissions all emphasized photography, despite the fact that there were serious questions about the flexible emulsions used on both wet and dry plates becoming distorted between the exposure and processing. There was also the recognition that precise positional measurements had never before been attempted from photographs! To support their photographic efforts, the American teams were equipped with heliostats, or horizontal refractors based on a model already in use at Harvard College Observatory. The photographic plate was enclosed by a 4-

meter (12-foot) tube and a wooden structure called the "Photographic House," which also contained a darkroom. Instruments of the type used at the 1874 transit gave a solar image 112 mm (4.4 in) in diameter upon the plateholder, which superimposed a reseau (a grid of fine lines) on the photographic plate. The French adopted a heliostat similar to the American, but the British, German, Dutch, and Russian photographers instead used an equatorially mounted telescope known as a "photoheliograph" (see fig. 65). Regardless of how they were obtained, the purpose of the photographs was the same as that of heliometer measurements: the determination of the distance between Venus and the limb of the sun.

The most innovative approach to photographing the transit was developed by French astronomer Pierre Jules César Janssen (1824–1907; see fig. 66), a member of the Bureau des Longitudes, and another early advocate of astronomical photography, Hervé Faye (1814–1902). As early as 1851, Faye had suggested that it would be less wasteful to expose a number of images of the sun on a single daguerrotype, by using a handle to move the sun by 2-millimeter intervals. The idea would later form the basis of Janssen's invention. Faye also hoped that photography would eventually prove capable of eliminating much of the error that appeared in measures of the positions of stars. For instance, by pressing a key, exposing a photographic plate and registering the time of the exposure by means of a Galvanic signal, he showed that it was possible to produce "a completely automatic observation" that could be produced by a "young apprentice who had no idea what he was doing."[24] Again, he noted, "[With photography] it is nature itself that appears under your eyes."[25]

Other methods of determining the earth-sun distance were being developed. In France, they included the method of physicist Léon Foucault (1819–68), who used a rapidly turning mirror apparatus to determine accurately the speed of light, which allowed the distances to the satellites of Jupiter to be determined from their eclipses by reversing the method used by Ole Rømer almost two centuries before. Rømer used the satellites' distance to measure the speed of light, whereas the Foucault method used the speed of light to measure the satellites' distances. And yet—though problems with the transit of

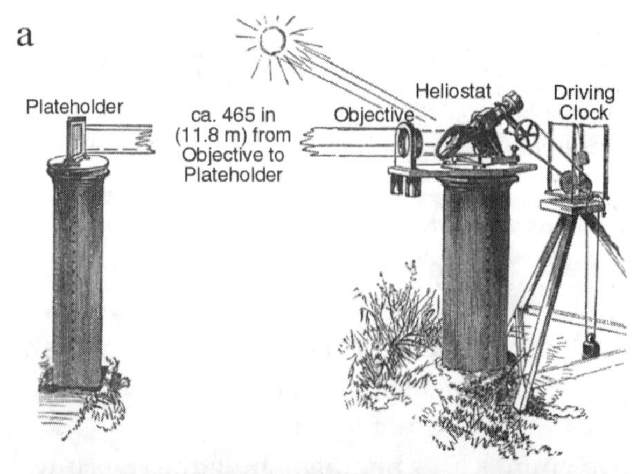

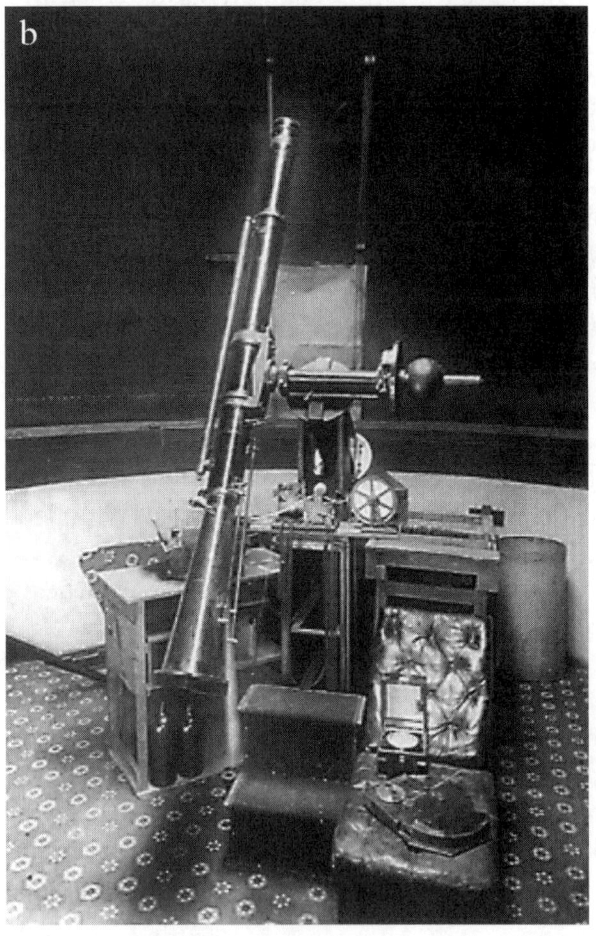

Figure 65. (*a*) Version of heliostat used for photography of transits of Venus by American parties in 1874 and 1882. (*b*) British-style photoheliograph, this one at the Woodford station in New South Wales, Australia, in 1874. (From Sydney Observatory, *Observations of the Transit of Venus*, 1892; courtesy of Richard Baum.)

Figure 66. Photograph of Pierre Jules César Janssen (1824–1907), taken in 1883. (Courtesy of the Mary Lea Shane Archives of the Lick Observatory, University of California–Santa Cruz.)

Venus method were widely being acknowledged—for many years, as Jimena Canales explains, "only Venus's transit method could . . . 'immediately convince the spirit' of the true value of the solar parallax."[26] Janssen, whom Faye once referred to as "the recognized expert on fugitive phenomena," took inspiration from Faye's idea of an automated observing machine built around the photographic plate as a means of recording the contacts at the transit of Venus. This led to his remarkable *revolver photographique* (photographic revolver; see fig. 67).

Janssen was a colorful figure—a true Parisian personality. Though unable to walk due to an accident in his youth, he led an adventurous life. An inveterate eclipse chaser, he observed the "Indian eclipse" of August 18, 1868, at Guntoor, and discovered (independently of Norman Lockyer), on the day after the eclipse, a spectroscopic method of observing the prominences outside of eclipses. In order to observe an eclipse that occurred while Paris was under siege by the Prussians in December 1870, he escaped—in a balloon—to Oran, Algeria, where he unfortunately was clouded out. He ascended Mount Etna to use the spectroscope in an attempt to detect water vapor in the atmosphere of Mars—he believed he had succeeded—and late in life established a temporary observatory on the summit of Mont Blanc.

At the heart of Janssen's photographic revolver, which in France became known simply as "the Janssen"—were two slotted disks. One disk, with twelve slots, made a continuous revolution in eighteen seconds; the other, a smaller one, was fixed and had only one slot, forming a window admitting only a small part of the solar

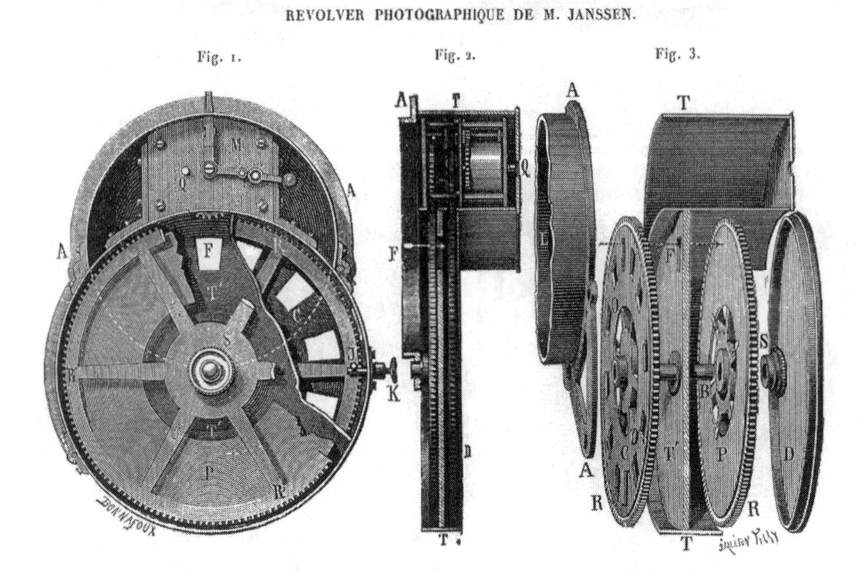

Figure 67. Internal mechanism of Jules Janssen's *revolver photographique*. (From Jules Janssen, *Œuvres Scientifiques*, ed. Hebri Dehérain [Paris: Société d'Éditions, 1930].)

image. Whenever the two slotted disks overlapped, an exposure was made on a curved daguerrotype plate located below.

The internal mechanism was based on that of the Colt revolver. The larger disk consisted of a toothed tray in contact with a small gear mechanism, which was operated by an electric current that, when interrupted by the beating of a pendulum clock, would advance the tray tooth by tooth beneath the single-slotted disk. A series of images of the limb of the sun would be produced on the plate, 48 exposures in all, at intervals of 1.5 seconds between each exposure. By starting the exposures just prior to the expected time of the first contact, "the photographic record of the contact would be comprehended (necessarily) in the set."[27] After removing the first plate, a new plate was inserted in order to get the second contact, and so on, until all four contacts had been recorded. Janssen hoped that the instant of each contact would be determined in the sequence of photographs, by selecting, for careful examination with a microscope, the one that was obtained at the "fortuitous moment."

In addition to its importance in French transit plans, the photographic revolver would be adopted and employed by American and British expeditions in 1874.

The "Compleat" Transit Station

A station's personnel couldn't control the weather, but in all other matters they were the key factor in determining an expedition's success or failure. First, most of the expeditions arrived at their destinations courtesy of their respective country's navy. Each party was dependent on the ship's captain, officers, and crew to deliver them to the correct location sufficiently ahead of the December 9, 1874, transit date in order to set up their station and make necessary preliminary observations. In addition, officers and crew often helped in establishing the transit camp and sometimes even aided in the observations.

Each transit expedition had a hierarchy as formal as the ship's complement.[28] At the top was the chief of the party, either a professional astronomer or a military officer with surveying, astronomical, or engineering experience. In inhabited areas, the party's leader was also responsible for relations with local officials and the general population, including negotiating for an observing site, supplies, and security. Often there would also be an assistant astronomer. Besides general management of the party, the astronomers were responsible for the visual observations, considered the most demanding, whether they took the form of timing contacts, making drawings, writing comments, or making micrometric, heliometric, or spectroscopic observations. Photographers were also employed by many of the parties. Photography was a more complicated pursuit than it is now, and the photographers had to manufacture and process their own plates, not simply expose them. They were regarded as tradesmen, technically proficient at their tasks but lacking in astronomical training. There were many others who were needed to support the transit expeditions—cooks, carpenters, mechanics, "artificers" (craftsmen), and now and then a professional hunter to help feed the group. Women were not generally part of these groups; the exceptions were the wives, daughters, or sisters of the astronomers, who often participated in making, recording, or timing observations.

In contrast to the expeditions of the eighteenth century, those of the nineteenth century appear to have had few problems in at least finding their destinations. Making a successful landing was another matter. Some expeditions went to well-equipped ports such as Nagasaki and Alexandria, where they could unload at a dock. With less-visited destinations, personnel, equipment, and supplies had to be brought ashore on lighters. At uninhabited places like Kerguélen, the ships' boats had to serve, landing fragile equipment and equally fragile party members through the surf.

Then a site had to be selected. In settled areas this meant gaining permission to use a suitable open space. This was rarely a problem, and usually the transit parties were welcome visitors. In uninhabited areas physical site factors were paramount. Most sites were more or less rectangular, although the French Campbell Island station was laid out parallel to the shoreline. Then the site had to be on flat, stable ground, as protected from wind as was practical, and had to have an unobstructed view of the sun throughout the transit.

Where security was an issue, fences were usually erected around the buildings to create a compound. Expeditions that went to uninhabited or "uncivilized" localities often brought prefabricated structures. If timber and other supplies were available locally, these could be used instead. However, even before the structures went up, the cement or brick-and-mortar instrument piers had to be constructed and allowed to set; only then would structures go up around them.

With standardized instruments, layout, and prefabricated buildings, the American stations were reminiscent of Roman camps. When housing and other facilities were locally available, their stations were completely devoted to observing.

Stations varied from basic to elaborate. A simple station that used local materials was the Italian station at Madhepur, India. The British station at Honolulu was more elaborate, with more instruments and buildings for them. They had the benefit of using an existing parcel of land owned and made available by Hawaii's last monarch, King Kaläkaua. The party themselves lived at a nearby cottage rented from

Princess Ruth, governess of Molokai, who also provided the site and housing for the satellite party at Kona on the Big Island.

The British stations were much less standardized than those of the Americans, adapting more to local conditions. At Mokattam Heights, Egypt, a few miles southeast of Cairo (now the Muqattam suburb), no fence was necessary. Located on a plateau, the station was protected on three sides by a cliff, while guards supplied by the Egyptian government protected it on the fourth side. Except for the more solid structures housing the transit, altazimuth, and 6-inch equatorial, tents served for most purposes.

In some rural areas local observers could be called upon, who could set up their own camp. If so, the station could be relatively open, perhaps set within a fenced field. This was the case with the bucolic setting at Woodford, New South Wales, Australia, whose observations were coordinated, along with other local stations, by Sydney Observatory (see fig. 68).

Yet another style was set by the camps of the French transit expeditions, for example, that at the Île St.-Paul. The island, a flooded volcanic caldera in the southern Indian Ocean, covers only 8 square kilo-

Figure 68. The locally built and manned 1874 transit of Venus observing station at Woodford, New South Wales, Australia. (From Sydney Observatory, *Observations of the Transit of Venus*, 1892.)

meters (3 square miles). Like most other small islands used by transit parties, there were few suitable places from which to choose, and the French station was placed near a penguin rookery (see fig. 69).

Every station possessed at least a basic set of instruments to view, and perhaps photograph, the transit. A well-equipped station also had instruments to measure its position accurately and to determine the local time. Given the "absolute longitude" and the local time, the time for a standard meridian (such as Greenwich or Paris) could be calculated and the station's results compared with those of any other transit station in the world. When the transit instrument ("transit" here referring to the passage of an object over the celestial meridian) was properly leveled and oriented north-south, the horizontal crosswire in the eyepiece and the vertical circle were used to measure the distance of a celestial object from the local zenith. In order to determine local sidereal time, an observer noted the moment a celestial object crossed the vertical wire. With this information, the latitude and longitude of the observing site could be calculated much more accurately than was done at the eighteenth-century transit stations. All that remained was the observation of the transit itself.

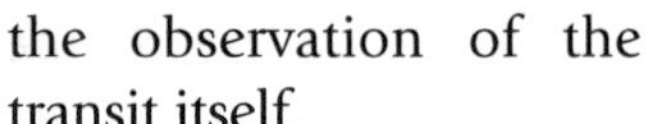

Figure 69. The French observing station at Île St.-Paul in 1874, shared with penguins. (From Alphonse Berget, *Le ciel* [Paris: Libraire Larouse, 1923].)

12

Upon the Flame-Cased Sun

She hangs upon the flame-cased sun,
And sucks the light as full as Gideon's fleece:
But then her tether calls her; she falls off,
And as she dwindles shreds her smock of gold
Amidst the sistering planets.

—Gerard Manley Hopkins,
fragment of speech for *Floris in Italy* (1864)

A Century's Wait Ends

The long wait for a transit of Venus finally ended at 3:06:22.3 PM Honolulu mean time, December 8, 1874 (1:45:49 on December 9, UT), when George Tupman became the first person in 105 years to see a transit of Venus. He had two advantages that gave him a head start: the Hawaiian stations were the closest in the world to the Delislean point of earliest ingress; and he was observing with a spectroscope that allowed him to spot Venus against the sun's inner atmosphere, the chromosphere, a full 39 seconds before it touched the visible solar limb.

The Hawaiian observers had clear skies at Honolulu on Oahu and Waimea on Kauai, but intermittent clouds blocked astronomer George Forbes's view of the first and second contacts at Kona on the Big Island. Forbes, a Cambridge graduate who had written a book on the transit, had arrived on the HMS *Scout* with an assistant named Barnacle, who proved to be more of an encumbrance than a help and was soon dismissed, as well as British consul Major Wodehouse. (Wodehouse's first order of business had been to travel eleven miles south to Kealakekua Bay, to restore the monument near the spot where the natives had killed Captain Cook in 1779. Apparently to prevent such unpleasantness recurring, the consul asked the ship's captain to furnish Forbes with a flag so that they would have "the protection of the British Flag here in case of any disturbance.")[1] Kona lacked today's luxury resorts, and Forbes complained bitterly of the accommodations, saying they were "unfit for Europeans." Then he almost drowned in the heavy surf trying unsuccessfully to save another man's life. At least his efforts to observe the transit were rewarded with partial success. In spite of the clouds, Forbes used a double-image micrometer to secure distances between the limbs of Venus and the sun. The pier on which he set up his telescope has survived, and a plaque marks the site. Following the transit Forbes went home by an indirect route via Beijing, the Gobi Desert, and then Siberia, where he negotiated the release from Irkutsk of two exiles. By 1877 he was a war correspondent with the Russian army in the Russo-Turkish War, and in the 1880s–90s he worked as an engineer on hydroelectric plans for Niagara Falls and the Nile River. He remained at least somewhat interested in astronomy and attempted to work out an orbit for a trans-Neptunian planet.

From Hawaii, where only the first two hours of the transit would be visible, the leading edge of the shadow cone of Venus passed obliquely across the earth, moving approximately southwestward at an average speed of 500 kilometers (300 miles) per minute. The next site where there were clear skies was the Russian station at Khabarovsk. The Russian stations had mixed results; thirteen had complete or partial success, but ten were completely clouded out, and the rest provided either no reports or were clouded out, if indeed they were occupied at all. Besides the Russian teams, an American party was situated at Vladivostok, on the Sea of Japan, headed by U.S. Naval Observatory astronomer Asaph Hall (1829–1907). Although the weather was frigid, and the observing site a full mile from the party's lodging, they had a clear horizon, and successful timings were taken of the first, second, and third contacts.

Several groups watched from Japan, including the American party at Nagasaki, led by George Davidson (1825–1911) of the U.S. Coast Survey. He was assisted by Captain Naranoyoshi Yanagi, head of the Japanese Hydrographic Bureau, and by Hikoma Ueno (1838–1904), Japan's first commercial photographer. At Yokohama, a Mexican party made observations under the direction of Don Francisco Diaz Covarrubias (1833–99).

Japan was all-important in the French transit plans. En route to Japan with the photographic revolver, Janssen and his team, which included the notable expert in celestial mechanics and future director of the Paris Observatory François Felix Tisserand (1845–96), had to weather a typhoon off Hong Kong. After arriving at Yokohama they established themselves at the ominously named Kompira-yama—the mountain of the god of typhoons—near Nagasaki. As a hedge against bad weather, Janssen sent a detachment of observers to Kobe. During the transit the weather turned out perfect at both sites, and after the transit Janssen telegraphed Dumas, president of the Transit Commission, as follows: "10 December. Transit observed at Nagasaki and Kobe, interior contacts obtained. No ligaments. Photographed with revolver [see fig. 70, for example]. . . . Venus seen over corona before transit, demonstrating the existence of the coronal atmosphere. Janssen."

In addition to his use of the photographic revolver, Janssen employed another telescope to place the slit of a spectrograph on the bright aureole of the planet's atmosphere as the planet stood in silhouette against the solar chromosphere (see fig. 71 for an illustration

Figure 70. A Janssen "revolver" in operation. The heliostat mirror, visible through the doorway, reflects the sun's image into the apparatus, while at right an exposed plate is being taken to the darkroom for processing. (From *La Nature* [1875].)

of this), and claimed to have established the existence of water vapor in the Venusian atmosphere. After 1882, he would retract this claim.[2]

The Nagasaki party, led by Janssen himself, was one of the few French expeditions to use his camera. The British, ironically, employed five "Janssen slides," as they called them, all made by the instrument-maker Dallmeyer. They imaged the sun with a half-inch solar disk, enlarged by projection to 4 inches, and obtained 50 frames at 1-second intervals on a circular glass plate 10¾ inches across. The British cameras were deployed at Honolulu, Rodrigues Island, Thebes, Roorkee, and Sydney. Only the Honolulu device failed, due to operator error when the planet's image was misplaced and bisected by the frame edges.

Far to the south, most of the parties sent to New Zealand reported clouds or even rain. A

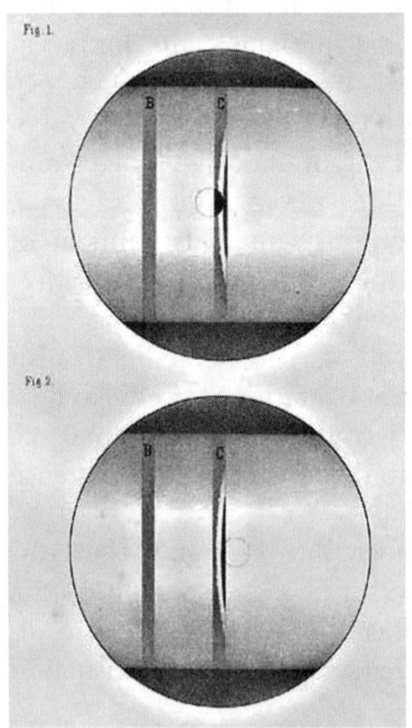

Figure 71. Venus captured in the slit of a spectroscope by P. Tacchini at Madhepur, India, in 1874. The upper drawing shows the first external, the lower the first internal, chromospheric contacts. (From Tacchini, *Il passagio di Venere sul Sol* . . . [Palermo: Stabilimento Tipografico Lao, 1875].)

German party in Bluff Harbor managed some timings and some photographs. The American team under Christian Heinrich Friederich Peters (1813–90; see fig. 72), which had continued on to New Zealand onboard the U.S. man-of-war *Swatara* after dropping off the Kerguélen Island and Hobart parties, was virtually the only New Zealand station to report even marginally good weather. Peters decided to press on from Bluff, where the *Swatara* had landed, into central Otago, which had better weather prospects, and set up a station at Queenstown on the shores of Lake Wakatipu. (The site is on Melbourne Street, the present site of the Millenium Hotel, with its "Observatory Restaurant.") Peters and his team captured the first and second contacts and exposed fifty-nine plates. After the transit, Peters gave a farewell dinner speech at the Queen Arm's Hotel, which was covered extensively in the *Lake Wakatip Mail.* He saw "in the stars a large city here in the future; railways converging upon it; people coming from many other parts of the earth to enjoy its beautiful climate, and behold its grand scenery (Loud Cheers)."[3]

Among the other New Zealand transit parties, the British central station at Burnham obtained just thirteen photographs, while only nine were taken at Christchurch. Better luck was had on the occupied Pacific islands near New Zealand. The Germans made successful timings and heliometric measurements on Auckland Island. The French obtained about 100 daguerreotype and collodion photographs at Nouméa, and they also observed at Campbell Island under Bouquet de la Grye. The Americans at Chatham Island under Edwin Smith struggled with clouds but still managed to take eight plates.

In Australia, the Americans in Hobart under William Harkness had clouds and could time only the first

Figure 72. Photograph of Christian Heinrich Friederich Peters (1813–90). (Courtesy of the Mary Lea Shane Archives of the Lick Observatory, University of California–Santa Cruz.)

contact, while those in Campbell Town could time only the third, although both exposed photographs (39 plates at Hobart and 55 at Campbell Town). Not all plates exposed were usable, but the attempt was made if there was even a marginal chance of recording the sun.

Fortunately, the rest of the Australian stations reported either completely clear skies, only occasional clouds, or thin clouds through which the sun remained visible. The last was the case at Mornington, about thirty miles from Melbourne, where William Parkinson Wilson, a professor of mathematics at the University of Melbourne, aged about forty-eight, excitedly observed the transit. The excitement was apparently too much for him, as he died only two days later. Visual observers made some of the most interesting observations. R. L. J. Ellery, using the Melbourne 8-inch Troughton & Simms refractor, reported an "occasional flicker of light between the limbs" some 43 seconds before his timing of the second contact ("sudden disappearance or rupture of the very thin dark thread joining the limbs"), along with an "appearance of smoky junction of limbs" 25 seconds before the reported contact.[4] The Government Observatory at Sydney was particularly active. Sixteen plates, containing 560 images, were obtained with a Janssen camera that can still be seen in the observatory museum. In addition, Henry Chamberlain Russell (1836–1907) used the observatory's 11½-inch refractor, stopped down to an aperture of only 5 inches, to make annotated drawings of the transit events, the originals of which were in color, including a series of views of egress showing both the black drop and the ring of light around the planet's limb, silhouetted against the sky (see fig. 73). The Australian observations were especially useful, as the continent lay near the center of the zone within which the entire transit could be seen.

As the transit continued to make its progress across the globe, the next concentration of observers to encounter the spectacle were those based in British India, who mainly enjoyed clear skies and a view of the entire transit. The Italian astronomers at Madhepur in Bengal made visual ingress and egress timings and used a spectroscope to watch the planet silhouetted against the solar chromosphere. In Vizhakhapatnam, a native of India, Ankitam Venkata

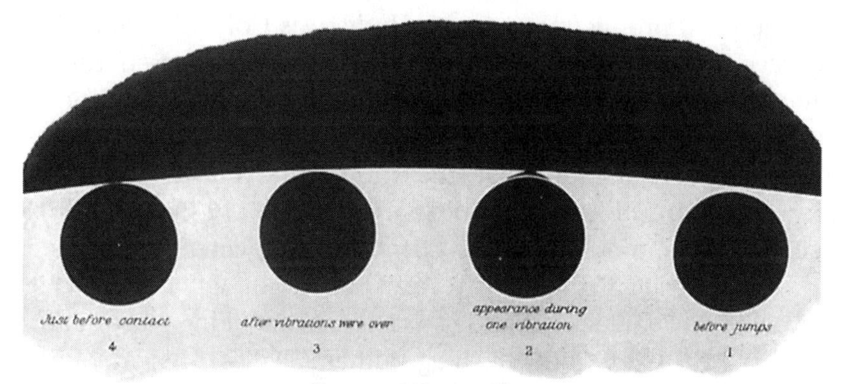

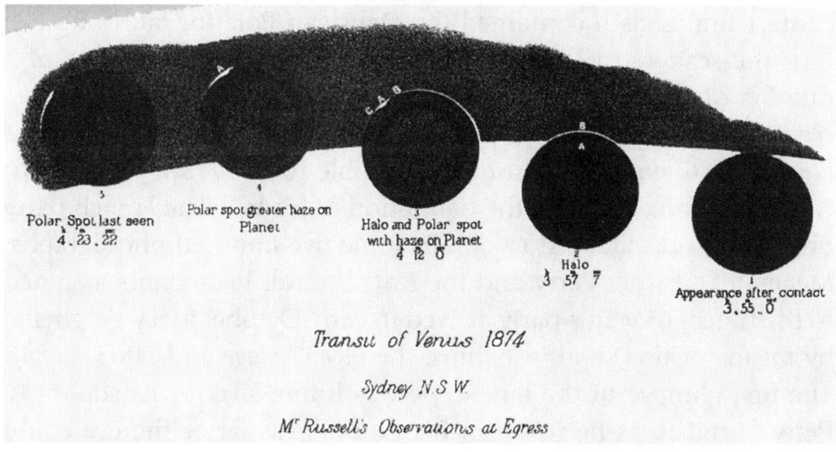

Figure 73. Drawings by H. C. Russell, director of the Sydney Observatory, made with the Observatory's 11.4-inch refractor, stopped down to 5 inches, showing the egress of Venus in 1874. The upper sequence shows an intermittent ("vibration") black-drop effect. The lower series shows the formation of a "halo" and polar light spot. Although north is at the top, these images are reversed because Russell viewed with a prism diagonal. (From Sydney Observatory, *Observations of the Transit of Venus*, 1892.)

Nursing Row (1827–92), used a 6-inch refractor to time egress. Largely self-taught in astronomy but elected a fellow of the Royal Astronomical Society of London three years earlier, Row equipped an observatory on his estate with the refractor, a transit circle and an equatorial clock.

The most productive station in India was Colonel Tennant's in Roorkee, the only Indian station armed with a Janssen camera, which he used to obtain 420 frames of the transit. In addition, he photographed with his 6-inch refractor and made visual timings of ingress and egress, noting that "the whole of the time the tremor was very great."[5] In Lahore, Captain George Strahan (1839–1911) noted that, about the time of the third contact, "the planet's edge was, in fact, encircled by a ring of light nearly as bright as the Sun, which prevented any contact, properly so called, from taking place at all."[6] Captain Archibald Cuthbert Bigg-Wither (1844–1913) at Multan also reported that Venus had "a ring of light concentric with her disk; there was absolutely no internal contact, whatever."[7]

The southern Indian Ocean was another strategic location for contact timing as it contained the Delislean Pole for latest ingress; thus the scattered islands in the region were the destinations of a number of transit-parties. Agnes M. Clerke later described their desolate character in her work *A Popular History of Astronomy during the Nineteenth Century* as "all but inaccessible rocks . . . swept by hurricanes, and fitted only for the habitation of birds."[8] The French party on Île St.-Paul managed to obtain some five hundred photographs. Meanwhile, Father Perry and the British naval lieutenants assigned to the transit of Venus party arrived in early October at icy Kerguélen by means of two small warships, the HMS *Volage* and HMS *Supply*. The first glimpse of the landscape was impressive for its starkness. Perry found it, as he wrote in his journal, "as far as the eye could reach . . . completely buried in snow. This was the end of spring for the southern hemisphere, so we had pleasant prospects of rambles in snow-shoes over rugged hills and half-frozen marshes and bogs" (see fig. 74 for a depiction of the island).[9] Perry's observations of the transit itself, with a 6-inch refractor at Observatory Bay, were partly spoiled by clouds. A large cloud came over the sun before ingress was complete and did not leave until Venus was well advanced on the sun. Later the murk returned and became so thick that the telescope did not show any trace of the planet at all. Fortunately, the sky cleared and allowed timings of the third and fourth contacts. Observers at the satellite station at Supply Bay had the opposite

Figure 74. The German observing station at Betsy Cove, Kerguélen Island, in 1874. From left to right are the sheds containing a visual refractor, a heliometer, a camera obscura, and a photographic refractor. (From A. von Schweiger-Lerchenfeld, *Atlas der Himmelskunde*, 1898)

experience, enjoying clear skies at ingress but being clouded out at egress (see fig. 75).

The other important observing point in the Indian Ocean was Mauritius, located between Rodrigues Island and Madagascar. At Port Louis, David Gill of Lord Lindsay's party used the world's only private heliometer. The other members of the group observed with 4- and 6-inch refractors and other instruments, taking 271 plates, of which about 100 were "of value."[10] Unfortunately, they did not observe ingress because of clouds. Gill noted that, shortly before the third contact, "a light grey shade appeared between the limbs of the Sun and planet," while his colleague Ralph Copeland (1837–1905), at the same location using a 6-inch refractor, felt that the ligament formed almost simultaneously with the contact.[11]

Probably the last observers to see the transit were those at the Russian station at Kharkiv in the Ukraine, the station closest to the Delislean point of latest egress. Soon after crossing over their loca-

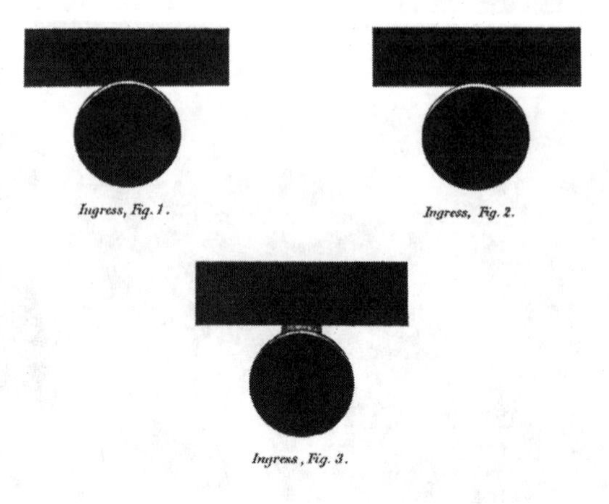

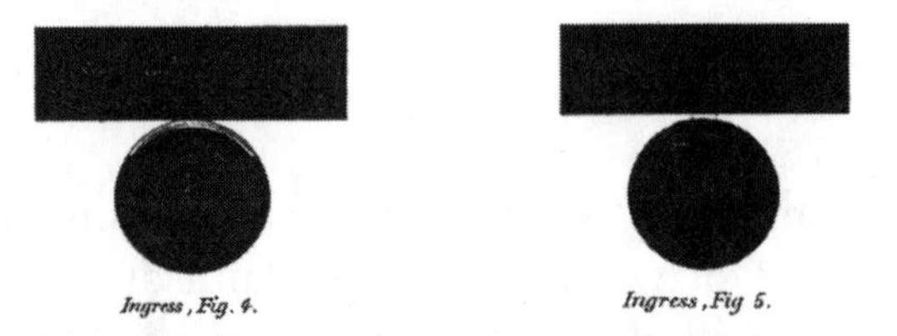

Figure 75. The ingress phase of the 1874 transit of Venus as drawn by the British party at Supply Bay, Kerguélen Island. (From George Biddell Airy, *Account of Observations of the Transit of Venus, 1874, December 8* [1881].)

tion, the trailing edge of Venus's shadow cone swept off into space, not to intersect the earth for another eight years, and then not again until our own time.

THE AFTERMATH OF THE TRANSIT

The 1874 transit had been widely observed from four continents and from numerous islands scattered across two oceans. But how successfully observed? It was not at first easy to say. The reduction of the observations was a long, drawn-out affair. Though preliminary results from some stations had begun to arrive by telegraph within hours of the transit, reports from isolated stations, such as mid-oceanic islands, took weeks or months to arrive. The contact times, reported sometimes as uncorrected readings from clocks or chronometers, or in local sidereal or solar time, had to be adjusted to standard time. This depended on the laborious reduction of hordes of lunar culminations, occultations, and other measurements in order to determine the longitudes of the observing stations. Photographic plates had to be carefully transported from remote places for measurement.

The plate measurements were themselves very laborious. Even when magnified only a few times, the photographs invariably showed the same fuzzy edge to Venus and the even fuzzier limb to the sun that had so long frustrated visual observers. An assistant at the Royal Greenwich Observatory, Charles Edward Burton (1846–82), had initially received the assignment of reducing the British plates. He applied himself diligently to the task until his eyesight—and health—began to fail. Then George Tupman (1832–1922) took over. Tupman commented on how tough he found the task: "When the negatives are placed under the microscope with an amplification of only 5 or 6 diameters, the limbs of both planet and Sun, even those which are pretty sharp to the unaided eye, become extremely indistinct, and the act of bisecting the limb with the wire or cross of the micrometer is mere guesswork."[12] Tupman added that there was only one fully satisfactory image in the whole collection.

Some nations did a little better in their photographic results. The French and American photographers, who used the simple optical system of a heliostat mirror and long-focus lens, had sharper images than those of countries whose parties used an additional enlarging lens. Nonetheless, almost all the photographs of all the parties were unsharp. Even the images obtained with the revolver cameras—the "cinematographic method"—suffered from the same difficulties as all the rest. Nevertheless, Janssen's device was influential. "Created first for the transit," writes historian Jimena Canales, "Janssen's apparatus was soon modified and moved to other areas of science and culture, most famously to Étienne Jules Marey's (1830–1904) physiological laboratory providing his famous sequential photographs, and then to Lumière's studio where it was gradually transformed into what would soon be named the cinematographic camera."[13]

Janssen has been considered in some quarters a founder of cinema. In 1895, only two years after the Lumière brothers (Louis Jean [1864–1948] and Auguste Marie Louis Nicolas [1862–1954]) produced their first motion picture camera, Janssen was featured in one of the first films to be shown publicly. However, the influence of his photographic revolver on future transit observations would be negligible. Indeed, almost all the transit committees, including the French, agreed that visual observations of the coming transit of 1882 would be better than photographs. Moreover, many astronomers had become convinced, even before the reductions of the 1874 transit had been completed, that the solar parallax would be more easily—and more accurately—achieved through other methods. Gill, for instance, despite enjoying a relationship "entirely without friction" with Lord Lindsay, left the latter's employ after his return from Mauritius.[14] Disillusioned with his experience with the vagaries of determining Venus's exact position relative to the sun during the transit, Gill had decided to concentrate his efforts on a different method of measuring the sun's distance—that of using Mars's changing position relative to the stars at a close opposition. In time for the apparition of 1877, he found himself, with his wife Isobel, on the desolate island of Ascension, in the mid-Atlantic. Isobel charmingly described her first sight of the skies from that remote island:

> Somehow, it all came right, and sitting that first evening after sunset in the verandah . . . we could speak of nothing, think of nothing, but the beauty of the heavens. Though Ascension was barren, desolate, formless, flowerless, yet with such a sky she could never be unlovely. The stars shone forth boldly, each like a living

> fire. Mars was yet behind Cross Hill, but Jupiter literally blazed in the intense blue sky now guiltless of cloud from horizon to zenith; and, thrown across in graceful splendour, the Milky Way seemed like a great streaming veil woven of golden threads and sparkling with gems.[15]

Gill, using the same heliometer he had used as a member of Lord Lindsay's expedition, determined the position of Mars relative to twenty-two comparison stars, on thirty-two evenings and mornings. From this set of data, Gill arrived at a value of the solar parallax of 8".780, making for an earth-sun distance of 149,840,000 kilometers (93,110,000 miles)—within 0.2 percent of the modern value. Gill's success with Mars underscored the fact that the geometric methods of Halley and Delisle were becoming passé. Henceforth, the trigonometric approach to the sun's distance would use Mars—or better yet, one of the asteroids—as an intermediary. There were also the methods noted earlier. Indeed, as a practical means of measuring the scale of the solar system, the 1882 transit would be both the Halleyan and Delislean methods' last hurrah.[16]

13

1882: THE LAST HURRAH

But were it told to me, to-day,
That I might have the sky
For mine, I tell you that my heart
Would split, for size of me.

. . . .

So, safer, guess, with just my soul
Upon the window-pane
Where other creatures put their eyes
Incautious of the sun.

—Emily Dickinson, untitled (Poem 327, c. 1862)

EIGHT EVENTFUL YEARS: 1874–1882

The 1874 transit of Venus had been largely a matter of scientific concern. In part, this was owing to the fact that no part of the transit was visible either from the United States or from most of

Europe. In contrast, the 1882 transit would be seen, at least in part, throughout the United States and the greater part of Europe. Public interest ran very high, stimulated by newspapers and periodicals and by a piquant awareness of the rarity of the event. It is probable that attention to the transit approached the notoriously high levels of interest associated with the returns of Halley's comet.

The year of the transit, 1882, fell within the long Victorian era—Queen Victoria was still seated securely on the British throne, and the British Empire was at its height. George Gordon was in Mauritius, soon to be recalled to rescue Egyptian garrisons in the Sudan; the British, interested in keeping control of the recently built Suez Canal, occupied Cairo. A year earlier, the Boers had repulsed the British at Laing's Nek and defeated them at Majuba Hill, forcing them to recognize the independent Transvaal Republic. These were skirmishes at the far-flung reaches of the British Empire. Other European countries were also involved in the colonial sweepstakes. For instance, Tahiti, from which Captain Cook had observed the transit of Venus in 1769, had changed hands once again; it was now a possession of France, a country resurgent, in its Third Republic, from the disaster of defeat in the Franco-Prussian War.

The clouds of another European war were not yet clearly visible on the horizon, even though it was in the same year that Italy joined Prussia and Austria in forming the Triple Alliance, which would be destined one day to struggle against France and Britain. Some inkling of the horror of that future war would have been grasped by anyone who recognized, in the devilish ingenuity of Hiram Maxim's recently invented recoil-operated machine gun, a gigantic leap forward in murderous efficiency. None could have foreseen the long twilight struggle of the trenches.

In the United States, Chester Arthur sat in the presidency. Charles Guiteau, the disappointed office-seeker who had shot Arthur's predecessor, James Garfield, in a railroad station the previous July, was hanged in Washington.

Far from the minds of most Europeans was the disastrous drought that ravaged much of the subtropical world as a result of the series of El Niño–Southern Oscillation events occurring in 1876–79.

Though millions died in India, China, and Brazil, historians rarely mention these events. Their plight certainly did not interfere with the plans of the Great Powers to observe the 1882 transit of Venus.

The spell of anticipation of the upcoming transit was upstaged, to some degree, by the appearance of a spectacular sun-grazing comet in September, the so-called Great September Comet. It was one of several sun-grazing comets; another, "Comet Tewfik," named after the Egyptian Khedive, was the first sun-grazing comet discovered (and photographed) during a solar eclipse, in May 1882, and was never seen again (see fig. 76).

Novelists took up the transit of Venus theme, including, somewhat improbably, composer John Philip Sousa, whose only venture into novel writing was *The Transit of Venus* (published in 1920, long after the 1882 transit—it must be one of the most atrocious novels ever written!). Sousa also wrote a "Transit of Venus March," performed in 1883 at the dedication of the Joseph Henry statue in front of the Smithsonian Institution, in Washington, DC, before an audience of five thousand.

The transit of Venus was taken up by an author far more gifted in letters, Thomas Hardy, already acclaimed for *Far from the Madding Crowd* and *The Return of the Native*. Between May and December 1882, he published, in serial form in the *Atlantic Monthly*, *Two on a Tower*, one of the finest astronomical novels ever written. It describes the May-December romance between the wealthy but aging Viviette, Lady Constantine, and young Swithin St. Cleeve, of Welland Bottom, the orphaned son of a curate hoping to win success as an astronomer.

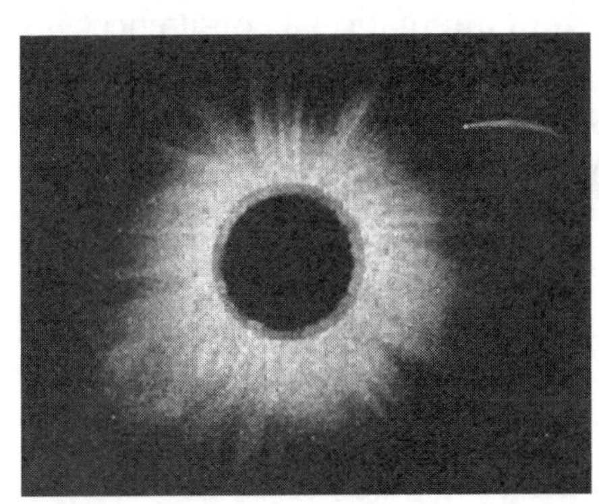

Figure 76. Drawing of the sun-grazing "Comet Tewfik" (*upper right*), from a photograph taken at Sohag, Egypt, during the total solar eclipse of May 17, 1882. (From William de Wiveleslie Abney and Arthur Schuster, "On the Total Solar Eclipse of May 17, 1882," *Philosophical Transactions of the Royal Society of London* 175 [1884]: 261.)

Hardy's interest in astronomy dated back to boyhood and his experiences with the bright stars of his native Dorset's skies and with a "big brass telescope" owned by his family. Though the idea of the novel was apparently inspired by the appearance of yet another comet (Tebbutt's) in June 1881—Hardy and his wife saw it from their house in Wimborne, eastern Dorset—the novel builds toward a climax involving the transit of Venus. That event occurred, of course, in the month when the last installment of Hardy's serialization appeared.

At the same time, real-world amateur and professional astronomers had also been busy. Those concerned with the upcoming transits had the opportunity to hone their transit-watching skills with transits of Mercury in 1878 and 1881. Many of those persons observed two or three transits in the 1874–1882 period, furnishing evidence that transit chasing, like solar eclipse chasing, may be addictive. Indeed, many of these observers also watched one or more of the total or annular solar eclipses that took place between the two transits of Venus. After all, the instruments and techniques used to observe solar eclipses are similar to those needed to observe transits.

BUILDING ON THE 1874 EXPERIENCE

As the 1882 transit approached, planners were able to take advantage of the experience of the 1874 transit, in contrast to 1874, when no one living had actually witnessed a transit of Venus. Most of the national transit commissions established for the earlier transit were still in place, and several were slowly analyzing the 1874 transit observations. To coordinate efforts, an international conference was held in Paris in October 1881. The French dominated: the honorary president was Jules Ferry (1832–93), the president of the Council of the Ministry of State Instruction and Fine Arts and the president was Jean Baptiste Dumas (1800–84), who had played a leading role in planning for the 1874 transit as permanent secretary of the French Academy of Sciences and president of the French Transit Commission. Austria-Hungary, Chile, and Norway sent representatives to the meeting but did not actually mount expeditions to observe the transit. Russia and the United States "abstained" from the meeting (see Table 5). Transit photography was among the topics discussed at length at the Paris conference. Its failure at the 1874 transit was

Table 5. International Transit of Venus Conference, Paris, October 5–13, 1881: Participating and selected nonparticipating countries

Country	Number of delegates	Government-sponsored* stations (and total number of parties) Planned for 1881	Actual
Argentina	1	(2)	Córdoba, Argentina (1)
Austria-Hungary	1	?	—
Brazil	1	Itapeva, Pernambuco, Rio de Janeiro, Antilles(?), Strait of Magellan(?) (5)	Punta Arenas, Chile; St. Thomas; Itapeva, Pernambuco, Rio de Janeiro, Brazil; (5)
Chile	1	Santiago (1)	—
Denmark	1	St. Thomas or St. Croix (1)	St. Thomas (1)
France	16	Cuba, Martinique, Florida, Mexico, Chile, Argentina (Santa Cruz, Chubut, Rio Negro) (8)	Havana, Cuba; Puebla, Mexico; St. Augustine, FL; Pétitionville, Haiti; Cerro Negro, Orange Bay, Chile; Santa Cruz, Bragado, Chubut, Rio Negro, Argentina; Fort-de-France, Martinique; Oran, Algiers, Algeria (13)
Germany	1	Argentina, Straits of Magellan or Falklands; 2 in United States (8)	Aiken, SC, Hartford, CN, United States; Punta Arenas, Chile; Bahia Blanca, La Plata, Argentina; South Georgia I.; Berlin, Dresden, Munich, Hamburg, Germany (10)
Great Britain	1	Bermuda, Jamaica, Barbados, 3 at Cape of Good Hope, Madagascar, New Zealand, Falklands, Sydney, Melbourne (11)	Kempshot, Jamaica; Gibb's Hill, Bermuda; Barbados; Greenwich, Great Britain; Cape Town, Aberdeen-Road, Durban, Touws River, South Africa; Nosy Vé, Madagascar; Melbourne, Sydney, Australia; Burnham, New Zealand (12)
Italy	1	?	Turin, Milan, Palermo, Naples, Italy (4)
Netherlands	2	Curaçao or St. Martin (1)	Curaçao (1)
Norway	1	?	—
Portugal	1	Lorenco Marques (1)	Lisbon (1)
Spain	2	2 in Cuba, Puerto Rico (3)	Havana, Cuba; Puerto Rico (2)
Switzerland	1	?	Geneva (1)

Countries not participating			
Belgium	—	—	San Antonio, TX; Cerro Negro, Chile (2)
Greece	—	—	Athens (1)
Mexico	—	Chapultepec (1)	Chapultepec (1)
United States	—	?	Tepusquet, CA, Cedar Keys, FL, Cerro Roblero, NM, Lehman's Ranch, NV, San Antonio, TX, Washington, DC; Santiago, Chile; Santa Cruz, Argentina; Wellington, South Africa; Hobart, Australia; Auckland, New Zealand (11)

*Includes less formal stations occupied by government staff, apparently on their own

generally acknowledged; indeed, it was revealed that neither the Germans nor the British planned to rely on photography at all in 1882.[1] The British position was summed up by Sir George Biddell Airy, who noted that, at least in terms of the British results, even plates taken with the Janssen revolver could not be used and "the ardour of the observers had been much cooled by the apparent general failure of the photographic principle and they were unwilling to spend further time on . . . reductions."[2]

Given the poor publication record of their transit observations in 1874, it appears that Russia had lost interest in transits as a means of determining the value of the solar parallax. The United States would be favored by geographic conditions of the transit's visibility in 1882 (see fig. 77). They were similar to those of the 1639 transit, at which Horrocks had lamented the transit's being "squandered" on the Americas. That once-barbarous region was now the site of some of the world's major observatories and some of its best-known astronomers. American observers would be able to see the entire transit without leaving home; indeed, the United States would sponsor within its borders foreign transit parties from Belgium, France, and Germany, who were eager to travel to points where the entire transit, not just the ingress-only view from Europe, could be seen.

In contrast to the British and the Germans, the Americans planned to use photography on a large scale at the 1882 transit. However, instead of wet bromo-iodide plates, which had been used

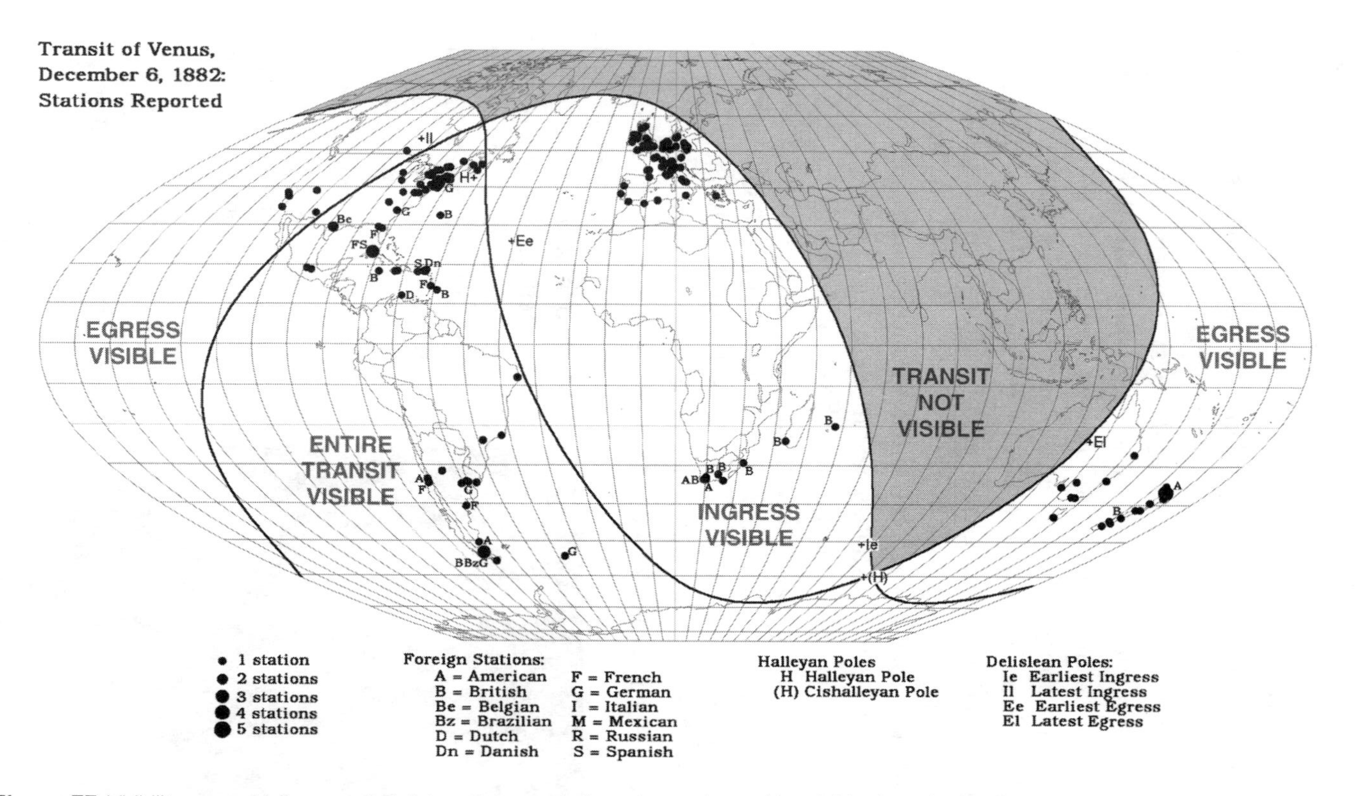

Figure 77. Visibility zones, Halleyan and Delislean Poles, and observing stations with published results for the transit of Venus of December 6, 1882. (Map by John Westfall.)

in 1874, they generally switched to dry-collodion plates, which could be manufactured in quantity before the transit and were more convenient to expose.

Despite Venus's obvious partiality in bestowing her graces on Americans at the 1882 transit, overseas expeditions were still needed, even by the Americans, to achieve the best geometry for a solar-parallax determination, whether using the Halleyan method or that of Delisle. The northern Halleyan Pole was conveniently placed just off the eastern seaboard of the United States, but the southern (actually Cishalleyan) pole lay in the southwesternmost Indian Ocean, off Antarctica, perhaps arguing the need for a return visit to Kerguélen. Of the four Delislean Poles, one ingress pole was in southern Canada, the other ingress pole was in the southwestern Indian Ocean; the egress poles were in central Australia (near modern Alice Springs) and in the mid-Atlantic between Bermuda and the Cape Verde Islands.

The U.S. Transit Commission had been funded with an $85,000 appropriation from Congress. Simon Newcomb (1835–1909) was nominally the leading figure of the U.S. Naval Observatory's Transit of Venus Commission and a leader in preparations for the 1874 transit (see fig. 78). However, he had in fact been unenthusiastic about the appropriation. By 1882 Newcomb had lost faith in the transit method for determining the solar parallax, though he would lead an overseas expedition himself to a site outside Cape Town, South Africa.

Indeed, preparations for the 1882 transit took place even as the task of reducing the 1874 transit observations had bogged down. Actual solar-parallax determinations based on the 1874

Figure 78. Photograph of Simon Newcomb (1835–1909). (Courtesy of the Mary Lea Shane Archives of the Lick Observatory, University of California–Santa Cruz.)

transit did not begin to appear until late in 1875, when Charles André, Charles Wolf's assistant, published values of 8.88 and 8.82 arc-seconds based on French timings made at Nouméa and Île St.-Paul. Victor Puiseux, using data from these stations together with Beijing, published values of 8.89 and 8.879 arc-seconds. A number of other results trickled out over succeeding years (see Table 6); they showed a good deal of scatter. In fact, only by discarding the disappointing British photographic results do the transit-derived values show greater mutual agreement than the visual timings of the eighteenth century.

Table 6. Solar parallax values from the 1874 transit of Venus

Value (arc-seconds)	Date published	Method	Authority (notes)
8.85 ± 0.02	1875	Contacts	André (mean of 8.82 and 8.88 arc-seconds)
8.89	1875	Contacts	Puiseux
8.879	1875	Contacts	Puiseux (Bejing and Île St.-Paul observations)
8.760	1877	Contacts	Airy
8.93	1877	Contacts	Tennant
8.754	1878	Contacts	Airy
8.884 ± 0.0369	1878	Contacts	Stone (1st group of stations)
8.88 ± 0.02	1878	Contacts	Stone (2nd group of stations)
8.699	1878	Contacts	Tupman (1st group of stations)
8.813 ± 0.033	1878	Contacts	Tupman (2nd group of stations)
8.846	1878	Contacts	Tupman (3rd group of stations)
8.165	1878	Photography	Tupman (British photographs)
8.8455	1878	Contacts	Tupman (range of 8.82–8.88 arc-seconds)
8.08	1878	Photography	Tupman (British photographs)
8.25	1878	Photography	Burton (same photographs as Tupman)
9.003	1881	Contacts	Puiseux
9.096	1881	Heliometer	Puiseux
8.88	1881	Contacts	Stone
8.883 ± 0.034	1881	Photography	Todd (American photographs)
8.80	1882	Combined	Proctor
		Post–1882 results from 1874 transit	
8.81 ± 0.06	1885	Photography	Obrecht (French photographs)
8.888 ± 0.040	1895	Photography	Newcomb (limb distances)
8.873 ± 0.060	1895	Photography	Newcomb (position angles)
8.876 ± 0.042	1895	Heliometer	Newcomb

There were two problems with the parallaxes based on the 1874 transit observations. The first was an embarrassment of riches: contact times, micrometer and heliometer measurements, and photographs from scores of stations. A complete solution based on all the observations was never attempted; had anyone wanted to do so they would have had to wait three-quarters of a century for digital computers to come into existence. Instead, the calculators of parallaxes chose small subsets of the observations, often subjectively weighted. Thus it was all too easy to give low weights to, or even discard, observations that differed from what was considered "the norm," raising doubts about the objectivity of the results. Often, the reducers of the observations used only the observations made by their countrymen, even if the geographic distribution of their observing stations was not ideal. It almost began to look as if the purpose of the expeditions was to find the distance from the sun to one's home country, rather than to the earth as a whole! Finally, there was increasing competition from a host of other methods of determining the sun's distance. There were Mars parallax determinations, such as was used by Gill at Ascension in 1877; asteroid parallax determinations, first employed by Johann Galle using the minor planet Flora in 1873; planetary-perturbation analyses; detailed studies of the moon's motion; analyses of the aberration of starlight; and methods based on determinations by physicists of the speed of light. Between 1874 and 1882 the solar parallax was found repeatedly by each of these six different methods. Completely independently of each other, they all seemed to be focusing on a "true" value in the range of 8.85 ± 0.05 arc-seconds. Although admittedly a mathematical shortcut, it is interesting simply to compare the unweighted means of the determinations of each of these methods, as is done in Table 7. In 1882, the *British Nautical Almanac* revised the "official" solar parallax to 8.848 seconds, the value first adopted by the *American Ephemeris* in 1870 from Simon Newcomb's review of parallaxes found by different methods up to then.[3] Thus by 1882, the overall uncertainty seemed about 0.5 percent.[4] The question was, could observations of the 1882 transit of Venus further reduce this uncertainty?

Table 7. Means of solar parallaxes found by six independent methods, 1874–82

Method	Number of determinations	Unweighted mean (arc-seconds)
Transit of Venus, 1874	15	8.867
Parallaxes of Mars, 1877	6	8.869
Parallaxes of asteroids	3	8.839
Planetary equation	3	8.915
Parallactic inequality	9	8.824
Annual aberration	3	8.810

The Last Great Attempt

The 1874 and 1882 transits had just two things in common: both were December transits, so they could best be seen from the Southern Hemisphere; and Venus's track across the face of the sun was almost parallel between the two transits (see the diagram of transit tracks in appendix B). Otherwise the geometry of the December 6, 1882, transit was radically different from its predecessor. To start, in 1882 Venus crossed the sun's southern hemisphere, rather than its northern. Then, in 1882 Venus crossed the sun's disk nearer to its center, making for a longer transit. For a geocentric observer, the duration from the first to the fourth contact was about 6.3 hours in 1882, as opposed to just 4.6 hours in 1874. This meant that, in 1882, the terrestrial zone where *all* of the transit could be seen was a little narrower than in 1874, but the area where *some* of the event was visible was noticeably larger.

It is often said that, due to disappointment with the results from the transit of Venus of 1874, there was less official interest in observing the 1882 transit. On the other hand, after 1882 there would be no second chance; the 2004 transit was a far-future event and was undoubtedly perceived much as we perceive the transit of 2117—the next such event after those of 2004 and 2012. Though the transit's visibility from most of Europe and from the Americas made travel unnecessary for many groups, no less than forty-five expeditions voyaged overseas in 1882. Their destinations fell within mainly four general areas: the Caribbean (British, Danish, Dutch, French, and Spanish), South Africa–Madagascar–Mauritius (American and British), Australia–New Zealand (American and British), and Patagonia–South Georgia Island (American, Argentine, Belgian, Brazilian, British, French, and German).

Besides the far-flung scientific attention, the 1882 transit was the first to lie within reach of the common man. As noted in the *New York Times*, December 7, 1882, "This is the first time within the memory of man that unlearned common people have been permitted to observe a transit, and it is the first revelation of the fact that a transit can be seen through smoked glass." The *Times* further documented the public reaction after the event: "The transit of Venus was a popular exhibition, and as it had been widely advertised in advance as the last performance for 122 years everybody who could possibly get a sight of the show embraced the opportunity. A very satisfactory view was obtained through smoked glasses, but the speck that was made on the disk of the sun by the planet was so small that it required some time of close application to the glass before it was recognizable. The dark spot appeared no larger than a small-sized dried pea."

Children, like the group depicted in *Harper's Weekly* (see fig. 79), were encouraged to experience an event that they would never see again.

At Meriden, Connecticut, schools were closed for the day, and the fire-alarm bell rang when the transit began. A five-year-old named Henry Norris Russell viewed the event from Oyster Bay, Long Island, and the sight of Venus against the sun inspired a lifelong interest in astronomy. He went on to become the leading American astrophysicist of his time, the director of the Princeton University Observatory, and a world-renowned expert on stars.

Enterprising proprietors of telescopes took advantage of the planet's rare rendezvous in front of the sun and stationed all sorts of instruments in any place likely to be frequented by a crowd, reaping a small fortune by exhibiting the planet at the price of ten cents a view. In City Hall Park in New York City, a telescope was erected, as described in the same *New York Times* article: "so great was the rush of people to take a look through that the services of Park policemen were required to keep them in line awaiting their turn. Once at the

Figure 79. A group of children viewing the 1882 transit of Venus (from *Harper's Weekly*, April 28, 1883). Their method of observation is *not* recommended.

telescope a view of a few seconds only was allowed, and by actual count 20 men peered through the glass in 5 minutes." One particularly enterprising individual set up a telescope on Broad Street, near the Stock Exchange, and sold peeks to the brokers.

Presumably every astronomical telescope in America favored by clear skies was trained on the sun on December 6, 1882. These included telescopes operated by amateurs and those in service of the small college observatories that had multiplied since the Civil War, as well as those at large professional observatories.

The amateurs included Lewis Morris Rutherfurd (1816–92), the pioneer American astrophotographer, who watched the transit with his 13-inch telescope on Second Avenue in New York City. Famous discoverer of comets E. E. Barnard (1857–1923) observed the transit at Vanderbilt University's observatory in Nashville, Tennessee. Pioneering woman astronomer Maria Mitchell (1818–89) photographed the transit at Vassar College with a smaller version of the "official" photoheliostat used by the American expeditions. The observing station at Poughkeepsie happened to be Mitchell's second choice. She had applied to join a transit party but was turned down because it was felt that women could not be accommodated on an overseas expedition (unless, of course, they were closely related to an expedition leader).

In New York City, John Krom Rees (1851–1907), using Columbia College's Alvan Clark refractor, commented about the arc of light surrounding Venus between the first and second contacts: "The light shining through Venus's atmosphere was a fine sight. I should say it first appeared to my eye when the planet was a little more than half-way on the sun. . . . This line of light, marking out the portion of Venus's disk not on the sun, changed its appearance considerably while my attention was fixed upon it."[5]

Though the Allegheny Observatory in Pittsburgh was plagued with clouds, like many locations in the eastern United States, Samuel P. Langley (1834–1906) enjoyed a view of a "very notable gathering of brightness" along the limb of Venus when it had not yet advanced upon the disk of the sun (see fig. 80). In addition, there were a number of official expeditions within the United States sponsored by either the U.S. Naval Observatory (USNO) or its chief rival, the Coast and Geodetic Survey (USCGS).[6] One USNO station was near San Antonio, Texas; another was at Cedar Keys, on Florida's Gulf coast. The USCGS utilized a private observatory just southwest of the Capitol in Washington as well as four stations in the western states and territories, the most successful of which was led by George Davidson (1825–1911), who had observed the 1874 transit from Nagasaki. Davidson traveled to Cerro Roblero (now called Cerro Robledo), near Fort Selden in New Mexico Territory (see fig. 81). Conditions were good, enabling the visual observers to time all four contacts. Cerro Roblero was the only complete transit station established by the USCGS in 1882, for which tons of prefabricated buildings, equipment piers, telescopes, and supplies had to be carried by mule up the steep mountainside.[7]

The parties farther west did not establish full-scale observing stations. They included an unidentified USCGS

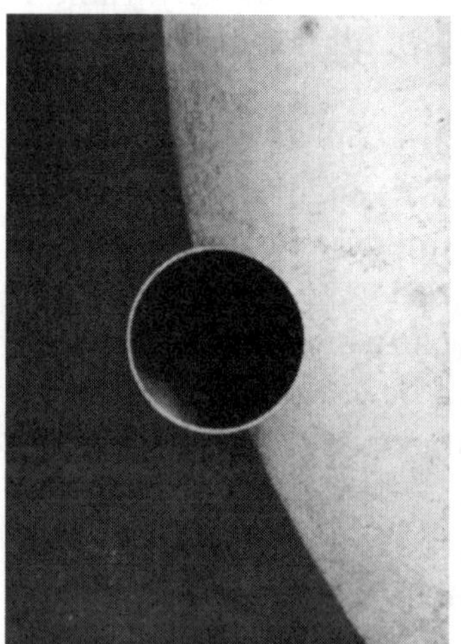

Figure 80. Drawing by S. P. Langley of the ring of light and "gathering of brightness" during ingress at the 1882 transit of Venus. Langley used the 13-inch (33 cm) Alleghany Observatory refractor at 244X. (From S. P. Langley, "Observation of the Transit of Venus, 1882, December 6, made at the Alleghany Observatory," *Monthly Notices of the Royal Astronomical Society* 43, no. 3 [January 1883]: 73.)

Figure 81. Photograph of the longitude-determination camp used by George Davidson's U.S. Coast and Geodetic Survey party for observations of the 1882 transit of Venus. The peak right of center is Cerro Roblero, New Mexico Territory; the view looks south and shows the camp alongside the railroad and telegraph line at Fort Selden. (Courtesy of the George Davidson Collection, Archives of the California Academy of Sciences.)

party that used George Davidson's private observatory in San Francisco, another California party that occupied the geodetic station on Tepusquet Peak in Santa Barbara County, and a group of USCGS observers in eastern Nevada, at Lehman's Ranch, not far from the present-day tourist attraction of Lehman's Caverns.

Across the Atlantic, the beginning phases of the transit would have been visible to hundreds of European amateur and professional astronomers, had the skies allowed it. Sadly, clouds, rain, and snow prevailed over much of the continent and the British Isles. Most stations in England were unable to see the event, including the observatories at Cambridge, Oxford, and Greenwich; even the solar observatory at Kew was deprived of a view of the sight not to be seen again until June 8, 2004. Scattered areas, however, were clear. For example, British solar observer Frederick Brodie (1823–96) was able clearly to see the ring of light around the part of Venus's limb projected on the sky, describing it as having "a soft white light, assuming a ruddy hue next to the edge of the planet's disk."[8] At Crowborough, Sussex, Charles Leeson Prince (1821–99) spotted Venus when half on the sun's disk, noting "that portion of the planet still outside the Sun's disk . . . appeared to be illuminated by a brilliant silver line of light, which most distinctly marked the limb of that portion of the planet . . . doubtless produced by the refraction of sunlight through the planet's atmosphere." Prince recorded the second contact, when the "shadow, or ligament, or whatever it was, *suddenly* left the Sun's limb in less than a second of time."[9]

France also suffered from inclement weather. Paris, Bordeaux, St.-Genis-Laval, and Lyon reported clouds and rain. On the other hand, in Nice, an observer named Michaud obtained five photographs, while André Puiseux watched the event projected on a screen (see fig. 82). At Orgères-en-Beauce in central France, an unnamed observer produced the evocative drawing shown in figure 83.

The sun was low in the sky during the transit in Germany, but some of it was seen from places such as the Potsdam Astrophysical Observatory, where Hermann Karl Vogel (1841–1907) drew the formation and disappearance of the black drop during ingress (see fig. 84).

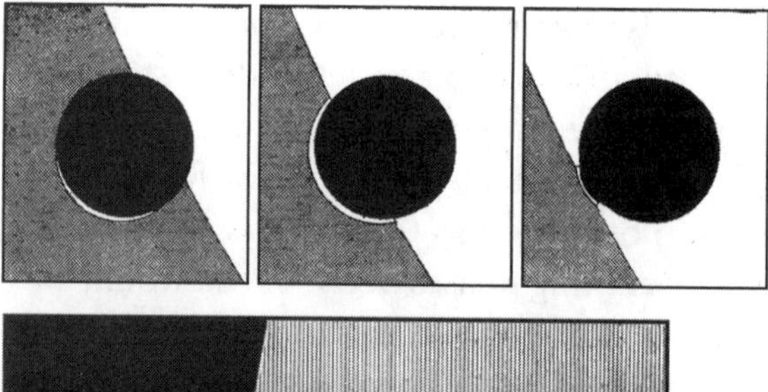

Figure 82. The aureole of the atmosphere of Venus, observed from Nice on December 6, 1882. (From Camille Flammarion, *Les terres du ciel*, 1877.)

Figure 83. Anonymous drawing, made at Orgères-en-Beauce in central France, of the 1882 transit of Venus, showing the aureole completely surrounding the planet. (From the reproduction in Camille Flammarion, *Les terres du ciel*.)

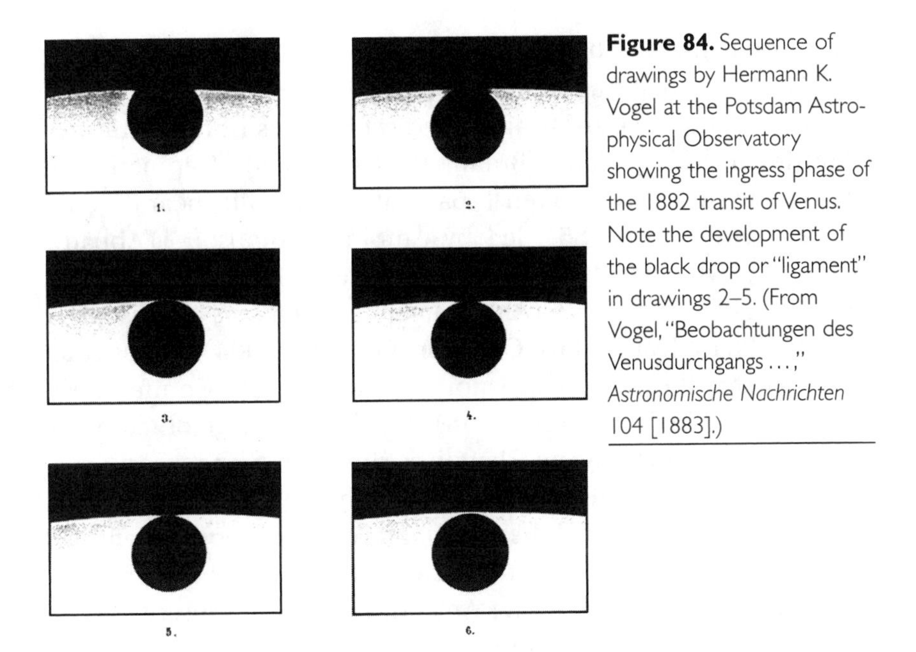

Figure 84. Sequence of drawings by Hermann K. Vogel at the Potsdam Astrophysical Observatory showing the ingress phase of the 1882 transit of Venus. Note the development of the black drop or "ligament" in drawings 2–5. (From Vogel, "Beobachtungen des Venusdurchgangs . . . ," *Astronomische Nachrichten* 104 [1883].)

Conditions were better south of the Alps, with successful observations reported in Italy from Milan to Palermo. The easternmost transit observers in Europe probably were those at the Athens Observatory in Greece.

In northern Africa, renowned French astronomer Pierre Jules César Janssen (1824–1907) observed the 1882 event from Oran in French Algeria. At the same time, Charles Trépied (1844–1907) watched at the Observatory of Algiers.

The USNO sent an expedition to the opposite end of the continent—Wellington, South Africa, about forty miles northeast of Cape Town. They observed from the Huguenot Seminary for Girls, where three of the female instructors and students joined the Americans in timing the ingress contacts. Conditions were good, and Simon Newcomb remarked, "I think that during the hours of the transit, which commenced about three o'clock, the definition of the Sun was as fine as I have ever seen it anywhere in my life."[10]

Unlike the case in 1874, there were just two parties on islands in the Indian Ocean; both were British and both observed ingress successfully. One, on the tiny island of Nosy Vé off the southwestern coast of Madagascar, was headed by the veteran of Kerguélen, Father Perry.

The second, under Charles Meldrum (1821–1901), watched from Lord Lindsay's 1874 station, the island of Mauritius.

Mexico and the Caribbean hosted expeditions from a variety of countries, including France, Britain, the Netherlands, Denmark, and Brazil. One such was the French party at Pétitionville, near Port-au-Prince, Haiti (see fig. 85), led by Antoine Thompson D'Abbadie (1810–97), who had traveled to solar eclipses in 1851 (Norway), 1860 (Spain), and 1867 (Algeria). Other members of his party included Pierre Jean Octave Callandreau (1852–1904), an assistant astronomer at the Paris Observatory, and a volunteer, Comte Aymar de la Baume-Pluvinel (1860–1938). A heliostat image obtained by expedition member Eugène Chapuis is shown in figure 86.

At Pernambuco (Recife), the site of one of three observatories within Brazil that watched the transit, one of the observers identified himself as "Pedro Alcanatara"; he was actually Brazilian emperor Dom Pedro II (1825–91). The Argentine National Observatory in Córdoba was the northernmost of the eighteen transit observing stations in southern South America, extending southward through Patagonia to Tierra del Fuego. One of the several French parties in this area was led by Lt. O. de Bernardières (d. 1900), located at Cerro Negro, Chile, south of Santiago. De Bernardières used the

Figure 85. Housing of the 16 cm (6.3 in) refractor at the French observing station in Pétitionville, Haiti, in 1882. (Courtesy of the Bibliothèque de l'Observatoire de Paris.)

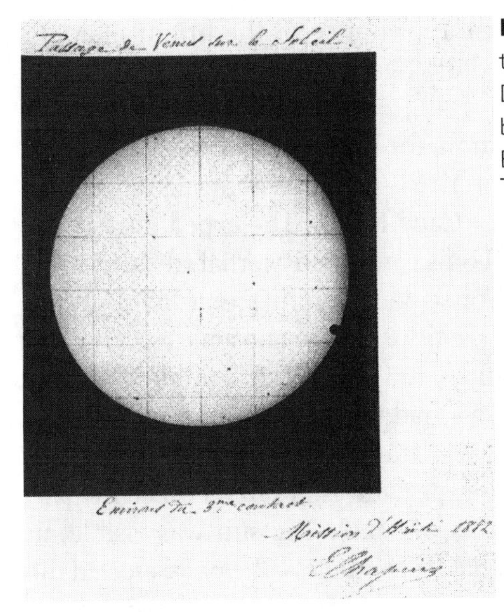

Figure 86. Photograph of Venus near the third contact of its transit on December 6, 1882, taken at Pétitionville by Eugène Chapuis. (Courtesy of the Bibliothèque de l'Observatoire de Paris.)

standard 8-inch (21 cm) refractor reserved for the leaders of the French parties and noted the following six events (in Paris mean time) associated with the second contact (translated by John Westfall; italics are in the original):

2h 24m 58s: The exterior limb of Venus is surrounded by a luminous aureole as clearly visible as the portion of the limb of the planet already on the sun's disk.

2h 25m 36s: The two disks still very clearly overlap, distinctly affecting the appearance of the solar limb.

2h 26m 08s: A small black band forms on the limb of the sun, as dark as the planet and wider than the geometric chord joining the points of contact of the two limbs. This band, which is very thin, appears frequently in the artificial transit observations.

2h 26m 28s: *First appearance of a bright filament, which appears intermittently.*

2h 26m 41s: *The ligament reappears, although the bright filament has a perceptible width and remains definitely visible.*

2h 26m 53s: The ligament has completely disappeared after passing through less and less definite shades; it has, as it were, faded away.[11]

These notes show how hard it was to define the instant of the second contact, or indeed any of the contacts. De Bernardières chose a time midway between the first appearance of the bright filament and the reappearance of the dark ligament, 2h 26h 34s, as the moment of this contact.

France sent one joint expedition, for both the transit of Venus and the 1882–83 International Polar Year—the "Mission scientifique du Cap Horn" (Scientific mission of Cape Horn). The expedition's ship, the *Romanche*, was captained by Louis Ferdinand Martial (d. 1885) and served as the base for a variety of scientific studies that involved the party for one year and one day (see fig. 87). The chosen observing station, Orange Bay on Hoste Island—at 55°.5 south latitude and only 44 miles northwest of Cape Horn—had the debatable honor of being the southernmost of all the 1882 transit stations. The notorious weather of the region lived up to expectations: it rained on eighty-two days during the last three months of 1882. The sun was visible an average of only three hours each day. Nonetheless, the party erected the prefabricated structures used by most international expeditions in 1882, and by October 22 they were ready. On the day of the transit, clouds and even rain interfered with observations; despite these difficulties, though, the party timed the second, third, and fourth contacts.

Figure 87. Model of the *Romanche*, the vessel that transported the French Cape Horn scientific expedition in 1882–83. Captained by Louis Ferdinand Martial, the ship could be powered either by wind or steam and is typical of the ships that served the nineteenth-century transit of Venus expeditions. (On display in the Museo Maritimo de Ushuaia, Tierra del Fuego, Argentina; photographed by John Westfall in 2002.)

The remaining cluster of observers was in Australia and New Zealand. In Auckland, New Zealand, the American Coast and Geodetic Survey party under Edwin Smith (1851–1912), who had led the Chatham Island party in 1874, timed the egress contacts and obtained seventy-four photographs. Counting independent local observers, about a half-dozen sites in Auckland saw transit egress. In Burnham, on the South Island, the British group was under the command of one of their most experienced transit observers, George Tupman (1838–1922), who had been the overall leader of the British parties in 1874. In contrast to the situation in 1874, the weather in New Zealand was generally favorable in 1882.

Results were mixed in Australia. Brisbane and Sydney were clouded out, but observations were made from Melbourne, Adelaide, and Hobart. It is likely that the successful groups in Australia were the last to watch the fourth contact in 1882, thus the last persons to see a transit of Venus until some observer in the same part of the world makes the first sighting of the first contact on June 8, 2004.

David Peck Todd and Lick Observatory

The transit of Venus of December 6, 1882, figured at least as a backdrop to one of the most celebrated passions of the century: a May-December romance between Austin Dickinson (1829–95), town leader of Amherst, Massachusetts, and Mabel Loomis Todd (1856–1932), wife of Amherst College's professor of astronomy, David Peck Todd (1855–1939). Austin was the brother of Emily Dickinson, the poet, who had once written: "For what are stars but Asterisks / To point a human life?"

In the years leading up to the transit in 1882, David Peck Todd appeared to be one of the most promising young astronomers in America (see fig. 88). After his graduation from Amherst in 1875, Todd was noticed by the doyen of American astronomers at the time, Simon Newcomb, then superintendent of the *Nautical Almanac* office in Washington, DC, a master of gravitational theory and tables of the motions of the planets, as well as the leading figure of the U.S. Naval Observatory's Transit of Venus Commission. Thus Todd's life would

Figure 88. Photograph of David Peck Todd (1855–1939), taken in middle age. (From Charles J. Hudson, "David Todd, 1855–1939: An Appreciation," *Popular Astronomy* 47 [1939]: 472.)

be fatally intertwined with the forthcoming transit of Venus. Indeed, Todd was employed at the U.S. Naval Observatory, still located in malarial "Foggy Bottom," along the Potomac, from 1877 to 1881. Though he was especially interested in the planet Jupiter, he had been the first to knowingly recognize a point of light near Mars as its inner satellite, Phobos, during the 1877 opposition of the planet.[12] Newcomb also assigned him the unenviable task of reducing the 1874 transit observations for the purpose of deriving a solar parallax.

In 1881, David and Mabel moved to Amherst, the little college town nestled between the Pelham and Mount Holyoke ranges. David had accepted a position as professor of astronomy at his alma mater, over Mabel's misgivings ("it seems to me you should hardly dare to give up the Office place entirely," she wrote to him[13]). Professionally for David, the move would eventually prove a mistake: Amherst was still poorly equipped for astronomy, boasting only the small observatory he had used as a student in timing the eclipses of Jupiter's satellites.

He was hoping to observe the transit of Venus, but at the moment his plans were very much up in the air. Newcomb had not chosen him to accompany any of the eight official U.S. government-sponsored expeditions to various points in the United States, South America, and New Zealand to observe the transit of Venus. But another opportunity arose: Capt. Richard S. Floyd (1843–90), president of the trustees of the observatory then being planned for Mount Hamilton, California, had hoped to lure the prestigious Newcomb himself to Mount Hamilton for the event. The observatory on the peak, 4,200 feet above sea level, was funded out of the munificent estate of the eccentric benefactor James Lick.[14] Floyd had even set up, at his recommendation, a special instrument, a "horizontal photoheliograph," for the express purpose of photographing the transit, of the type used by the American transit field parties in 1874 and 1882. It was a splendid instrument, with a 5-inch (13 cm) objective lens fashioned by the celebrated telescope makers Alvan Clark and Sons, but arrived on Mount Hamilton with one end missing and without information about the lens's focal length (it was 40 feet [12 meters]). In the end, Newcomb worried about Mount Hamilton's suitability after meteorological records showed that unfavorable conditions had frequently prevailed on or about the date of the transit to come.[15] Also undoubtedly aware that only the transit egress would be visible from California, he opted for the site described earlier, near Cape Town. Since the Transit Commission in Washington was unwilling to fund any other expeditions than those already under way, Lick Observatory seemed to have been left out in the cold.

At almost the last minute, however, Floyd engaged Todd and his assistant, Amherst photographer John L. Lovell, to come to Mount Hamilton for the transit rather than allow the site and all the preparations made there to go to waste. Floyd generously agreed to pay all of their expenses out of his own pocket. Todd accepted, although members of Newcomb's commission had warned him that there would be nothing on Mount Hamilton with which to work, and he was advised, sarcastically, to bring a faucet with him to "put in the bowl for a tank in his Photo Room."[16] Nonetheless, Floyd and another member of the Lick Trust, Captain Thomas Fraser (1843/44–91), were ceaseless in their efforts at getting everything ready for the transit. Their preparations included taking a series of test photographs to determine the exact focal length of the photoheliostat's lens. Later, Fraser would grouse that Todd, in his official report, did not do enough to credit Floyd's efforts, but Todd may have been too preoccupied to notice, for as Fraser assisted Todd with the final adjustments of the instruments, he found him "bang crazy with his responsibility."[17] There was reason for the tension: it would be the last chance to photograph a transit of Venus in his lifetime.

Everything turned critically on the weather, the main reason Newcomb had decided to abandon the idea of observing from Mount Hamilton. The attempts to photograph the transit of 1874 had largely been frustrated by poor seeing from the various sites, notably Asaph Hall's at Vladivostok. Clouds were always a possibility, and their effects on an astronomer's psyche when they presumed to interfere with an event as rare as a transit or a total eclipse could be devastating. Near the end of his life, Todd was to confess to his daughter that "what saddened his life were three cloudy eclipses of the Sun."[18]

Even apart from the possibility of clouds, the steadiness of daytime seeing at Mount Hamilton could not always be trusted at this time of year, so that even with perfect transparency, the results might still be ruined by tremulousness of the image. There were three or four heart-stopping days of cloudy weather at the end of November, but then the sky cleared—from November 30 to the afternoon of December 7 "we saw no cloud, day or night, which could interfere with any observation." There were "fitful gusts night and day" on the third and fourth as well as on the morning of the fifth, but they subsided about noon; "for the next fifty or sixty hours the utmost tranquility prevailed."[19]

As the sun rose, close to 7 AM, December 6, Venus was already on its disk. Floyd observed the transit with the 12-inch (30 cm) refractor that had been shipped out to Mount Hamilton a year earlier for the November 7, 1881, transit of Mercury, while Todd and Lovell worked the photoheliostat. In their first image, the sun's disk was still noticeably flattened and distorted on the bottom due to its low elevation above the horizon. In preparing for the transit, Todd had opted to employ fine-grained gelatin wet plates rather than the newfangled dry plates favored by the other expeditions, including those of the Transit of Venus Commission, and the results were to bear out his judgment. He kept up the photographic work until the planet left the sun (with a particularly feverish flurry taken between the two egress contacts). There were 147 plates in all, of which 125 were deemed good enough for micrometric measurement.

Todd was delighted with the results, telegraphing Mabel: "Splendid day. Splendid success." His diary was even more exuberant: "Day as perfect as a June day in New England—sky perfectly cloudless. A day built only for the Gods and Mt. Hamilton astronomers. . . . We saw things as plain as was ever seen with any glass in the world."[20] Euphoric as he felt, when he tried to make micrometric measures of the best plates he found that the results fell short of complete success, for reasons that are only too apparent: the images of Venus, instead of being crisp and perfectly well-defined, were fuzzy around the edges, owing to diffraction produced by the small lens and the inevitable vagaries of the seeing). And yet for all their obvious shortcomings, the Mount Hamilton plates were generally regarded as the sharpest and most perfect obtained anywhere, in either 1874 or 1882.

One of Todd's transit of Venus plates is shown in figure 89. His series may well be the best surviving set of photographs of a transit of Venus, and this is one of the best of the series, but it illustrates the problems encountered as soon as one attempts to measure from such photographs. An enlargement of the portion of the plate showing Venus and the sun's limb is shown as an inset; both have indistinct borders, brought out by the brightness-contour plot in the inset.

To Newcomb's great disappointment, Todd would never complete the task of reducing the 1882 observations to a form that would yield a value for the solar parallax, as he had done with the 1874 observations. The two men, who had been close in earlier years, became increasingly alienated. David and Mabel remained married—involved in a ménage à trois, with Austin Dickinson as the third apex of the triangle—until Austin's death in 1895, and then the couple chased eclipses together. In all, Todd went on thirteen eclipse expeditions, making him one of the most prolific eclipse chasers of his time; many of them, including far-flung expeditions to Japan and Angola, were made with Mabel. Pursuing his passion, he also applied his talents in mechanical engineering to devise a "pneumatic commutator," which permitted automated photography of eclipses. On an eclipse expedition to Algeria in 1900, the globetrotting couple was accompanied by Percival Lowell, who had become celebrated for his observations of the alleged Martian canals. Lowell funded David to take the Amherst 18-inch refractor and a photographic camera of his design on an expedition to Alianza, Chile, in 1907, for the purpose of photographing the red planet. That year David and Mabel posed together in front of Observatory House in Amherst. Mabel appears care-worn; David has a

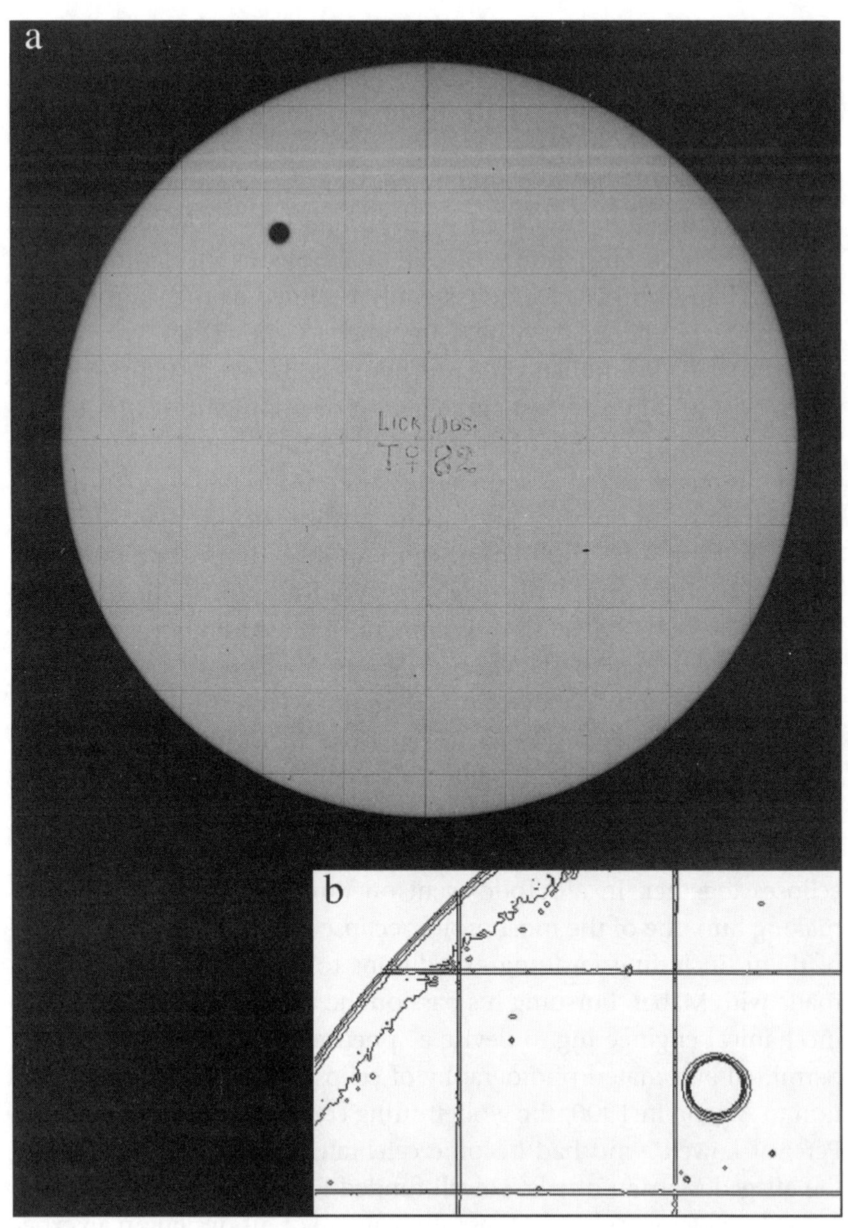

Figure 89. (*a*) One of the photoheliostat plates of the transit of Venus taken at Lick Observatory, California, December 6, 1882, by David P. Todd. (*b*) An enlarged brightness-contour plot of Venus and the solar limb. (Courtesy of Tony Misch, Lick Observatory.)

rather wild look in his eyes, and from this point forward he became increasingly wild in his ideas. He planned, in 1909, to take a balloon above the earth in order to detect Hertzian waves from Mars. Though he was eased into early retirement from Amherst in 1917, he continued to make feverish efforts to contact Mars until he was institutionalized in 1922.[21] He spent the last years of his long life—he died in 1939—in hospitals and nursing homes. Mabel had died in 1932, long after she accomplished the one thing for which she is most fondly remembered today: the task of discovering and beginning the process of preserving from oblivion Emily Dickinson's poems and letters.

The Closing Act of the Transits

Todd's transit of Venus plates had been deposited in the plate vault at Mount Hamilton, forming "the first batch of strictly original scientific data stored in the vaults of the Lick Observatory."[22] There—in the aftermath of the transit, as astronomers realized that even with photography the Venus-transit method of obtaining the solar parallax remained unsatisfactory and as they turned to other methods—the plates were forgotten. A reference to them in one of Todd's letters in the Mary Lea Shane Archives of the Lick Observatory, discovered 120 years later, led William Sheehan, in collaboration with Anthony Misch, a resident astronomer on Mount Hamilton, to search for them. They were still in the vault, in mint condition, enough of them to put together into a motion picture; there was Venus, strangely flickering as it marched across the sun (see fig. 90), which we could not resist accompanying with the seldom-heard music of John Philip Sousa's "Transit of Venus March."

We could not help thinking of the long-lost world these images (and music) brought, in a manner of speaking, back to life again. We could not help thinking of Todd and Captain Fraser, keeping vigil on Mount Hamilton as astronomers have done ever since; of Mabel and Austin, standing under the leaden skies of a severe New England winter's day, perched on the verge of their great romance; of the expeditions far and wide over the vast surface of the earth, to such

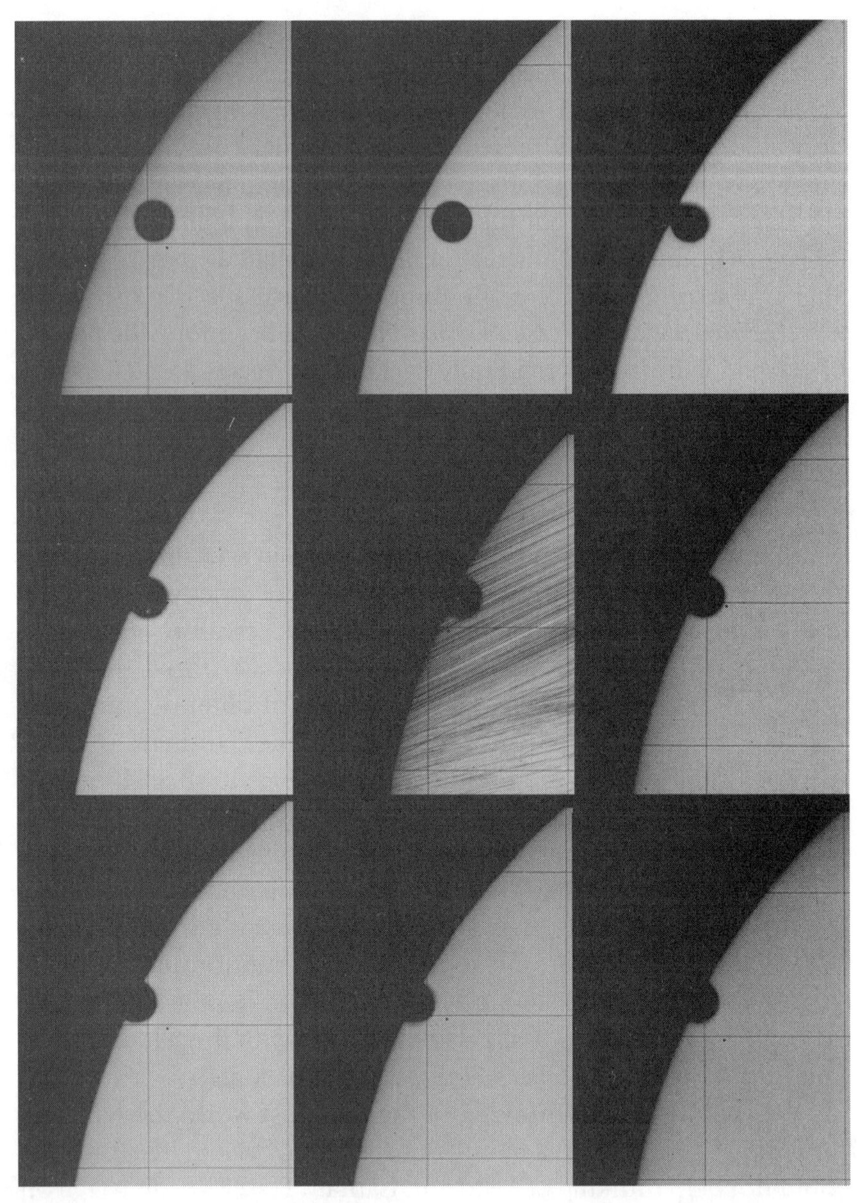

Figure 90. Sections of nine plates of the egress phase of the 1882 transit of Venus, taken by D. P. Todd at Lick Observatory. (Courtesy of Tony Misch, Lick Observatory.)

far-flung places as Argentina, South Africa, New Zealand; of all the upcast eyes, whirring clock-drives, and snapping shutters.

We cannot estimate how many people were thrilled by the transit of Venus across the sun on that December day in 1882. In contrast to only two, Horrocks and Crabtree, who witnessed Venus against the sun in 1639, millions must have been held spellbound by this six-hour celestial shadow show. These millions will, however, be a fraction of those who will view the transit of 2004 and, eight years later, that of 2012. Those who observed the transit of 1882 are all gone; they have joined the "innumerable caravan."[23]

And so, in turn, will we prospective observers of the 2004 and 2012 transits. For us, as for them,

> Heaven [is] but the Vision of fulfilled Desire,
> And Hell the Shadow from a Soul on fire
> Cast on the Darkness into which Ourselves,
> So late emerged from, shall so soon expire.
>
> We are no more than a moving row
> Of Magic Shadow-shapes that come and go
> Round with the Sun-illumined Lantern held
> In Midnight by the Master of the Show.

(From *The Rubáiyát of Omar Khayyám*, LXVII–LXVIII,
translated by Edward FitzGerald, 1859)

14

A "Compleat" Guide to Our Transits: 2004 and 2012

We are now traversing distances beside which the immense line stretching from the earth to the sun is but an invisible point.

—Thomas Hardy, *Two on a Tower* (1882)

The Grail Achieved

The December 6, 1882, transit came and went. The field parties returned home. The observers, who had watched or photographed the now-familiar phenomena of ingress and egress, attempted to time the four contacts, and witnessed the black drop and the "ring of light," waited for mathematical astronomers to assemble the data they had obtained and convert them into a solar parallax.

Curiously, fewer parallax results were published for the 1882 transit than for the one in 1874. In part, the reason for this is that most astronomers had already become fully aware that there were better methods of working out the earth-sun distance. One of these was based on the observation of asteroids that approached relatively close to the earth. Several asteroids were used in this way, but a special opportunity was presented by Eros—a cigar-shaped body discovered photographically in 1898 by Gustav Witt at the Urania public observatory in Berlin and independently by Auguste Charlois (1864–1910) of the Nice Observatory. It is a member of the class of Earth-approaching asteroids later known as Amor Objects, named after another asteroid of the class, Amor, discovered in 1932. Eros's highly elliptical orbit brings it much closer to the earth at some oppositions than at others, and in 1900–1901 it came within only 48 million kilometers (30 million miles) of the earth, nearer than it would approach until 1931. This is much closer than most asteroids, which keep to better-behaved and less alarming orbits between Mars and Jupiter. It was carefully measured in relation to neighboring stars by eyepiece micrometer and by photography at about two dozen observatories around the world. From these observations, astronomers derived a number of values of the solar parallax; the mean of six published figures was 8.8006 ± 0.0022 arc-seconds, which gave an astronomical unit equal to 149,488,000 ± 38,000 kilometers (92,888,000 ± 24,000 miles).

In 1931, Eros approached the earth within only 26 million kilometers (16 million miles). This event was again widely observed, with 2,847 plates, exposed at twenty-four observatories, being measured and serving as the basis for the British Astronomer Royal, Sir Harold Spencer Jones (1900–60), to work out a revised solar parallax, which he published in 1941: 8.7904 ± 0.0010 arc-seconds. This implied a mean distance from the earth to the sun of 149,675,000 ± 17,000 kilometers (93,005,000 ± 11,000 miles).

Writing a popular account in a magazine, Jones concluded that he had reached "the goal for which astronomers have so long been striving" and predicted that it would be the last word on the subject for many years to come.[1]

It certainly was good enough for all practical purposes, at least in 1941, and a vast improvement over any values of the solar parallax published on the basis of the transit of Venus method. This assured that at the next transit of Venus, "when the June flowers are blooming in 2004," as U.S. Naval Observatory astronomer William Harkness had put it on the eve of the 1882 transit, the event would have no role as a method of determining the solar parallax. In the race to determine the value of the solar parallax, Eros had trumped Venus, as it were. But even Eros was not the last word.

Indeed, by the time Spencer Jones died in 1960, the Space Age had begun. Radio waves had been reflected off the moon in 1946 and their echo detected back on the earth. Then, when Venus came to inferior conjunction in March 1961, radio astronomers using the large-dish radio antenna at the future Goldstone Tracking Station, on the Mojave Desert, beamed a signal at the planet and received its echo in return. The 6½ minutes needed to reach Venus and return implied an astronomical unit equal to 149,599,000 kilometers (92,957,000 miles). After this success, Venus during the early 1960s was bombarded with radar signals, which yielded the publication of numerous independent measures of the astronomical unit that agreed, typically, to within a few hundred kilometers, or to one or two parts per million. Remember, Halley had dreamed of using the transit of Venus method to determine the value of the earth-sun distance to one part in five hundred! Even the method of bouncing radio signals off Venus—a large, extended object, after all, with surface irregularities—would be superseded by a method using the time-lag of radio communications sent to and received from interplanetary spacecraft.

Currently the most accurate interplanetary range is to the Viking Lander on Mars, whose daily mean distances show a scatter of only about six meters.[2] At this level of accuracy, the astronomical unit itself is no longer a direct measure of the earth-sun distance, since the latter varies by several thousand kilometers from one orbit to the next owing to oscillations around the center of mass (barycenter) of the earth-moon system and gravitational perturbations produced by the planets. Such is the level of accuracy attained at the present time that orbital models include the small perturbations produced by the three hundred or so largest asteroids!

The definition of the astronomical unit is now, in fact, rather abstract. It is a "derived constant," specified in terms of more fundamental constants. Its value is equal to the radius an ideal body—not the actual earth—would follow, entirely free of perturbations, in an idealized one-year circular orbit around a specified mass approximating that of the sun. The current best estimate of this quantity, adopted by the Solar System Dynamics Group of NASA's Jet Propulsion Laboratory: 1 astronomical unit equals 149,597,870.691 km, with an uncertainty of 3 meters, while the light time for 1 astronomical unit equals 499.004783806 ± 0.00000001 seconds.[3] Using this value of the astronomical unit, the distance from the earth to Venus (center to center) at the midpoints of the two upcoming transits of Venus, will be (at the times indicated, calculated to the nearest kilometer)

June 8, 2004, at 08h 19m 40s UT	43,216,312 km (26,853,371 mi)
June 6, 2012, at 01h 29m 28s UT	43,189,411 km (26,836,656 mi)

DISTANCES TO STARS AND GALAXIES

Obviously, from what we have said above, the transits of Venus of our time will no longer be viewed as important scientifically in providing an accurate value of the astronomical unit. However, the astronomical unit itself remains the basis of all the highly accurate calculations needed to work out, precisely, the distances and positions of the moon and the planets. In addition, since its value is equal to half of the diameter of the long baseline the earth traverses once every year in its orbit around the sun, it furnishes the first rung in the so-called cosmic distance ladder. An astronomical unit is equal to one-half the length of the baseline used in making trigonometric measures of the parallaxes of the stars; thus we can use it to measure the distances to the stars.

The first stellar parallaxes—for the stars 61 Cygni, Alpha Centauri, and Vega—were obtained by Friedrich Wilhelm Bessel, Thomas Henderson, and Wilhelm Struve in the late 1830s.[4] They were remarkable achievements at the time, since the measures involved are extremely delicate. The angles of displacement produced even across a baseline as great as that of the earth's orbit are very tiny, since even the star nearest to us after the sun—the Alpha Centauri triple system—is some 4.3 light-years away from us. Given that each light-year is almost 10 trillion kilometers (6 trillion miles), Alpha Centauri lies a million times farther away from us than Venus does when it lies near inferior conjunction.

Parallaxes, obtained from measures by the satellite Hipparcos, are now available for stars out to a distance of some three thousand light-years. These stars include, significantly, the members of the Hyades cluster, a stellar grouping that forms the horns of the constellation of Taurus, the Bull, which has been important in calibrating the relationship of the color (spectral class) of a star to its intrinsic brightness. Once one has worked out this relation, one can take the next step and estimate a star's brightness—hence its distance—directly from its spectrum.

Eventually, astronomers managed to grope their way out to pulsating giant stars known as the Cepheids (so called because the prototype of these stars is Delta Cephei), whose pulsation periods are related directly to their intrinsic brightness. Cepheids, being very brilliant stars, can be identified in other galaxies, those massive and majestic conurbations of stars that used to be referred to as "island universes" lying at distances of tens and even hundreds of millions of light-years. In this way astronomers use the overlapping methods of estimating distances from one group of celestial objects to the next until they have leapfrogged their way out to the deepest reaches of the cosmos.

The most brilliant "standard candles" of all are the exploding stars known as type I supernovae. They have been used to reveal the distances to very remote galaxies—10 billion light-years away, near the edge of the observable universe—and have revealed that the expansion of the universe at those remote distances seems to be accelerating.

Venus Revealed

Although Venus appears as nothing more than a black silhouette during transit—round as a billiard ball—we must remind ourselves that it is a planet, roughly the size of the earth. If the earth and its companion, the moon, were observed in transit from a planet more distant from the sun, such as Mars, they, too, would appear just as smooth and perfectly serene. The earth would appear as a round, black pearl dropped upon the sun out of the surrounding ocean of space. As a matter of fact, there were indeed transits of the earth and moon observable from Mars on May 8, 1905, and May 11, 1984 (see fig. 91), and the next will occur on November 10, 2084—perhaps humans will be able to view that spectacle by then.

For the first time since transits of Venus have begun to be observed, we who witness the forthcoming events will have some comprehension of its nature as a world of its own. It is no longer completely veiled and utterly mysterious.

Because for visual observers the planet is always shrouded in an impenetrable and notoriously deadpan mask of clouds, its surface was for a long time inaccessible. Sketches of its dazzling white ball showed only amorphous cloud features. In at least one case, that of Percival Lowell, an observer recorded what he confidently supposed to be surface features; they turned out to have been nothing more

Figure 91. Artist's rendition of the transit of the earth and moon as seen from Mars on earth date May 11, 1984. The sun, earth, and moon are highly magnified but remain at equal scales with each other; the earth is near the sun's upper left limb, whereas the smaller moon is toward the lower left limb. (Drawn by John Westfall.)

than the blood vessels of his own retina—his large telescope having been turned, effectively, into an ophthalmoscope![5]

Until even the early 1960s, the planet's rotation period was unknown; the results offered by early observers such as Cassini and Bianchini were no better than guesses. All that changed soon after the beginning of the Space Age. By bouncing radio waves off the planet's surface and studying the detailed structure of the resulting echo—essentially, the same method that led to the refined value of the astronomical unit—it was possible to produce rough maps of the surface. Though most of the planet seemed to consist of rolling plains, several mountainous regions were revealed and given names such as Alpha and Beta Regio. The rotation period turned out to be slow—indeed, it occurs in a backward, or retrograde, direction. The period, 243.0 days, is longer than the planet's 224.7-day orbital period around the sun.

The massive atmosphere—so dense it is more like an ocean—consists mostly of carbon dioxide, with a lacing of sulfuric acid droplets that form the opaque death shroud of clouds. The atmosphere is a very efficient heat trap; radiation coming in is not allowed to escape. What this means is that Venus is subject to a runaway greenhouse effect. At the surface the temperature is around 500 degrees Centigrade (over 900° F)—higher than the melting point of lead—with conditions resembling nothing so much as the medieval vision of hell.

The Russian Venera and American Pioneer, Venus, and Magellan missions have produced detailed radar maps of the surface (see fig. 92); the rolling plains prove to be volcanic and are dotted with impact craters and, in places, rise into towering shield volcanoes.[6]

In short, the Evening Star that glares in the twilight sky or darkens in silhouette during its rare passages across the sun would be unpleasant in the extreme to visit. It is a searing, hot, inhospitable world. Its volcano-dotted landscape—massively oppressive beneath its deep, sluggish, oceanlike depth of carbon dioxide, starless beneath a relentless smog of sulfuric acid droplets—is no place for man or woman. The earth's sister in space is no sister of ours. It is instead a sinister realm—sans lakes, sans streams, sans winds, sans everything.

Figure 92. Simulated view of the surface of Venus, with topography as revealed by Magellan orbiting spacecraft. (NASA Jet Propulsion Laboratory image P40176.)

Still, our generation's discovery of the true nature of Venus's surface will not diminish its allure when we see that brilliant, beautiful speck dominate the western evening sky in the spring of 2004 and watch it gradually descend to its appointment with the sun on June 8. On that day the future will merge with the past, and the historical narrative of the transits and their observers, our theme up to this point, will yield to our own sense impressions and instrumental records of the reality of a transit.

THE 2004 TRANSIT APPROACHES

We now turn to the main events—the two transits of Venus that will occur during our own era.

Since not every generation can see a transit of Venus, we will be fortunate to be able to watch the two coming transits of June 8, 2004, and June 5–6, 2012. Except for Antarctica and southern Argentina and

Chile, there is no place on the earth that will not be able see at least one of these events. For the millions of people who will watch them, it will hardly matter that the transits will not be used to measure the distance from the earth to the sun, except possibly as an intellectual exercise. They will be enjoying a rare spectacle, identifying with astronomers of centuries past, who also watched their transits with awe.

In the spring of 2004 Venus will be high in the evening sky, at its farthest from the sun in the sky on March 29, when it will be at half phase. Then, in April, May, and early June, the brilliant planet will appear to approach the sun, accelerating in its course toward the western horizon from day to day (see fig. 93). At the same time, its disk will grow in size as its phase narrows to a crescent (see Table 8). The changes in Venus's disk size and phase will be easily visible with a small telescope.

A transit of Venus also means an unusually close inferior conjunction, one allowing us to view the planet in its narrow crescent phase when near the sun. When the phase shrinks to 5–10 percent sunlit, a moderate-size telescope may begin to show evidence of Venus's atmosphere, in the effect known as the cusp-extension. At such times, the sunlit crescent stretches more than 180 degrees and sometimes creates a light-ring all the way around the planet, owing to sunlight scattered in the haze layer above the planet's cloud tops. The effect is best seen either without a filter or sometimes with a red filter. As the planet approaches the sun yet more closely in the sky, the effect becomes greater, but then the planet must be viewed in the daytime. *Due to the danger of pointing the telescope accidentally at the sun without adequate filtration, viewing Venus when within a few degrees of the sun is recommended only for the expert and requires special precautions.* In brief, one must shield the aperture of the telescope from direct sunlight, either by means of a lengthy tube or by a screen suspended in front of the instrument; even so, great care must be taken to avoid accidentally shifting the telescope to point directly at the sun.[7]

Appendix C gives the Universal Times (the standard time at the Greenwich Meridian) for the transit contacts visible for major cities throughout the visibility zone for 2004. (To obtain daylight savings time from UT, subtract 3 hours for Atlantic DT, 4 hours for Eastern

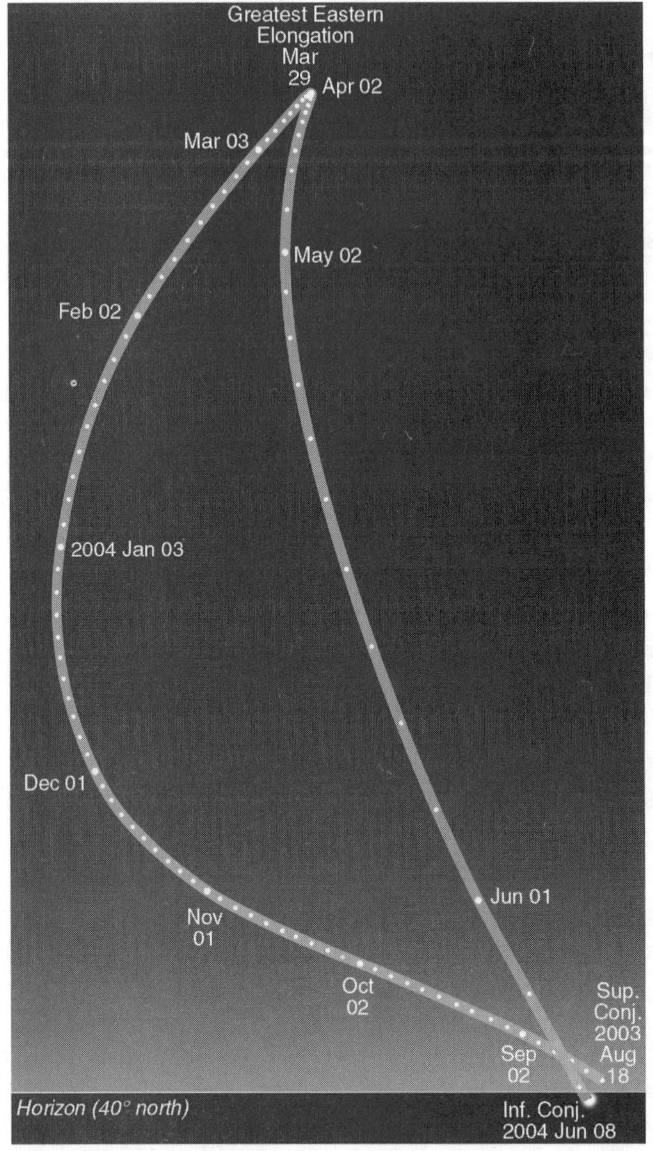

Figure 93. Venus's position in the sky at sunset in relation to the setting sun (lower right), from its superior conjunction on August 18, 2003, to its inferior conjunction (transit) on June 8, 2004. The viewpoint is at latitude 40° north. (Diagram by John Westfall.)

DT, 5 hours for Central DT, 6 hours for Mountain DT, 7 hours for Pacific DT, 9 hours for Alaska DT, and 10 hours for Hawaii-Aleutian DT.) The table also gives the altitude of the sun above the horizon for each contact and the long-term probability of actual sunshine during daylight hours in June.

Table 8. Venus near the sun in the evening and morning skies, March 29–August 18, 2004

Date (2004)	Solar elongation (°)	Phase (%)	Diameter (")	Evening altitude (°)	Date (2004)	Solar elongation (°)	Phase (%)	Diameter (")	Morning altitude (°)
MAR 29	46	50.9	24	44	JUN 10	3	0.02	58	1
APR 02	46	49.2	24	44	13	7	0.8	57	4
05	46	47.5	25	44	16	12	2.1	55	6
08	46	45.6	26	43	19	16	4.0	54	9
11	45	43.7	27	43	22	20	6.2	52	12
14	45	41.7	28	43	25	24	8.8	50	14
17	45	39.7	30	42	28	27	11.5	48	17
20	44	37.6	31	41	JUL 01	30	14.3	46	19
23	43	34.6	32	41	04	33	17.0	44	21
26	42	33.0	34	40	07	35	20.0	42	23
29	41	30.5	35	38	10	37	22.7	40	25
MAY 02	40	28.1	37	37	13	39	25.4	38	27
05	38	25.4	38	36	16	40	27.9	36	28
08	37	22.7	40	34	19	41	30.5	34	30
11	35	20.1	42	32	22	42	32.8	33	31
14	32	17.1	44	29	25	43	35.0	31	33
17	30	14.4	46	27	28	44	37.2	30	34
20	27	11.3	49	24	31	44	39.3	29	35
23	24	8.6	51	21	AUG 03	45	41.3	28	36
26	20	6.0	53	17	06	45	43.3	27	37
29	16	3.8	55	13	09	46	45.1	26	38
JUN 01	12	2.0	56	9	12	46	46.9	25	38
04	7	0.7	57	5	15	46	48.6	24	39
07	2	0.1	58	1	18	46	50.3	23	40

Note: "Solar Elongation" is the angular Venus-sun distance in the sky; "Phase" is the fraction of Venus's disk that is sunlit; "Diameter" is the diameter of Venus's disk in arc-seconds; and "Altitude" is the height of Venus in degrees above the horizon at sunset (evening) or sunrise (morning) at latitude 40°N. Data are for 0 hours UT for each date.

On June 8, 2004, Venus's disk will be 58 arc-seconds across. The planet will take 19 minutes to cross the limb of the sun and 6.2 hours to complete the entire transit. (See the diagram of 1639–2012 transit-tracks across the sun in appendix B.) About a quarter of the earth will see the entire transit from ingress to egress; another half will see part of the transit, including either ingress or egress. The zone of visibility of the 2004 transit is centered on the Old World (see fig. 94). Almost all of Europe and most of Africa and Asia will be able to view the entire transit. Ingress, but not egress, will be visible from easternmost Asia, Australia, westernmost New Zealand, and islands in the western Pacific. Conversely, egress, but not ingress, will be seen from northwest Africa and eastern North and South America (see fig. 95). Since the 2004 and 2012 transits occur in early June, the entirety of both transits will be visible above the horizon in the Arctic, north of latitude 67 degrees north.

If you are fortunate enough to reside in either or both of the transit visibility zones, you still may wish to travel to another part of the zone that has better weather prospects than does your locality. And, naturally, if you must travel to have a chance of seeing the transit at all, it only follows that you should travel to a place that has a good record of clear daytime skies in June. The right-hand columns of the 2004 and 2012 "local circumstances" tables in appendices C and D give June sunshine statistics for the cities selected; you may also wish to consult the global June cloudiness map in appendix E. (Cloudiness is not quite the converse of sunshine, but either statistic is useful for planning purposes. Recall the thwarted astronomers of yore, and remember that the tables and the map give long-term means and are no guarantee as to what will actually occur on transit day!)

OBSERVING THE TRANSITS

A transit itself must be viewed during daylight when the planet is silhouetted against the (literally) blinding disk of the sun. This is very like observing sunspots or the partial phases of solar eclipses.

> **WARNING: Looking directly at the sun with the naked eye, binoculars, or a telescope is extremely dangerous, if done incorrectly, and may cause serious damage to the eyes or even blindness.**

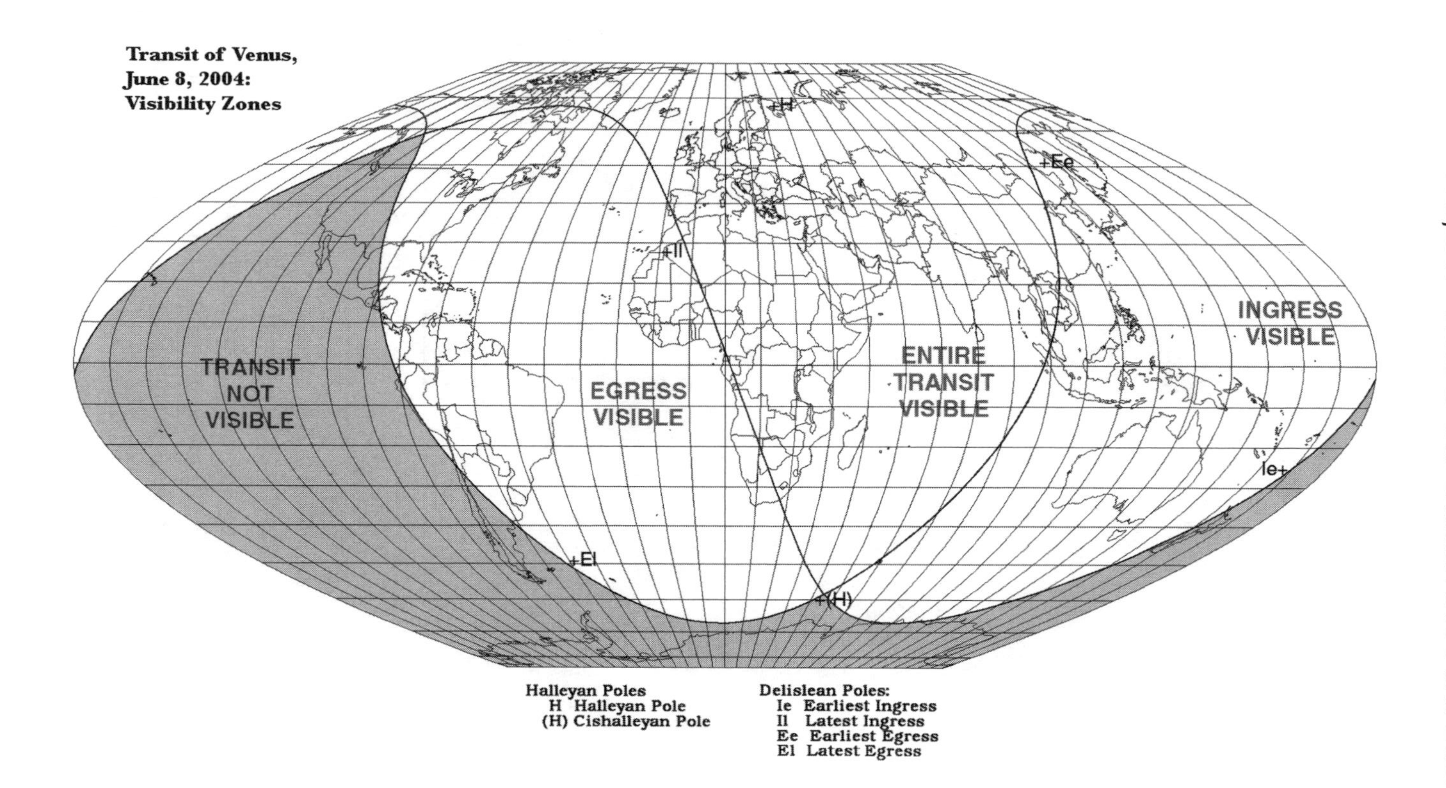

Figure 94. Visibility zones, with Halleyan and Delislean Poles, for the transit of Venus of June 8, 2004. (Map by John Westfall.)

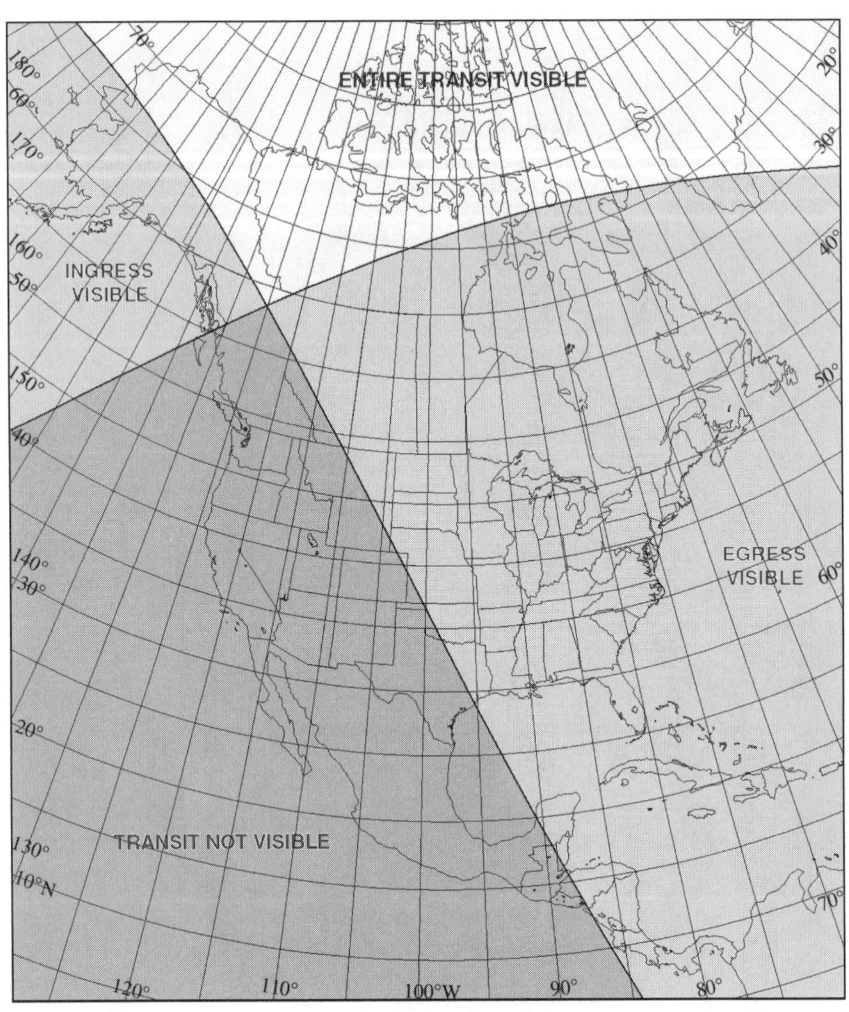

Figure 95. Visibility of the June 8, 2004, transit of Venus from North America. (Map by John Westfall.)

Fortunately, there are safe ways of viewing a transit. The naked-eye observer must avoid looking at the sun without an adequate filter, such as those used to observe solar eclipses. A welder's filter of shade 13 or 14 is safe; unfortunately, its optical quality is not good enough to use with binoculars or a telescope. With the naked eye, a dark speck will be all that you will be able to see of the planet. A pair of binoculars will show Venus as a disk, and powerful binoculars

(15× or more) may be used to time the ingress and egress contacts. Of course, with binoculars one must cover *both* lenses with safe filters or place a safe filter over one lens and an opaque cover over the other. Make sure the cover is securely fitted and that it doesn't get bumped or jarred loose while observing the sun!

Serious transit observing will require a telescope with a full-aperture solar filter, either metal-coated glass or metal-coated mylar (a durable polyester film). The filter must be securely placed over the telescope aperture (the "front end") and must transmit no more than 1/100,000 of the sun's radiation, not just of visible light but also in the dangerous ultraviolet and infrared bands.[8] If you purchase a glass solar filter it will most likely be available in a cell to fit your telescope size. However, some observers have concluded that metal-coated mylar filters give clearer images than glass ones. With mylar, you may have to make a cell yourself to fit the filter to your telescope.[9]

Some solar filters impart false color to the sun's disk, often either a blue or yellow tint. If this bothers you, the coloration is usually weak enough to be corrected by a color-compensation filter for film photography or by image-manipulation software (e.g., Photoshop) for digital images.

Check any metal-coated solar filter initially, and periodically thereafter, for scratches and pinholes. These will let in unwanted direct sunlight that will reduce contrast in your image and can endanger your eyesight, particularly if the defects are large. A limited number of small defects can be painted over, but overdoing this will degrade the image, which means that the filter will have to be replaced.

Metal-coated solar filters are shiny on both sides, so if used with a refractor or a catadioptric reflector with a corrector lens, they can produce a ghost image that reduces contrast. The problem is that the sunlight passing through the filter is reflected from the lens onto the rear face of the filter, which then reflects the unwanted light back into the telescope. The solution is to tilt the filter cell to the optical axis by a degree or so. (Filter cells that are threaded to screw onto the telescope should be avoided, since they can't be tilted.)

Another method involves projecting the sun's image through an eyepiece. The projection method can be used for individual or group viewing either with tripod-mounted binoculars or with a telescope. Although the image may be thrown on any surface, even the ground or a wall, results are much better if it is cast on a white surface shaded from direct sunlight.

Even here there is one caveat: anything placed too near the eyepiece will receive a concentrated solar image and may melt, shatter, or, if inflammable, catch fire. Projection is best done with a small telescope (6-inch aperture or less) because with larger instruments there is danger of the eyepiece itself overheating and shattering. A reflector's secondary mirror—especially in catadioptric systems, like the popular Schmidt-Cassegrains—can also overheat.

The size of the image can be adjusted by changing the eyepiece—the shorter the focal length of the eyepiece, the larger the image—or by moving the projection screen toward or away from the eyepiece.

One disadvantage of the projection method is that it provides relatively low magnification and low contrast, so it may not show fine detail; another is that the projected image is difficult to photograph without distortion.

No matter which method of viewing the sun you use, remember to cap the finder telescope's lens during a transit. Otherwise, the intense sunlight may burn out the crosswires during the several hours of a transit, or worse, someone may try to peek through the finder and damage his or her eyesight.

For daytime observing, the seeing is rarely steady enough to allow a telescope of more than 10–20 cm (4–8 in) aperture to be used to advantage. Magnifications in the range of 50× to 200× are best. With such magnifications, an equatorial mounting and clock drive are a virtual necessity for tracking Venus during the several hours of a transit. In general, the atmosphere is more turbulent during the daytime than at night, but it usually helps to place the telescope on a lawn, rather than on pavement, and if possible the line of sight should project over vegetation or a body of water rather than, say, a parking lot, which will generate heat currents.

Finally, it is worth mentioning that you will find a transit a more enjoyable experience if you avoid standing out in the heat and the glare of direct sunlight. It helps to place a sunshade around the telescope in order to shade the eyepiece area—even a cardboard disk will help.

Specialized Equipment

Except for the high-density filter, a transit may be observed using standard binoculars, telescopes, eyepieces, and mountings. However, the advanced observer may also wish to utilize more specialized gear.

Hydrogen-alpha filters permit the sun to be viewed in the narrow spectral region of 6563 angstroms (deep red), which is especially effective in bringing out its chromosphere and prominences. With such a filter it should be possible to view Venus when it is entirely off the visible disk of the sun—before the first contact or after the fourth contact.

Though the great majority of people who will view a transit of Venus will probably be satisfied with just that—looking at the event and carrying away the memory of having been privileged to watch something that does not happen very often—some may wish for a more permanent record. By whatever means a transit is recorded, the transitional phases—ingress and egress—are obviously the most dramatic parts. However, it is always possible that at some point during the portion of a transit when Venus is entirely upon the sun that the planet will pass near, or even across, a sunspot or sunspot group.

In drawing Venus, the same techniques used in drawing sunspots can be used. The projection method facilitates sketching because you can fasten a piece of paper on the projection screen and simply trace the sun's limb, Venus, and any other features on the sun's disk. By this means it is possible on the same sheet to trace the outline of Venus at regular intervals throughout a transit (assuming you are using a telescope on an equatorial mounting; otherwise the image will rotate during the course of the transit).

A transit can also be photographed by various means, either on film or digitally, and with still images or video. In attempting to do so, you will have a greater chance of success if you practice photographing the sun before the transit with the same film (or digital light-sensitivity setting), optical system, and filter that you plan to use during the transit. This will help you find the correct exposure. (Practice also helps those who plan to sketch a transit.) The less automatic your camera, the better. If it has an autofocus function, turn it off because it may become confused when used with a telescope and a solar filter. An autoexposure function may fail for the same reason. Remember to treat your camera as you do your eyes and always employ a safe solar filter. (If sunlight—visible light, infrared, and ultraviolet—is not adequately filtered the camera can be damaged; worse, your eyes may be injured if the camera's viewfinder gives a direct view of the sun.) If you are using film, slower films usually give higher resolution and better tonal reproduction than high-speed films; with plenty of light, we recommend films in the ISO 25–100 range. The same principle works with digital cameras that allow the light sensitivity to be adjusted; images obtained with "slow" settings have less noise and better quality than those with "fast" settings. If your exposure time is 1/30 second or shorter, the image should not trail even if your telescope has no clock drive (although a clock drive will be very useful in keeping Venus centered during the course of a transit). Finally, if you use a digital camera whose resolution (number of rows and columns of pixels) can be chosen, set it to the highest resolution possible—a transit of Venus deserves this!

The simplest way to photograph a transit is to use the camera by itself, without a telescope. Whether on film or silicon chip, the disk of Venus will be only 1/3500 as wide as the effective focal length of your optical system. (The sun will be much larger, 1/109 of the focal length.) Thus a telephoto lens will be needed—for example, a 300 mm telephoto will show the sun as 2.75 mm across with a 1/12 mm spot that represents Venus. You can further increase your telephoto lens's magnification by adding a teleconverter lens, which typically magnifies the image by an additional 1.3–5×.

Using the camera with a telescope will enlarge Venus from a tiny spot to a distinct circle. How large it appears depends on the effective focal length, which in turn depends not only on your telescope but also on how you use your camera with it. If your camera's lens is not removable, your only choice is the afocal method. Here, you focus your telescope at infinity, which is the focal setting that is sharpest to your eye (either without glasses if you don't wear them, or with them, if you need them for distance vision). Then focus your camera at infinity and hold it to the eyepiece, preferably with a bracket to keep the camera in place. If you have a digital camera you can preview the image for centering, exposure, and focus. To find the effective focal length for the afocal method, divide the camera lens's focal length by the eyepiece's focal length and then multiply the result by the telescope's focal length.

You will probably obtain better results if you can remove the lens from your camera, as is the case with single-lens reflex (SLR) cameras. Such cameras have two advantages: they allow you to view through the lens, seeing exactly what the film (or chip) will see, and they also allow you to use your telescope as the lens. Using only the direct focus of the telescope is called the prime-focus method. You also have the option of enlarging the prime-focus image with a Barlow (negative) lens or by projection with an eyepiece. Compared with 35 mm film, the silicon chips used in consumer and even "prosumer" digital still cameras and videocameras are very small. Even so, you should be able to fit Venus in the frame, but with high magnification it may be difficult to find the tiny portion of the sun's disk where Venus is silhouetted.

Another option is to use a CCD (charge-coupled device) camera. These cameras are specialized digital imaging devices that have low noise levels because they have cooling units. In addition, they allow one to correct the raw images for variable pixel response by using a flat-field frame, and for noise, with a dark frame. The ability to make these corrections makes CCD frames the best form of image for photometric measurements. Their chief drawback is that most CCD cameras must be used with a computer. As with more conventional film or digital photography, it is up to you to determine by experiment the best solar exposure time for your telescope and solar filter. As with other forms of photography and imaging, this should be done by photographing or imaging the sun before the transit day, using the same equipment that you plan to use for the transit itself.

The above forms of still photography usually require several seconds between successive images (although some cameras can operate in "burst" mode for short periods). This limitation might cause you to miss recording rapidly changing phenomena such as the formation or dissolution of the black drop during ingress and egress. Using a camcorder to videotape the transit, or a "webcam" to save images directly to a computer, avoids this problem.

Some camcorders have considerable "zoom" range, so that their longest optical focal length is likely to show Venus on the face of the sun. (Use optical zoom, rather than digital zoom; the latter simply enlarges the image without improving the resolution.) You may increase the magnification yet more by using a teleconverter lens. Naturally, you should place a safe solar filter on the camcorder's lens before turning the unit toward the sun.

Camcorders can be used in the afocal mode, similarly to film or digital still cameras. However, probably the most flexible way to obtain video through a telescope is to use one of the small lensless video cameras, supplied by several vendors, that are designed to fit a standard eyepiece tube. You can then use the videocamera as you would an SLR camera: either at prime focus or with a Barlow lens or eyepiece projection to enlarge the image. A standard video cable is needed to connect the video output from the eyepiece camera to the video input of a camcorder used to record the video. Conveniently, the audio channel of the camcorder can be used to record shortwave time signals.

As a transit of Venus is a very rare event, you will probably want to aim for the best recording quality that is practical. At the present time this is digital video. (At the time of writing, HDTV [high-definition television] camcorders are about to enter the market at prices that are expensive [around $3,000], but no longer staggering.) If your camcorder has a progressive scan function, you should use it. However, based on results recently received from the May 7, 2003, transit of Mercury, inexpensive webcams can also give high-quality images.[10] As with CCD cameras, webcams cannot be used by themselves but need to input their images into a computer.

An advantage of digital images, whether originally so or from scanned photographs, is that they may be "enhanced" by computer.

One useful enhancement is to "stretch" an image's tonal range to achieve greater contrast. Naturally, any such technique can be overdone, resulting in a stark almost black-and-white image where faint tones have become black and bright areas are saturated. Also, the widely used unsharp-masking technique, if overdone, can create artifacts, including a light area within, and a bright ring surrounding, the disk of the transiting planet (interestingly, both illusions have sometimes been reported by visual observers). Finally, the quality of digital images can be greatly improved by creating a composite image by "stacking" (averaging) from a few to a few hundred individual frames. However, if too great a time span is compressed into a single final image, the motion of the planet relative to the sun will blur either or both bodies. Venus's relative motion will be about 4 arc-seconds per minute of time, so "stacking" more than a few seconds worth of digital images (usually recorded at 30 frames per second) will create noticeable blurring.

Needless to say, in order to maximize their value, your drawings, photographs, or other images of the transit should include pertinent information about your equipment and the observing conditions. If for no other reason, this information may help you better to record the next transit! Information about your equipment should include (where applicable) the type of filter used; the telescope or camera aperture, f-ratio, and magnification; film type or digital camera light-sensitivity setting; and exposure time. The Universal Time (UT) should be recorded for all observations, and this should be to 1-second accuracy for contact timings and photographs and videos of the ingress and egress phases. Finally, you should note relevant atmospheric conditions, such as "seeing," haze, and clouds. Don't forget to record your longitude and latitude (to 0.01 degree or 1 arc-minute of precision), as the transit contacts will occur at different times at different locations.

What to Look For

In a sense, we have striven to provide throughout this book an orientation to observers of the 2004 and 2012 transits, since the expectation of what to look for is founded on the experience of past observers. For example, observers may wish to make accurate timings of the contacts and compare them with the results of distant colleagues in order to calculate the value of the solar parallax. Indeed, the European Southern Observatory's Educational Office and other groups are planning to coordinate just such programs.[11] Nonetheless, remember that any solar parallax value found by the transit method cannot compete with the current 3-meter precision of the definition of the astronomical unit! Outside of parallax measurement, there is still useful information to be gained from observing the transit, especially regarding the several optical phenomena associated with these events:

- *A bright spot reported centrally or eccentrically located within the disk of Venus when in transit.*[12] This phenomenon is most likely produced by scattered light, whether atmospheric, telescopic, or ocular, or by reflections in the telescope or eyepiece.
- *A halo of light seen around Venus when fully on the sun's disk (more often reported).* This is illusory and appears to be a contrast effect.[13] It is most emphatically *not* produced by the atmosphere of Venus, though this is still sometimes alleged.
- *The black drop.* This is, of course, the most famous of all transit phenomena, but alas, it is no longer very mysterious. It is thoroughly explained by blurring of the limbs of Venus and the sun. The fuzziness of their disks is in turn due to several factors, including atmospheric seeing, limited telescope resolution, and blurring in the retina of the human eye. It is also recorded in photographs and CCD images; film emulsions and silicon chips also have their limitations. The black drop has been photographed during transits of Mercury but never (yet!) for a transit of Venus. To help understand historical reports of the black drop, it would be useful to videotape Venus, when near the second and third contacts, with a number of telescopes of different apertures and designs, utilizing a variety of colored filters.
- *A light ring or aureole that forms around the planet during ingress and egress.* To be distinguished from the illusory brightening

around the planet upon the disk of the sun, this appearance was noted and frequently drawn during the eighteenth- and nineteenth-century transits but never photographed. The aureole is caused by refraction of sunlight in the upper atmosphere of Venus.[14] It should not be confused with the cusp-extension described earlier, which is due to sunlight scattered in the planet's haze layer and is orders of magnitude fainter. Different observers in 1874 and 1882 reported differences in the aureole's appearance, such as asymmetry in its location and brightness distribution (for instance, one 1882 observer recorded a "bright bead"). Some observers also reported fluctuations in the aureole's extent and brightness. Observers in 2004 may record the aureole from its first appearance to its disappearance during ingress (and the reverse sequence during egress) on videotape, on film, or with digital photographs taken at regular intervals. Unfortunately, we cannot give exposure guidelines, except to say that somewhat longer exposures will probably be required for the aureole than are suitable for the sun's disk itself. Also, because the "ring of light" is very narrow, good seeing with a moderate-sized instrument (say, 10–20 cm) will be essential to resolve it adequately.

Other projects will no doubt come to mind, such as spectroscopy, near-infrared photography or imaging, or constructing stereoscopic views of Venus using simultaneous photographs from different locations (or pseudo-stereoscopic views using images obtained at intervals during the transit). We leave other projects to the observer's ingenuity and imagination.

At the time of writing (late 2003), the number of Web sites devoted to the 2004 transit of Venus is rapidly growing. They include over twenty commercial tours to the zone of visibility. Some give useful background information, particularly that created by Fred Espenak of NASA's Goddard Spaceflight Center: http://sun earth.gsfc.nasa.gov/eclipse/transit/venus0412.html. Another useful Web site is http://www.transitofvenus.org, which has links to many other transit-related Web sites.

2012 IS NOT THAT FAR AWAY!

In the spring of 2012, Venus will dominate the western evening sky, gradually approaching the sun much as it did in 2004. (Refer to fig. 9 in chap. 3 to see its movement in the sky.) Once again it will eventually cross the sun itself; this time on June 5 and 6, 2012. Rather providentially, almost anyone unfortunate enough to live in a place where the 2004 transit is unobservable will see at least part of the transit of 2012 (see fig. 96). At the transit of 2012, Venus will take about 18 minutes to traverse the sun's limb at ingress and egress, and the entire transit will take 6.7 hours. In 2012, Venus crosses the sun's disk closer to its center than in 2004, resulting in a longer duration of transit and causing Venus to cross the sun's limb at a less oblique angle, which means a briefer ingress and egress (see fig. B1 in appendix B.) Note also that Venus crosses the northern half of the sun in 2012, as opposed to the southern half in 2004.

As in 2004, the great majority of the human race will be able to see some or all of the 2012 transit of Venus, given clear skies of course. Of the major land masses, only southern and eastern South America, western Africa, and the southern and western Iberian Peninsula will miss out entirely.

The event will be centered on the Pacific Basin, so the land areas where the entire transit can be seen will be limited. They will include northern and eastern Asia, including Japan and the most populated portions of China, as well as the Philippines, New Zealand, most of Australia, Alaska, and northwest Canada. Many Pacific Islands, such as Fiji and Hawaii, will fall in this zone of optimum visibility. Finally, as is always the case for a June transit, the whole transit will in theory be visible from the lands of the "midnight sun"—the Arctic, including northern Alaska and Canada.

If you plan to travel to the zone of complete visibility in 2012, the question arises as to where in this area the long-term prospects for clear skies are highest. The rightmost column in the table in appendix D gives the June probability of sunshine for selected cities throughout the 2012 transit visibility zone. Referring to it, we see that sunshine occurs for at most half the daytime hours for most

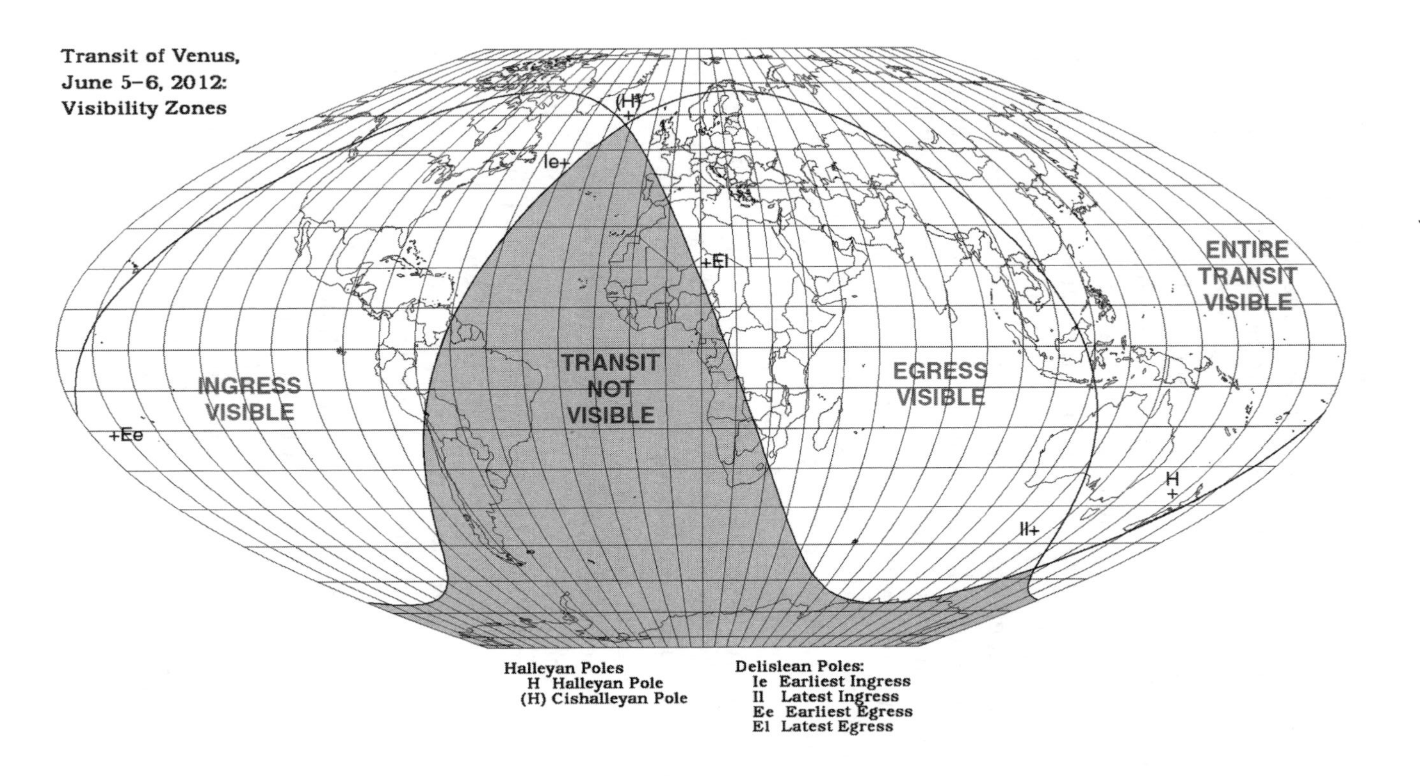

Figure 96. Visibility zones, with Halleyan and Delislean Poles, for the transit of Venus of June 5–6, 2012. (Map by John Westfall.)

places in densely inhabited eastern Asia that could see the entire transit. Beijing and presumably nearby areas in northern China are an exception. Cloudiness is also frequent in the Arctic, even in June. However, June is in the dry season for tropical and subtropical Australia, with over 80 percent chance of sunshine in Alice Springs and Darwin. Here the would-be transit chaser will have to make a difficult decision, because the sun will be low in the sky during transit ingress in this region. Hawaii and Samoa are in a similar situation, but with the sun very low at the end of the transit. A low solar elevation means a high chance of poor seeing; and even of cloudiness, because if there are even just a few clouds in the sky, the effect of foreshortening will make them more frequent near one's horizon. For a higher solar altitude throughout the 2012 transit, it will be necessary to locate nearer the center of the Pacific Basin, which means an island location, and perhaps some inconvenience, such as infrequent flights, in simply reaching one's intended location. Still, based on the table in appendix D, Pacific-island stations such as Yap, Johnston Island, Guam, New Caledonia, and New Guinea might be good compromises between a relatively high solar elevation throughout the transit and a fairly good prospect of sunshine for the event.

The remainder of Asia, almost all of Europe, and eastern and northeastern Africa will witness the end of the transit but will miss its beginning. The chance of clear skies will be fairly good, 50 percent or better, throughout Mediterranean Europe and the northern coast of Africa, but the sun's elevation will be low. The probability of clear skies, and also the elevation of the sun, will go up as one moves eastward, and both criteria are favorable throughout the Middle East.

Unlike the case in 2004, the entire United States will have a chance to see part of the transit of Venus in 2012, in this case the ingress phase, but the sun will set with the transit under way. This will be true also for most of Canada, Mexico, Central America, and the Caribbean, together with northwest South America.

Figure 97 focuses on the North American visibility zone. Almost all the cities listed in the United States, and half of those given for Canada, have a better than even chance of clear skies. In most of the

American southwest, the sunshine probability rises over 70 percent, and some places in that region have sunshine percentages in the 80s and even the 90s. Also, the farther westward one goes in North America, the higher the sun's altitude will be during transit ingress.

Besides favoring different parts of the world than were favored in 2004, there will be several other differences between the two events. In 2012 we will have profited from the observations of 2004, which may raise new questions about the ring of light and the black drop—and may perhaps even reveal unexpected new phenomena. The ring of light will have been photographed and electronically imaged for the first time in 2004, so in 2012 we will know from recent experience what exposure times and filters will best capture it.

The equipment used in 2012 may differ from that used in 2004. Technological forecasting is risky, but it seems very likely that high-definition digital camcorders will be relatively inexpensive by 2012 and will quite likely record directly onto DVDs or some similar dig-

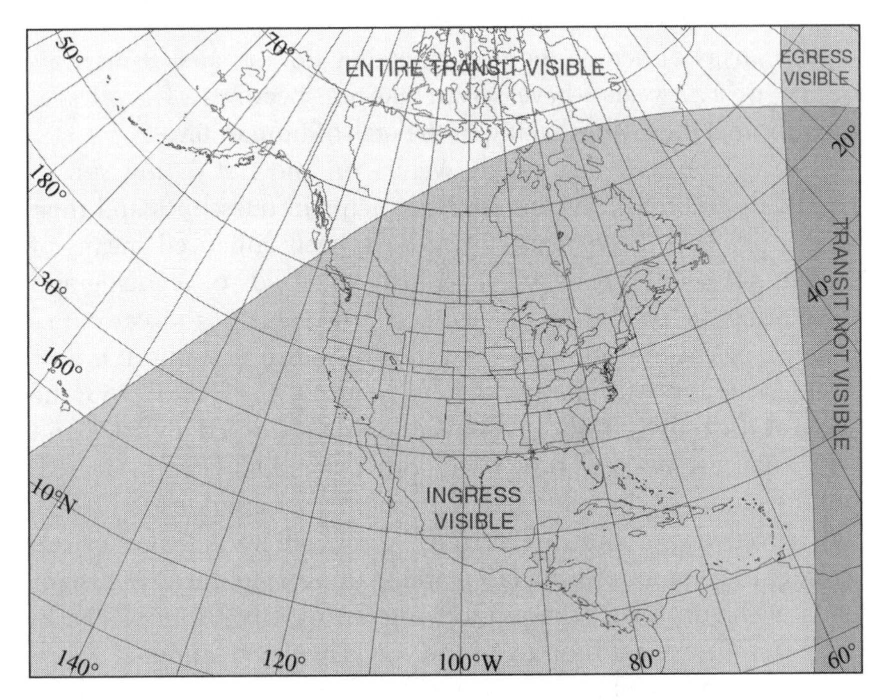

Figure 97. Visibility of the June 5–6, 2012, transit of Venus from North America. (Map by John Westfall.)

ital random-access medium. Also, even present-day films and electronic imaging devices can, with a suitable filter, take pictures in the near-infrared band. At present, "thermographic" cameras that image in the far-infrared band are very expensive, but their price has been dropping. By 2012, perhaps it will be practical to image Venus during the transit using "heat radiation." One wonders what the ring of light, or Venus's nightside, will look like in the thermal band. Then there is the prospect of adaptive optics, now confined to large telescopes and nighttime imaging, being used to alleviate the blurring effects of atmospheric seeing for the transit.

But the greatest difference from 2004 will be psychological, even existential. One can perhaps justify putting off an expensive trip to see the 2004 transit, knowing that the 2012 event will be visible from where one lives. But observers who miss the 2012 transit will not have another chance, since Venus will not cross the face of the sun again until 2117!

That realization, in turn, must encourage some reflections. Indeed, transits of Venus—occurring, as they do, in pairs, at intervals of 105 or 122 years—have served like the beatings of a celestial metronome, the swings taking in the eras of human life.

Since Horrocks first caught Venus on the fast-setting sun at Hoole, in 1639, astonishing improvements in telescopes and their accessories have been made; there have been enhanced means of making astronomical records and better methods of reducing and calculating the results. As we have seen, the value of the astronomical unit has been confined within steadily narrower limits; it is now straitjacketed to within a few meters. In 1639, 1761, and 1769, the sun was the only star whose distance could be grasped. By 1874 and 1882, the distances of some of the other stars had been measured, but the nature of the spiral nebulae was still a matter of speculation and effectively unknown. We now know full well that they are galaxies of stars, and we are fast working our way to the most distant parts of the universe. Since, in looking farther one also looks backward in time, we are probing to the very dawn of creation.

One wonders what the world will be like when the next cycle of transits begins in 2117. It is a sobering reflection. Despite all the advances noted in human knowledge, other, less edifying themes have occurred over the history of the transits with all too depressing regularity. Above all, war has been the theme that has been inextricably involved with the series of transits. Kepler first tackled the movements of the planets during the Thirty Years' War, fueled by religious differences. Horrocks observed just before the beginning of the English civil wars. The transits of the eighteenth century occurred against the backdrop of the Seven Years' War. During the period of the nineteenth-century transits, the world seemed relatively at peace—the transits occurred against the backdrop of international fairs and exhibitions, in a world increasingly tied together by telegraphs and railways, international financial institutions, and flush with the fruits of a global economy. However, in retrospect one sees that the clouds of future wars were also gathering: the blocks and alliances that would struggle to the death during the disastrous global war of 1914–1918 and whose potential for destruction is still not spent to this day.

The ensuing twentieth century would prove to be perhaps the most maniacal and blood-soaked in all human history.

Writing on the eve of the transit of 2004, with wars recent and imminent, it is impossible to look comfortably ahead as far as the transits of 2117 and 2125. One hopes for peace in their time, if not in ours, but at the same time suspects that the odds are equally good for another "war to end all wars," perhaps even a war to end all humanity.

In 1945, atomic bombs were exploded at Hiroshima and Nagasaki. It is sobering to reflect that the latter city's "ground zero" occurred not far from where European astronomers—the exemplars of the highest science and the cream of the culture of their day, people like Jules Janssen and George Davidson—had witnessed the transit of Venus only seventy-one years before. After the bombs were dropped and the frightening magnitude of the destruction they wrought had become manifest, H. G. Wells, the most famous futurist of the time, uttered a somber verdict: "Man should face his culminating destiny with dignity and mutual aid and charity, without hysteria, meanness, and idiotic misrepresentations of each other's motives."[15]

We should do so, but we have not done so yet; it only remains to be seen whether we shall have done so by 2117, when Venus—the classic planet of love—in perfect tranquility, oblivious to our loves, our wars, our science, or our madness, glides once more across the face of the sun.[16]

EPILOGUE

Muse, tell me the deeds of golden Aphrodite,
the Cyprian, who stirs up sweet passions in the gods
and subdues the tribes of mortal men and birds that fly in air
as well as all manner of creatures that the dry land rears,
and those that inhabit the sea:
all these love the deeds of rich-crowned Cytherea.

—Homer, "Hymn to Aphrodite" (ca. 850 BCE)

The zone of visibility of the June 8, 2004, transit of Venus will be centered on the Old World. Almost all of the population of Europe, and of most of Africa and Asia, will be able to see the entire transit. Some of the best locations in terms of visibility, elevation of the sun above the horizon, and weather conditions will be found in the Middle East. Much of this book has been concerned with the scientific effort to use Venus to measure the distance from the earth to the sun. This quest for a number and an endeavor has, for sociolog-

"Hymn to Aphrodite" from Homer, *Homeric Hymns*, trans. Apostolus N. Athanassakis (Baltimore: Johns Hopkins University Press, 1976).

ical reasons, and perhaps for psychological reasons, mainly been the enterprise of males. However, the 2004 transit of Venus will be scientifically important chiefly for the study of transit phenomena themselves. It will be an event memorable as any previous transit for its romance—it is an occurrence to be etched in memory for the rest of one's lifetime, recounted in detail to one's children and grandchildren. Savor the experience—after all, one may not live to see another; although there will be one more, in 2012, that one may be clouded out or not favored for other circumstances. Then the next will not be until 2117: a treat for one's grandchildren or great-grandchildren.

"Love that which will never be seen twice," says poet Alfred de Vigny. A transit of Venus is rarer than a total eclipse of the sun, or even a brilliant long-tailed comet. It is something to be savored for the sheer spectacle of it. It is something to go out of one's way to see: the intersection of two worlds, as romantic love is the intersection of two lives. There is no question: it has in it some element of romance, even apart from the fact that Venus (or Aphrodite, as she is known in Greek) is the namesake of the goddess of love:

> Sing, Muse, and remind us of the deeds of golden Aphrodite!
>
> Sing, Muse, and tell us where—all other things being equal—
> we may be for the transit.

The easy answer is that one should be wherever on the earth the transit is visible. One would decidedly not wish to be, on June 8, 2004, on that side of the globe in which the entire transit occurs at night, when the sun is hidden. Most of the zone from which the transit is a lost cause—not even the briefest glimpse of ingress or egress being visible—is in the Pacific Ocean basin. Land areas to be avoided include: Tierra del Fuego, almost all of Mexico, British Columbia, and the Western United States, Hawaii, Antarctica, and in all but the most extreme westernmost tip of New Zealand.

Anyone wishing to emulate earlier transit observers involved in the quest for solar parallax might try to get close to one of the Halleyan Poles as possible, where the transit lasts as long, or as briefly, as possible. In the Southern Hemisphere, Kerguélen Island, destination of so many nineteenth-century transit expeditions, would be such a choice; in the Northern Hemisphere, Russia near the Kara Strait or the island of Novaya Zemlya. For the Delislean Poles, for as early or as late a transit as possible, see the 2004 visibility map (refer to fig. 94). In 2004, it would admittedly be solely one's desire to recreate history to cause one to attempt to reach such remote and isolated places.

Admirers of Horrocks might choose Hoole; those of Halley, London or Oxford; of Delisle, Paris; of Chappe, Siberia; or of Le Gentil, Pondicherry.

All things being equal, Iraq might make a good observing site. After all, here, within the Fertile Crescent between the Tigris and Euphrates, lay ancient Mesopotamia, where Venus was once worshiped as Inanna and more famously as Ishtar, the Queen of Heaven, the Evening Star who followed the sun into *kur*, the underworld. The poet John Milton, in his "Hymn on the Morning of Christ's Nativity" (1629), remembers "moonèd Ashtaroth, / Heaven's Queen and Mother both." Inanna-Ishtar-Astarte was worshiped throughout the ancient Middle East, and one recognizes in her the archetype of Aphrodite-Venus.

Milton's references to "moonèd" Ashtaroth is interesting, because there is an old tradition that the Babylonian astronomers glimpsed the planet's crescent phase. The Babylonian astronomers certainly followed the planet's movements; we know this from the famous Venus tablet found at Nineveh (near Mosul, northern Iraq), which consists of the copy of a record made during the reign of King Ammisaduqua, the penultimate king of the first Babylonian dynasty, who lived nearly four thousand years ago. At the moment, however, we can only say that sites within Iraq sadly are not, because of the present turmoil, ideal choices for observing the transit.

The Anatolian coast of Turkey would make a lovely destination; it is rich in lore of the goddess of love from Homeric times, and Aphrodite-Venus took part—on the Trojan side—in the battle for Troy, since the Trojan prince Paris voted her more beautiful than Hera or Athena. One could find compelling reasons to observe Venus's passage from one of the islands of the wine-dark Aegean.

> The isles of Greece, the isles of Greece!
> Where burning Sappho loved and sung
> . . . .
> Eternal summer gilds them yet.
>
> (Byron, *Don Juan*, canto 3 [1821])

Perhaps Samos itself, Pythagoras's isle, or Lesbos (Mytilene), from which the favorite poet of feminists often invoked the goddess of love. Sappho's heartfelt utterances might have been written yesterday: "Sweet mother, I cannot ply the loom, vanquished by desire for a youth through the work of soft Aphrodite" (Wharton Fragment 114, ca. 590 BCE).

There is Cythera—or Kythera, as the Greeks spell it—the southernmost of the Ionian Isles, just south of Cape Malea (Peloponnese). It would be an appropriate destination, since it lies close to the spot where lovely Aphrodite (Cytherea) is supposed to have arisen naked from the foam of the sea, which was impregnated by the genitals of Uranus severed by Saturn in one of the primal mythological wars. When retelling this story, it is almost impossible not to mention Botticelli's famous painting in the Uffizi.

Riding on a scallop shell, Aphrodite, whose name means "foam-born," stepped ashore at Kythera, but finding the island too small to suit her, left for the Peloponnese; as she walked, grass and flowers sprang from the soil. At last she took up residence at Paphos, in Cyprus—though according to others, she was born there.

In any case, the Cyprian groves became the center of the worship of the Paphian queen. And so Paphos—or anywhere at all in Cyprus—would be a perfect site from which to observe the transit, both from the point of view of astronomical data-collecting and romantic associations.

But wherever one happens to be—provided one is not at one of the unfortunate viewless sites—the goddess will make an appearance. In the company of a good telescope, or at least a safe filter for your unassisted eyes, with friends and a good bottle of wine, one may imagine oneself—for the duration of the transit—to be in a bower of bliss. There one may "a stately pleasure dome decree" and picture one's surroundings as the

Fairest Isle, all isles excelling,
Seat of pleasures, and of loves;
Venus here will choose her dwelling,
And forsake her Cyprian groves.

(Dryden, *King Arthur* [1691])

Savor the moment, for it will soon pass—to be emulated once again, on the other side of the world, in 2012. And for those anticipating 2012, it will come in time. Once Venus has rolled off the sun—once the culmination of so much effort, all the excitement, the cheers and hoots and shouts of wonder at the precision-timing of the heavenly pageant and the novelty of the appearance—is over there will be ample time to rest and recover and to meditate on the theme *sic gloria mundi transit* ("thus passes the glory of the world") or those invoked in Pindar's *Nemean* VII (467 BCE?):

From every task a pause is sweet;
Even honey cloys,
And Aphrodite's delicious flowers.
By nature each is different,
Each has a life that sets him apart.
To one person this is allotted, to others that,
And no one man can be happy in all things.

(Trans. William Sheehan)

Those who will have seen the transit will have enjoyed at least a few hours of happiness and can look forward to 2012—the last transit of our still-budding century and, barring medical miracles, of our lives. Yet, whatever triumphs or follies lie ahead for our descendants, the transits of Venus will faithfully continue their centuries-long cycles as long as the sun exists and Venus and our earth circle it.

Appendix A

Transits of Venus, −2970 to +7464

(2971 BCE to 7464 CE; Julian Calendar through +1526, Gregorian Calendar for +1631 and later)

Date	DT (hr)	Series
−2970 Nov 18	09.3	A
−2864 May 20	18.5	D
−2856 May 18	11.5	E
−2727 Nov 18	08.1	A
−2621 May 21	21.8	D
−2613 May 19	15.0	E
−2484 Nov 18	06.4	A
−2378 May 22	01.4	D
−2730 May 19	18.5	E
−2241 Nov 19	05.2	A
−2135 May 22	04.8	D
−2127 May 19	22.0	E
−1998 Nov 19	02.4	A
−1892 May 22	09.6	D
−1884 May 20	02.5	E
−1763 Nov 21	12.0	B
−1755 Nov 19	01.3	A
−1649 May 23	12.9	D
−1641 May 21	06.1	E
−1520 Nov 21	10.9	B
−1512 Nov 18	23.9	A
−1406 May 23	16.3	D
−1398 May 21	09.3	E
−1277 Nov 22	09.5	B
−1269 Nov 19	22.8	A
−1163 May 23	19.7	D
−1155 May 21	12.9	E
−1034 Nov 22	08.5	B
−1026 Nov 19	21.5	A
− 920 May 23	23.1	D
− 912 May 21	16.2	E
− 791 Nov 22	07.1	B
− 783 Nov 19	20.3	A
− 677 May 18	01.9	D
− 669 May 22	19.8	E
− 548 Nov 22	06.0	B
−540 Nov 19	*18.9*	*A*
−426 May 22	23.2	E
−305 Nov 23	04.4	B
−183 May 23	02.5	E
− 62 Nov 23	03.2	B
+ 60 May 23	05.9	E
181 Nov 23	01.8	B
303 May 24	09.2	E
424 Nov 23	00.5	B
546 May 24	12.7	E
554 May 22	06.0	F
667 Nov 23	23.3	B
789 May 24	16.1	E
797 May 22	09.2	F
910 Nov 23	22.0	B
1032 May 24	19.5	E
1040 May 22	12.8	F
1153 Nov 23	20.9	B

1275 May 25	22.8	E	3089 Dec 18	22.9	C	5763 Jan 10	17.0	C
1283 May 23	15.9	F	3219 Jun 20	00.0	E	5900 Jul 11	22.3	F
1396 Nov 23	19.5	B	3227 Jun 17	16.9	F	5908 Jul 09	14.6	H
1518 May 26	02.0	E	3332 Dec 20	22.2	C	5998 Jan 15	02.8	G
1526 May 23	19.2	F	3462 Jun 22	02.8	E	6006 Jan 12	16.6	C
1631 Dec 07	05.3	C	3470 Jun 19	19.8	F	6143 Jul 14	00.4	F
1639 Dec 04	18.4	B	3575 Dec 23	21.2	C	6151 Jul 11	16.7	H
1701 Jun 06	05.3	E	*3705 Jun 24*	*05.6*	*E*	6241 Jan 17	02.8	G
1769 Jun 03	22.4	F	3713 Jun 21	22.5	F	6249 Jan 14	16.7	C
1874 Dec 09	04.1	C	3818 Dec 25	20.5	C	6386 Jul 16	02.8	F
1882 Dec 06	17.1	B	3956 Jun 24	01.3	F	6394 Jul 13	18.9	H
2004 Jun 08	08.3	E	4061 Dec 26	19.6	C	6484 Jan 19	02.6	G
2012 Jun 06	01.5	F	4119 Jun 26	04.3	F	6492 Jan 16	16.5	C
2117 Dec 11	02.9	C	4304 Dec 29	19.1	C	6629 Jul 18	04.9	F
2125 Dec 08	16.1	B	4442 Jun 28	07.3	F	6637 Jul 15	21.2	H
2247 Jun 11	11.7	E	4547 Dec 31	18.5	C	6727 Jan 22	02.7	G
2255 Jun 09	04.8	F	4685 Jun 30	09.7	F	6735 Jan 19	16.9	C
2360 Dec 13	02.0	C	4791 Jun 02	18.1	C	6872 Jul 19	07.1	F
2368 Dec 10	15.0	B	4928 Jul 02	12.5	F	6880 Jul 16	23.1	H
2490 Jun 12	14.7	E	5026 Jun 07	04.1	G	6970 Jan 23	02.9	G
2498 Jun 10	07.8	F	5034 Jun 04	17.7	C	6978 Jan 20	17.1	C
2603 Dec 16	00.7	C	5171 Jul 05	14.8	F	7115 Jul 23	09.0	F
2611 Dec 13	14.1	B	5269 Jan 08	03.7	G	7123 Jul 21	01.2	H
2733 Jun 15	18.0	E	5277 Jan 05	17.4	C	7213 Jan 25	03.2	G
2741 Jun 13	11.0	F	5414 Jul 07	17.5	F	7221 Jan 22	17.5	C
2846 Dec 17	00.1	C	5512 Jan 12	03.3	G	7358 Jul 24	11.1	F
2854 Dec 14	13.2	B	5520 Jan 09	17.0	C	7366 Jul 22	02.9	H
2976 Jun 16	20.9	E	5657 Jul 08	19.8	F	7456 Jan 28	03.4	G
2984 Jun 14	14.0	F	5755 Jan 13	03.2	G	7464 Jan 25	17.7	C

Sources: For the period –1998 through +3956, the table is based on Jean Meeus, *Transits* (Richmond, VA: Willmann-Bell, 1989). Information on earlier and later transits is from computer simulations using the program SkyChart 2000.0 v2.5.1.

Note: Transits are grouped into series (lettered A–H in an extension of the system used by Meeus). Each member of a series occurs 243 years after the last member; each series lasts several thousand years. Sometimes the first or last transit in a series is a *partial transit,* where Venus grazes the limb of the sun; such events are listed in italics. The time of day is given in "Dynamic Time" (DT), or "atomic time." Due to unpredictable variations in the earth's rotation, the Universal Time (UT) of transits in the distant past or distant future are uncertain, so their terrestrial visibility zones cannot accurately be computed.

Appendix B

Data on Observed Transits, 1639–2012

The diagram below shows the tracks across the face of the sun of the transits of Venus occurring between 1639 and 2012. The tracks are *geocentric* (as seen from the center of the earth) and may shift north or south up to about one-half Venus-width, depending where the observer is actually located. Celestial north is at the top and celestial west is to the right. Astronomical telescopes usually invert the image, so this will be upside-down for a northern-hemisphere observer using such an instrument. In addition, if a star diagonal is used, creating an additional reflection, left and right will be reversed.

Below is a table summarizing the geometrical circumstances of the six transits of Venus occurring between 1761 and 2012. The duration information, particularly the difference between the minimum and maximum duration, is pertinent to the usefulness of the Halleyan method for determining the solar parallax for that transit. (The difference was greatest for the 1874 transit, indicating that the Halleyan method was the most efficient for that event, and was the

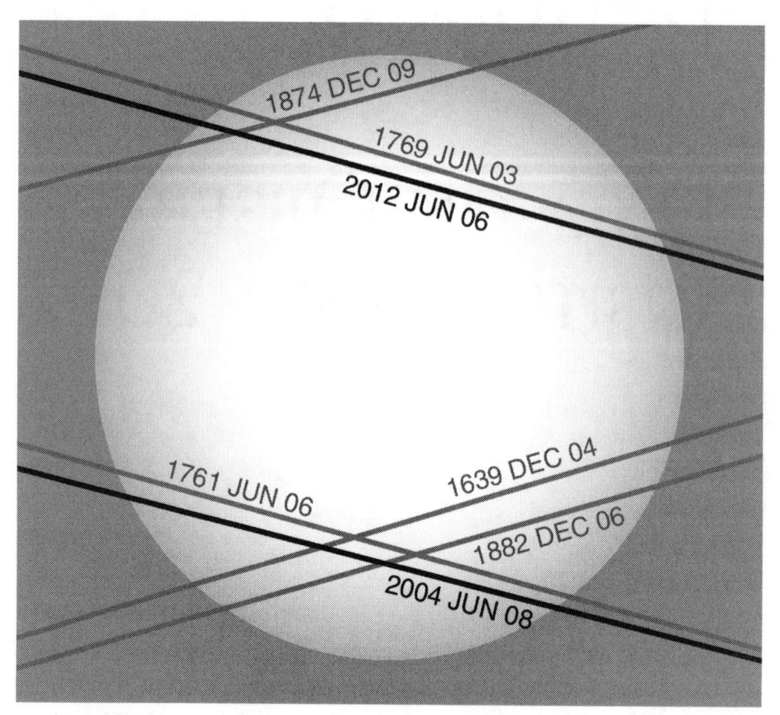

Figure B1. Tracks of Venus across the sun during the 1639–2012 transits. Venus's motion is consistently left to right (geocentric view; celestial north at top). (Diagram by John Westfall.)

least, and thus the least efficient, for the 1761 transit.) Similarly, the difference in time of earliest and latest ingress and egress describe the efficiency of the Delislean method for each transit—the larger the difference, the more applicable the method. (Thus the Delislean method was most efficient for the 1874 transit and will be least efficient for that of 2012.)

Table B1. Summary of transit of Venus geometrical circumstances

Transit date (UT)	1761 Jun 06	1769 Jun 03/04	1874 Dec 09	1882 Dec 06	2004 Jun 08	2012 Jun 05/06
Geocentric contact, UT (hours)						
Contact 1	02.031	19.256	01.816	13.943	05.224	22.158
Contact 2	02.335	19.573	02.311	14.284	05.548	22.457
Contact 3	08.306	25.271	05.937	19.916	11.108	26.525
Contact 4	08.610	25.589	06.432	20.257	11.432	26.824
Minimum separation (+ = Venus north of sun's center, – = Venus south of sun's center)						
(arc-seconds)	–570.84	+609.35	+829.95	–637.28	–626.87	+554.38
Semidiameters (arc-seconds)						
Sun	945.03	945.29	974.58	974.27	945.37	945.69
Venus	29.09	29.10	31.82	31.80	29.11	29.13
Duration (contact 2 to contact 3)						
Geocentric (hours)	5.971	5.689	3.626	5.632	5.560	6.068
Minimum (hours)	5.816	5.475	3.264	5.385	5.391	5.861
Maximum (hours)	6.032	5.901	3.910	5.856	5.652	6.255
Difference (minutes)	13.02	25.55	38.78	28.28	15.64	23.62
Point of minimum duration						
Latitude (°)	58.2 N	41.9 S	73.8 S	42.5 N	67.1 N	37.9 S
Longitude (°)	89.0 E	141.0 W	8.5 W	62.2 W	36.3 E	170.6 E
Solar Alt. Con. 2 (°)	+37.7	+21.6	+8.7	+21.3	+32.9	+22.9
Solar Alt. Con. 3 (°)	+47.4	+6.9	+19.0	+6.5	+43.6	+9.0
Point of maximum duration						
Latitude (°)	59.3 S	60.0 N	64.2 N	60.0 S	60.2 S	58.6 N
Longitude (°)	99.3 E	23.3 E	115.7 E	101.5 E	54.4 E	22.5 W
(Solar Alts. = 0°.0)						
Ingress (contact 2)						
Earliest (hours)	2.220	19.453	2.111	14.152	5.424	22.344
Latest (hours)	2.452	19.695	2.529	14.419	5.673	22.572
Difference (minutes)	13.90	14.54	25.05	16.04	14.90	13.63
Latitude of earliest (°)	20.8 S	50.2 N	40.0 N	50.8 S	25.7 S	45.8 N
Longitude of earliest (°)	132.9 W	7.4 E	144.2 W	86.1 E	176.7 E	40.1 W
Latitude of latest (°)	22.9 N	52.1 S	44.9 S	53.0 N	28.0 N	47.5 S
Longitude of latest (°)	42.6 E	174.0 W	25.4 E	95.0 W	8.3 W	138.2 E

Egress (contact 3)						
Earliest (hours)	8.1896	25.1489	5.7191	19.7805	10.9826	26.4105
Latest (hours)	8.4209	25.3915	6.1361	20.0483	11.2307	26.6380
Difference (minutes)	13.88	14.56	25.02	16.07	14.89	13.65
Latitude of earliest (°)	48.4 N	25.3 S	64.9 S	26.1 N	51.4 N	21.9 S
Longitude of earliest (°)	174.8 E	119.1 W	113.7 W	40.6 W	137.0 E	166.2 W
Latitude of latest (°)	46.6 S	23.1 N	62.2 N	23.7 S	49.4 S	20.1 N
Longitude of latest (°)	10.5 W	58.4 E	49.0 E	136.6 E	49.2 W	11.3 E
Duration of ingress (contact 1–contact 2) and egress (contact 3–contact 4)						
(minutes)	18.2	19.0	29.7	20.5	19.4	17.9

Source: Computed by John Westfall.

Appendix C

Local Circumstances for the Transit of Venus, June 8, 2004

	Contact 1		Contact 2		Contact 3		Contact 4		
Region/Place	UT (05h+)	Sun alt. (°)	UT (05h+)	Sun alt. (°)	UT (11h+)	Sun alt. (°)	UT (11h+)	Sun alt. (°)	June Potential Sunshine (%)
Geocentric	13.4m	—	32.8m	—	06.5m	—	25.9m	—	—
United States									
Albany, NY	—	—	—	—	05.6	17	25.6	21	59
Anchorage, AK	13.3	12	33.3	9	—	—	—	—	46
Atlanta, GA	—	—	—	—	06.3	7	26.4	10	67
Baltimore, MD	—	—	—	—	06.0	14	25.9	18	62
Birmingham, AL	—	—	—	—	06.2	5	26.3	8	66
Boston, MA	—	—	—	—	05.7	19	25.7	23	63
Buffalo, NY	—	—	—	—	05.4	14	25.4	17	68
Charleston, SC	—	—	—	—	06.6	10	26.6	14	62
Chicago, IL	—	—	—	—	05.2	7	25.2	11	67
Cincinnati, OH	—	—	—	—	05.6	8	25.7	12	62
Cleveland, OH	—	—	—	—	05.5	11	25.5	15	65
Columbus, OH	—	—	—	—	05.6	10	25.6	14	66
Dallas, TX	—	—	—	—	—	—	25.9	0	72
Dayton, OH	—	—	—	—	05.6	9	25.6	13	68

Des Moines, IA	—	—	—	—	05.0	3	25.1	6	66
Detroit, MI	—	—	—	—	05.3	10	25.4	14	67
Grand Rapids, MI	—	—	—	—	05.2	9	25.2	13	61
Harrisburg, PA	—	—	—	—	05.8	14	25.8	18	64
Hartford, CT	—	—	—	—	05.7	18	25.7	22	60
Houston, TX	—	—	—	—	—	—	26.3	0	73
Indianapolis, IN	—	—	—	—	05.5	8	25.6	11	66
Jackson, MS	—	—	—	—	06.2	2	26.3	6	70
Jacksonville, FL	—	—	—	—	06.9	8	26.8	12	59
Kansas City, MO	—	—	—	—	05.2	1	25.3	5	68
Knoxville, TN	—	—	—	—	06.1	8	26.1	12	64
Little Rock, AR	—	—	—	—	05.8	1	25.9	5	73
Louisville, KY	—	—	—	—	05.7	7	25.8	11	66
Madison, WI	—	—	—	—	05.0	7	25.1	10	64
Memphis, TN	—	—	—	—	05.9	3	26.0	7	74
Miami, FL	—	—	—	—	07.5	7	27.4	11	62
Milwaukee, WI	—	—	—	—	05.0	8	25.1	11	64
Minneapolis/ St. Paul, MN	—	—	—	—	04.6	5	24.7	8	64
Nashville, TN	—	—	—	—	05.9	6	26.0	10	68
New Orleans, LA	—	—	—	—	06.5	0	26.6	4	70
New York, NY	—	—	—	—	05.8	17	25.8	20	64
Norfolk, VA	—	—	—	—	06.2	14	26.2	18	68
Oklahoma City, OK	—	—	—	—	—	—	25.6	1	73
Peoria, IL	—	—	—	—	05.2	5	25.3	9	66
Philadelphia, PA	—	—	—	—	05.9	16	25.9	19	62
Pittsburgh, PA	—	—	—	—	05.6	12	25.7	16	57
Providence, RI	—	—	—	—	05.8	19	25.7	22	60
Raleigh, NC	—	—	—	—	06.3	12	26.3	16	61
Richmond, VA	—	—	—	—	06.1	13	26.1	17	68
Rochester, NY	—	—	—	—	05.4	15	25.4	18	66
St. Louis, MO	—	—	—	—	05.4	4	25.5	8	69
San Juan, PR	—	—	—	—	08.9	17	28.7	22	57
Savannah, GA	—	—	—	—	06.6	8	26.6	12	65
Scranton/Wilkes- Barre, PA	—	—	—	—	05.7	16	25.7	19	60
Shreveport, LA	—	—	—	—	—	—	26.1	3	71
Syracuse, NY	—	—	—	—	05.5	16	25.5	19	59
Tampa, FL	—	—	—	—	07.1	6	27.1	10	67
Toledo, OH	—	—	—	—	05.4	10	25.4	13	64
Tulsa, OK	—	—	—	—	—	—	25.6	3	65
Washington, DC	—	—	—	—	06.0	14	25.9	17	66
Wichita, KS	—	—	—	—	—	—	25.4	2	69
Canada									
Edmonton, AB	—	—	—	—	—	—	23.0	1	49
Montreal, PQ	—	—	—	—	05.2	18	25.2	21	47

Toronto, ON	—	—	—	—	05.3	14	25.3	17	56
Winnipeg, MB	—	—	—	—	03.9	5	24.0	8	51
Mexico									
Merida, Yucatan	—	—	—	—	—	—	27.6	2	52
West Indies									
Kindley AFB, Bermuda	—	—	—	—	07.2	22	26.9	26	60
Willemstad, Curaçao	—	—	—	—	09.6	13	29.3	17	64
Santo Domingo, Dominican Republic	—	—	—	—	10.8	7	30.5	12	47
Ft.-de-France, Martinique	—	—	—	—	09.5	21	29.1	25	57
South America									
Buenos Aires, Argentina	—	—	—	—	13.5	2	32.8	6	45
Sao Paulo, Brazil	—	—	—	—	13.1	17	32.3	20	56
Montevideo, Uruguay	—	—	—	—	13.6	4	32.8	7	55
Caracas, Venezuela	—	—	—	—	09.8	14	29.5	18	53
Europe									
Vienna, Austria	19.8	21	39.5	24	03.7	65	23.1	64	52
Brussels, Belgium	19.8	14	39.7	17	03.9	61	23.4	62	43
Prague, Czech Rep.	19.7	20	39.5	23	03.6	63	23.0	62	50
Copenhagen, Denmark	19.5	20	39.3	23	03.1	57	22.6	57	47
Helsinki, Finland	18.9	27	38.7	30	02.2	52	21.8	51	52
Marseilles, France	20.2	12	40.0	16	04.7	68	24.1	69	70
Paris, France	20.0	12	39.8	16	04.2	62	23.7	63	47
Berlin, Germany	19.6	20	39.4	23	03.4	60	22.9	60	48
Munich, Germany	19.9	18	39.7	21	03.9	62	23.3	65	42
Athens, Greece	19.9	25	39.5	28	04.3	73	23.5	70	73
Budapest, Hungary	19.8	23	39.5	26	03.6	65	23.0	64	57
Reykjavik, Iceland	18.8	8	38.8	10	03.1	43	22.9	44	30
Dublin, Ireland	19.7	9	39.7	12	04.0	56	23.6	57	39
Milan, Italy	20.0	16	39.8	19	04.3	67	23.7	67	53
Rome, Italy	20.1	17	39.9	20	04.5	71	23.9	71	60
Den Helder, Netherlands	19.7	15	39.6	18	03.7	59	23.2	60	46
Oslo, Norway	19.2	20	39.1	22	02.8	53	22.4	53	43
Warsaw, Poland	19.5	25	39.2	28	03.1	60	22.5	59	39
Lisbon, Portugal	20.3	1	40.2	4	05.8	65	25.2	68	74
Bucharest, Romania	19.7	27	39.3	31	03.5	66	22.8	64	58
Astrakhan, Russia	18.5	42	38.0	46	02.1	54	21.5	51	65

Moscow, Russia	18.8	34	38.4	37	02.0	53	21.5	51	49
St. Petersburg, Russia	18.8	30	38.5	32	02.1	51	21.6	50	55
Belgrade, Serbia	19.8	23	39.5	27	03.8	67	23.1	66	58
Madrid, Spain	20.3	5	40.2	9	05.4	67	24.8	70	74
Stockholm, Sweden	19.1	24	38.9	26	02.6	53	22.1	53	57
Zurich, Switzerland	20.0	16	39.8	19	04.1	65	23.5	66	46
Kiev, Ukraine	19.3	30	38.9	33	02.8	60	22.2	58	55
Kew, United Kingdom	19.8	12	39.7	15	04.0	59	23.5	61	43
Southwest Asia									
Kabul, Afghanistan	17.0	60	36.1	64	01.6	41	20.8	37	80
Nicosia, Cyprus	19.5	32	39.0	35	03.9	69	23.1	66	86
Shiraz, Iran	18.3	47	37.5	51	03.1	55	22.2	51	87
Tehran, Iran	18.4	46	37.7	50	02.7	55	21.9	51	80
Baghdad, Iraq	18.9	40	38.2	44	03.3	61	22.5	57	81
Jerusalem, Israel	19.4	33	38.8	37	04.1	69	23.2	65	97
Amman, Jordan	19.4	33	38.8	37	04.0	68	23.2	64	92
Shuwaikh, Kuwait	18.6	43	37.9	47	03.4	59	22.5	55	71
Beirut, Lebanon	19.4	33	38.8	37	03.9	68	23.1	64	84
Damascus, Syria	19.4	34	38.8	38	03.9	68	23.0	64	86
Istanbul, Turkey	19.6	29	39.2	33	03.7	68	22.9	64	70
Urfa, Turkey	19.2	36	38.6	40	03.4	64	22.6	61	88
South and East Asia									
Rangoon, Burma	13.7	82	32.5	84	01.3	14	20.5	10	27
Phom Penh, Cambodia	12.5	78	31.3	76	01.3	4	—	—	51
Beijing, China	13.1	69	32.3	66	59.3*	6	18.9	3	58
Hong Kong, China	12.2	78	31.1	74	00.2	0	—	—	39
Shanghai, China	12.1	71	31.2	67	—	—	—	—	33
Urumchi, China	15.7	66	34.9	68	00.2	28	19.6	24	58
Mumbai, India	16.2	64	35.1	68	02.5	35	21.6	31	41
Kolkata, India	14.7	78	33.7	83	01.3	23	20.5	18	36
New Delhi, India	16.1	68	35.1	72	01.5	34	20.7	30	46
Jakarta, Indonesia	11.4	61	30.1	59	—	—	—	—	62
Osaka, Japan	11.3	58	30.5	55	—	—	—	—	42
Tokyo, Japan	11.2	55	30.4	51	—	—	—	—	34
Vientiane, Laos	13.1	85	31.9	83	00.9	9	20.2	4	38
Kuala Lumpur, Malaysia	12.4	70	31.1	70	02.2	4	—	—	54
Karachi, Pakistan	16.9	59	36.0	64	02.4	42	21.5	38	57
Manilla, Philippines	11.0	70	29.9	66	—	—	—	—	42
Singapore, Singapore	12.1	68	30.8	67	02.3	0	—	—	50
Seoul, Rep. of Korea	12.2	64	31.4	60	—	—	—	—	50
Colombo, Sri Lanka	14.9	64	33.8	68	03.1	25	22.0	21	43
Taipei, Taiwan	11.7	72	30.7	68	—	—	—	—	41
Bangkok, Thailand	13.1	81	31.9	80	01.3	9	20.5	5	46
Ho Chi Minh, Vietnam	12.3	77	31.1	75	01.3	2	—	—	27

Africa									
Algiers, Algeria	20.4	9	40.2	12	05.5	74	24.8	75	70
Brazzaville, Congo	19.4	3	38.7	8	09.1	63	27.9	62	36
Cairo, Egypt	19.6	29	39.1	33	04.5	72	23.6	68	84
Addis Ababa, Ethiopia	18.8	30	38.0	34	06.0	62	24.9	58	47
Nairobi, Kenya	18.5	24	37.6	28	07.1	57	25.9	53	48
Tripoli, Libya	20.3	15	40.0	19	05.4	80	24.6	79	66
Tananarive, Madagas.	16.6	23	35.5	27	07.9	36	26.6	33	62
Marrakech, Morocco	—	—	40.4	2	06.6	69	26.0	73	75
Windhoek, Namibia	—	—	37.2	1	10.5	44	29.2	44	93
Lagos, Nigeria	—	—	39.7	1	08.8	71	27.8	73	30
Dakar, Senegal	—	—	—	—	09.0	61	28.2	65	65
Cape Town, So. Africa	—	—	—	—	11.1	33	29.8	32	61
Pretoria, South Africa	17.3	5	36.3	9	09.9	39	28.6	37	86
Khartoum, Sudan	19.4	27	38.7	31	05.8	70	24.8	66	75
Dar-es-Salaam, Tan.	18.0	23	37.1	27	07.5	51	26.2	48	64
Entebbe, Uganda	18.9	21	38.0	25	07.3	60	26.2	57	51
Harare, Zimbabwe	17.8	11	36.8	15	09.1	45	27.8	43	76
Oceania									
Alice Springs, Australia	08.4	34	27.1	32	—	—	—	—	81
Darwin, Australia	08.9	45	27.6	42	—	—	—	—	86
Melbourne, Australia	07.7	18	26.4	15	—	—	—	—	33
Perth, Australia	09.7	34	28.3	32	—	—	—	—	55
Sydney, Australia	07.3	17	26.1	14	—	—	—	—	53
Suva, Fiji	06.4	6	25.5	2	—	—	—	—	41
Nouméa, New Caledonia	06.6	14	25.6	10	—	—	—	—	52
Port Moresby, New Guinea	07.8	36	26.6	32	—	—	—	—	62

Source: Contact times computed by John Westfall from algorithms given in Meeus (1989). Sunshine data are from Rudloff (1981) and Ruffner and Bair (1987). The maximum differences from times independently computed by the U.S. Naval Observatory and by the program winOccult version 1.5.1 are 0.1–0.2 minutes.

Note: — = Sun below horizon at time of contact; UT = Universal Time, to 0.1 minute; Sun alt. = Altitude of center of the sun above the horizon (unrefracted); June potential sunshine = long-term percentage of June daylight hours with sunshine.

*Beijing, China, contact 3 UT = 10h 59.3m

Appendix D

Local Circumstances for the Transit of Venus, June 5–6, 2012

	June 05				June 06				
	Contact 1		Contact 2		Contact 3		Contact 4		
Region/Place	UT (22h+)	Sun alt. (°)	UT (22h+)	Sun alt. (°)	UT (04h+)	Sun alt. (°)	UT (04h+)	Sun alt. (°)	June Potential Sunshine (%)
Geocentric	0.95m	—	27.4m	—	31.5m	—	49.4m	—	—
United States									
Albany, NY	03.6	24	21.3	21	—	—	—	—	59
Albuquerque, NM	05.5	49	23.1	45	—	—	—	—	84
Anchorage, AK	06.3	51	23.9	51	30.7	15	48.4	13	46
Atlanta, GA	04.3	31	21.9	27	—	—	—	—	67
Baltimore, MD	03.8	26	21.4	22	—	—	—	—	62
Birmingham, AL	04.4	33	22.1	29	—	—	—	—	66
Boise, ID	05.7	54	23.3	51	—	—	—	—	76
Boston, MA	03.5	22	21.2	19	—	—	—	—	63
Buffalo, NY	03.8	28	21.4	25	—	—	—	—	68
Charleston, SC	04.1	27	21.8	24	—	—	—	—	62
Chicago, IL	04.2	34	21.8	31	—	—	—	—	67

Cincinnati, OH	04.1	32	21.8	28	—	—	—	—	62
Cleveland, OH	03.9	30	21.6	27	—	—	—	—	65
Columbus, OH	04.1	31	21.7	28	—	—	—	—	66
Dallas/Ft. Worth, TX	05.0	41	22.6	37	—	—	—	—	72
Dayton, OH	04.1	31	21.7	28	—	—	—	—	68
Denver, CO	05.2	47	22.8	44	—	—	—	—	70
Des Moines, IA	04.5	39	22.1	35	—	—	—	—	66
Detroit, MI	04.0	31	21.6	28	—	—	—	—	67
El Paso, TX	05.6	49	23.3	45	—	—	—	—	88
Fairbanks, AK	06.1	48	23.6	47	31.1	16	48.8	14	49
Grand Rapids, MI	04.1	33	21.7	29	—	—	—	—	61
Harrisburg, PA	03.8	26	21.4	23	—	—	—	—	64
Hartford, CT	03.6	23	21.2	20	—	—	—	—	60
Honolulu, HI	10.0	85	27.6	89	26.5	9	44.5	5	74
Houston, TX	05.0	40	22.7	36	—	—	—	—	73
Indianapolis, IN	04.2	33	21.8	30	—	—	—	—	66
Jackson, MS	04.6	35	22.3	32	—	—	—	—	70
Jacksonville, FL	04.3	28	21.9	24	—	—	—	—	59
Juneau, AK	05.9	52	23.5	51	30.7	8	48.5	5	34
Kansas City, MO	04.6	39	22.2	36	—	—	—	—	68
Knoxville, TN	04.2	31	21.8	27	—	—	—	—	64
Little Rock, AR	04.7	37	22.3	34	—	—	—	—	73
Los Angeles, CA	06.3	58	23.9	55	—	—	—	—	65
Louisville, KY	04.2	32	21.8	29	—	—	—	—	66
Madison, WI	04.2	35	21.9	32	—	—	—	—	64
Memphis, TN	04.5	36	22.2	32	—	—	—	—	74
Miami, FL	04.4	26	22.1	22	—	—	—	—	62
Milwaukee, WI	04.2	34	21.8	31	—	—	—	—	64
Minneap./St.Paul, MN	04.4	38	22.0	35	—	—	—	—	64
Nashville, TN	04.3	33	22.0	30	—	—	—	—	68
New Orleans, LA	04.7	35	22.4	31	—	—	—	—	70
New York, NY	03.6	24	21.3	21	—	—	—	—	64
Norfolk, VA	03.8	25	21.5	21	—	—	—	—	68
Oklahoma City, OK	04.9	42	22.6	38	—	—	—	—	73
Peoria, IL	04.3	36	22.0	32	—	—	—	—	66
Philadelphia, PA	03.7	25	21.4	21	—	—	—	—	62
Phoenix, AZ	05.9	54	23.5	50	—	—	—	—	94
Pittsburgh, PA	03.9	29	21.5	25	—	—	—	—	57
Portland, OR	05.9	57	23.5	54	—	—	—	—	48
Providence, RI	03.5	22	21.2	19	—	—	—	—	60
Raleigh, NC	03.9	27	21.6	23	—	—	—	—	61
Richmond, VA	03.8	26	21.5	22	—	—	—	—	68
Rochester, NY	03.7	27	21.4	24	—	—	—	—	66
St. Louis, MO	04.4	36	22.0	33	—	—	—	—	69
Salt Lake City, UT	05.5	52	23.1	49	—	—	—	—	78
San Francisco, CA	06.4	61	23.9	57	—	—	—	—	73
San Juan, PR	04.1	11	21.9	7	—	—	—	—	57

Savannah, GA	04.2	28	21.8	24	—	—	—	—	65
Scranton/ Wilkes-Barre, PA	03.7	25	21.4	22	—	—	—	—	60
Seattle, WA	05.8	56	23.4	53	—	—	—	—	54
Shreveport, LA	04.8	39	22.5	35	—	—	—	—	71
Syracuse, NY	03.7	26	21.3	23	—	—	—	—	59
Tampa, FL	04.4	28	22.1	24	—	—	—	—	67
Toledo, OH	04.0	31	21.7	28	—	—	—	—	64
Tucson, AZ	05.9	53	23.5	49	—	—	—	—	93
Tulsa, OK	04.8	40	22.4	37	—	—	—	—	65
Washington, DC	03.8	26	21.5	23	—	—	—	—	66
Wichita, KS	04.8	42	22.4	38	—	—	—	—	69
Canada									
Edmonton, AB	05.2	48	22.7	46	—	—	—	—	49
Montreal, PQ	03.6	25	21.2	21	—	—	—	—	47
Toronto, ON	03.8	28	21.5	25	—	—	—	—	56
Vancouver, BC	05.8	55	23.4	53	—	—	—	—	43
Whitehorse, YT	05.8	50	23.4	49	30.9	9	48.7	7	51
Winnipeg, MB	04.5	40	22.1	37	—	—	—	—	51
Mexico									
Mazatlan, Sinaloa	06.1	48	23.8	44	—	—	—	—	62
Merida, Yucatan	05.2	33	22.8	29	—	—	—	—	52
Mexico City, D.F.	05.9	41	23.5	37	—	—	—	—	43
Central America—Caribbean									
Kindley AFB, Bermuda	03.5	15	21.2	11	—	—	—	—	60
San Jose, Costa Rica	05.5	24	23.3	20	—	—	—	—	31
Willemstad, Curaçao	04.6	11	22.4	7	—	—	—	—	64
Sto.Domingo, Dom.R.	04.2	15	22.0	11	—	—	—	—	47
S. Salvador, El Sal.	05.6	30	23.3	26	—	—	—	—	47
Kingston, Jamaica	04.6	20	22.4	16	—	—	—	—	59
Ft.-de-France, Martin.	04.2	5	22.0	1	—	—	—	—	57
South America									
Bogota, Colombia	05.4	13	23.2	19	—	—	—	—	29
Quito, Ecuador	05.9	15	23.8	11	—	—	—	—	52
Lima, Peru	06.9	9	24.8	5	—	—	—	—	12
Caracas, Venezuela	04.6	9	22.4	5	—	—	—	—	53
Europe									
Vienna, Austria	—	—	—	—	37.5	14	55.1	17	52
Brussels, Belgium	—	—	—	—	37.3	8	54.9	10	43
Prague, Czech Rep.	—	—	—	—	37.4	14	55.0	16	50
Copenhagen, Denmark	—	—	—	—	37.0	14	54.6	17	47

Helsinki, Finland	—	—	—	—	36.6	22	54.2	24	52
Marseilles, France	—	—	—	—	37.7	5	55.4	8	70
Paris, France	—	—	—	—	37.4	6	55.0	9	47
Berlin, Germany	—	—	—	—	37.2	14	54.8	16	48
Munich, Germany	—	—	—	—	37.5	11	55.1	14	42
Athens, Greece	—	—	—	—	37.8	16	55.4	20	73
Budapest, Hungary	—	—	—	—	37.5	16	55.1	19	57
Reykjavik, Iceland	03.4	5	21.1	4	35.4	4	53.6	6	30
Dublin, Ireland	—	—	—	—	37.0	4	54.7	6	39
Milan, Italy	—	—	—	—	37.6	9	55.3	12	53
Rome, Italy	—	—	—	—	37.8	9	55.4	12	60
Den Helder, Nether.	—	—	—	—	37.2	9	54.8	12	46
Oslo, Norway	—	—	—	—	36.7	15	54.3	17	43
Warsaw, Poland	—	—	—	—	37.2	18	54.8	21	39
Bucharest, Romania	—	—	—	—	37.6	19	55.2	22	58
Astrakhan, Russia	—	—	—	—	36.7	35	54.2	38	65
Moscow, Russia	—	—	—	—	36.6	28	54.2	31	49
St. Petersburg, Russia	—	—	—	—	36.5	24	54.1	27	55
Belgrade, Serbia	—	—	—	—	37.6	16	55.2	19	58
Madrid, Spain	—	—	—	—	—	—	55.4	1	74
Stockholm, Sweden	—	—	—	—	36.7	18	54.3	20	57
Zurich, Switzerland	—	—	—	—	37.5	9	55.2	12	46
Kiev, Ukraine	—	—	—	—	37.1	24	54.7	26	55
Kew, United Kingdom	—	—	—	—	37.2	6	54.9	8	43

Southwest Asia

Kabul, Afghanistan	—	—	—	—	35.6	52	53.1	55	80
Nicosia, Cyprus	—	—	—	—	37.7	23	55.2	27	86
Shiraz, Iran	—	—	—	—	36.8	38	54.3	42	87
Tehran, Iran	—	—	—	—	36.8	38	54.3	41	80
Baghdad, Iraq	—	—	—	—	37.2	32	54.8	35	81
Jerusalem, Israel	—	—	—	—	37.7	24	55.2	28	97
Amman, Jordan	—	—	—	—	37.6	25	55.2	28	92
Shuwaikh, Kuwait	—	—	—	—	37.1	34	54.6	38	71
Beirut, Lebanon	—	—	—	—	37.6	25	55.2	28	84
Damascus, Syria	—	—	—	—	37.6	25	55.1	29	86
Istanbul, Turkey	—	—	—	—	37.6	21	55.2	24	70
Urfa, Turkey	—	—	—	—	37.4	28	54.9	31	88

South and East Asia

Rangoon, Burma	—	—	—	—	32.8	74	50.3	78	27
Phom Penh, Cambodia	—	—	—	—	31.7	77	49.2	79	51
Beijing, China	09.9	14	27.7	17	31.9	72	49.3	71	58
Hong Kong, China	11.7	6	29.6	10	31.2	88	48.7	84	39
Shanghai, China	11.0	15	28.8	19	31.0	78	48.5	75	33

Urumchi, China	—	—	—	—	34.3	61	51.7	64	58
Mumbai, India	—	—	—	—	35.2	54	52.7	58	41
Kolkata, India	—	—	—	—	33.8	69	51.3	73	36
New Delhi, India	—	—	—	—	35.0	59	52.4	63	46
Jakarta, Indonesia	—	—	—	—	30.7	61	48.4	61	62
Osaka, Japan	10.8	27	28.5	31	30.0	66	47.6	63	42
Tokyo, Japan	10.7	31	28.4	35	29.8	63	47.4	59	34
Vientiane, Laos	—	—	—	—	32.2	80	49.7	84	38
Kuala Lumpur, Malay.	—	—	—	—	31.7	68	49.3	70	54
Karachi, Pakistan	—	—	—	—	35.8	50	53.2	54	57
Manilla, Philippines	12.7	10	30.6	14	30.2	78	47.7	75	42
Singapore, Singapore	—	—	—	—	31.4	67	49.0	68	50
Seoul, Rep. of Korea	10.4	21	28.1	25	30.9	70	48.4	68	50
Colombo, Sri Lanka	—	—	—	—	34.1	56	51.7	60	43
Taipei, Taiwan	11.6	13	29.5	17	30.6	81	48.2	77	41
Bangkok, Thailand	—	—	—	—	32.2	76	49.8	79	46
Ho Chi Minh, Vietnam	—	—	—	—	31.5	77	49.0	78	27
Africa									
Algiers, Algeria	—	—	—	—	37.9	1	55.6	4	70
Cairo, Egypt	—	—	—	—	37.8	21	55.4	24	84
Addis Ababa, Ethiopia	—	—	—	—	37.4	20	55.0	24	47
Nairobi, Kenya	—	—	—	—	37.2	15	54.9	19	48
Tripoli, Libya	—	—	—	—	38.1	7	55.8	10	66
Tananarive, Madagas.	—	—	—	—	35.6	16	53.3	19	62
Pretoria, So. Africa	—	—	—	—	—	—	53.8	0	86
Khartoum, Sudan	—	—	—	—	37.8	17	55.5	21	75
Dar-es-Salaam, Tanzania	—	—	—	—	36.8	14	54.5	18	64
Entebbe, Uganda	—	—	—	—	37.4	11	55.1	15	51
Harare, Zimbabwe	—	—	—	—	36.5	2	54.3	6	76
Oceania									
Alice Springs, Aus.	15.6	6	33.7	10	27.5	39	45.3	37	81
Darwin, Australia	15.0	9	33.0	13	28.0	51	45.8	48	86
Melbourne, Australia	16.1	7	34.2	10	26.6	22	44.6	20	33
Perth, Australia	—	—	—	—	28.9	35	46.8	35	55
Sydney, Australia	15.9	13	34.0	16	26.2	23	44.2	21	53
Yap, Caroline Islands	13.4	24	31.3	28	28.3	62	45.9	58	52
Roratonga, Cook Is.	13.4	46	31.3	46	—	—	—	—	51
Suva, Fiji	14.5	41	32.4	43	24.9	15	42.9	11	41
Johnston Island	11.1	73	28.8	77	26.1	18	44.0	14	74
Guam, Mariana Islands	13.0	31	30.8	35	27.9	58	45.6	54	61
Nouméa, New Caled.	15.2	31	33.2	34	25.3	22	43.3	18	52
Port Moresby, N.Guin.	14.9	24	32.8	28	26.7	44	44.5	40	62
Auckland, New Zealand	15.4	24	33.5	25	25.1	7	43.3	4	42
Apia, Samoa	13.7	49	31.6	51	24.7	8	42.8	4	67

Source: Contact times computed by John Westfall from algorithms given in Meeus (1989). Sunshine data are from Rudloff (1981) and Ruffner and Bair (1987).

Note: — = Sun below horizon at time of contact; UT = Universal Time, to 0.1 minute; Sun Alt. = Altitude of center of the sun above the horizon (unrefracted); June potential sunshine = Long-term percentage of June daylight hours with sunshine.

Appendix E

June and December Sunshine Probability Maps

The two maps below, figures E1 and E2, show the mean long-term ratio of hours with sunshine to total daylight hours for the months of June and December—the months when transits of Venus occur in our era. The data plotted come from three sources: Rik Leemans and Wolfgang P. Cramer, *The IIASA Database for Mean Monthly Values of Temperature, Precipitation, and Cloudiness on a Global Terrestrial Grid,* IIASA Publication RR–91–18 (Laxenberg, Austria: International Institute for Applied Systems Analysis, 1991); Willy Rudloff, *World-Climates: With Tables of Climatic Data and Practical Suggestions* (Stuttgart: Wissenschaftliche Verlagsgesellschaft, 1981); and James A. Ruffner and Frank E. Bair, eds., *The Weather Almanac,* 5th ed. (Detroit: Gale Research, 1987).

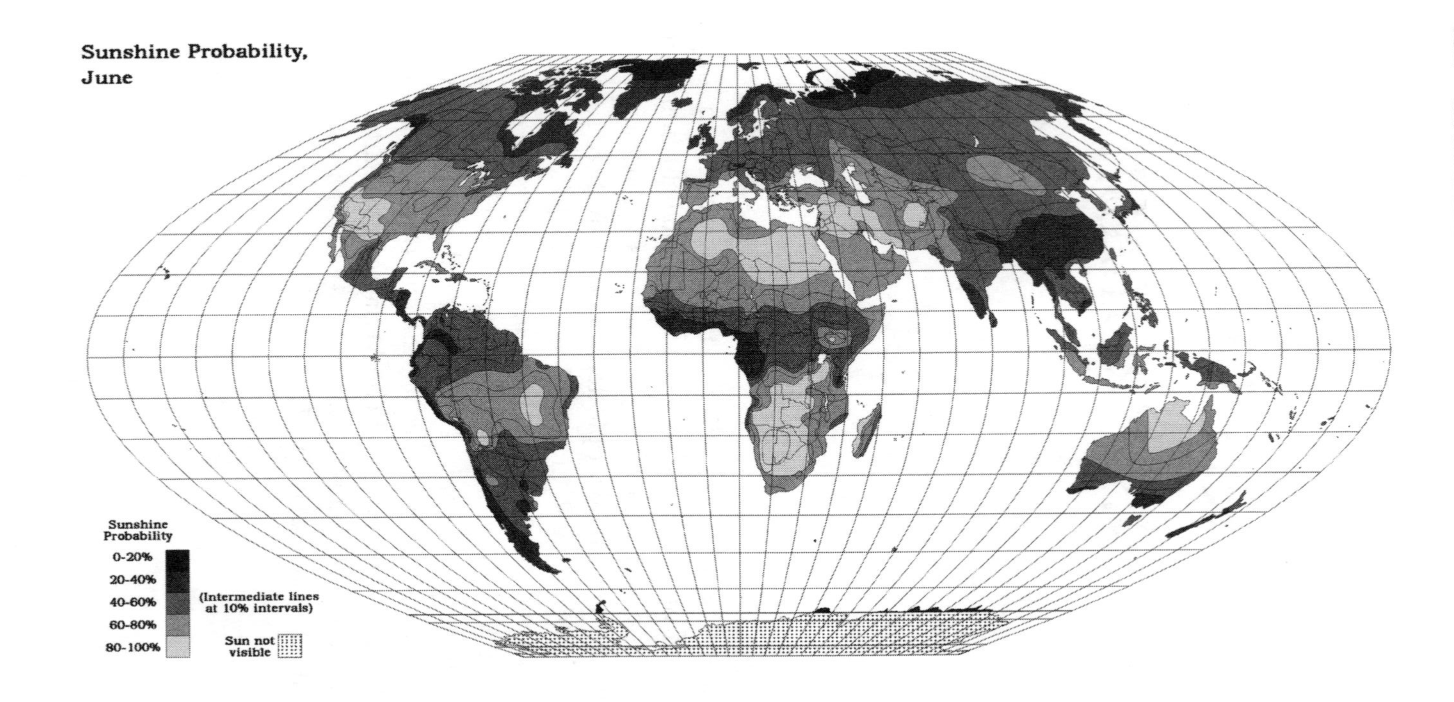

Figure E1. Mean probability of sunshine during the month of June, worldwide. (Map by John Westfall.)

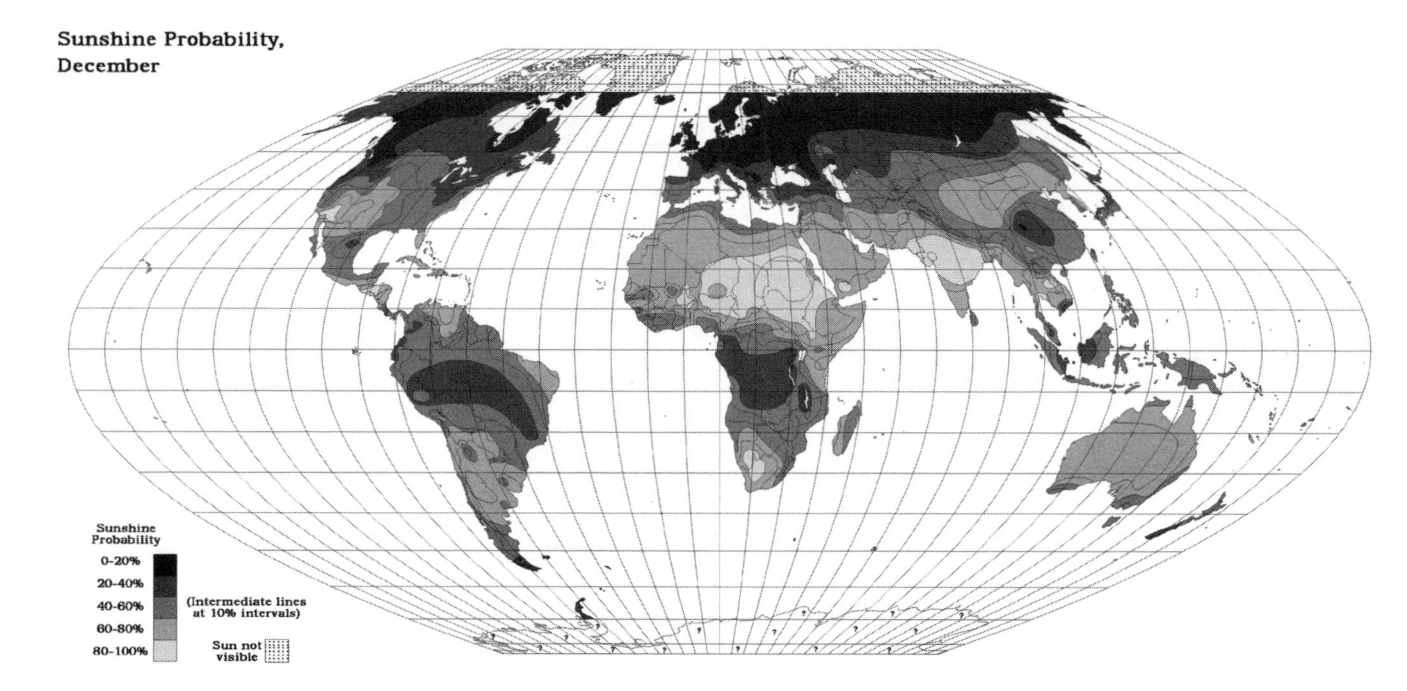

Figure E2. Mean probability of sunshine during the month of December, worldwide. (Map by John Westfall.)

NOTES

CHAPTER 1

1. James Jeans, *Physics and Philosophy* (London: Cambridge University Press, 1943); quoted in Paul D. MacLean, *The Triune Brain in Evolution: Role in Paleocerebral Functions* (New York: Plenum Press, 1990), p. 5.

2. Other useful sources relevant to chapter 1 are A. Pannekoek, *A History of Astronomy* (New York: Dover, 1989; reprint of 1961 edition); George Sarton, *A History of Science*, 2 vols. (New York: W. W. Norton, 1970); and Albert van Helden, *Measuring the Universe: Cosmic Dimensions from Aristarchus to Halley* (Chicago: University of Chicago Press, 1985).

CHAPTER 2

1. Lorna J. Marshall, *Nyae Nyae !Kung Beliefs and Rites* (Cambridge, MA: Peabody Museum of Archaeology and Ethnology, Harvard University, 1999), p. 251.

2. Akhenaten, "Hymn to the Aten," trans. James H. Breasted, in *A History of Egypt* (1909).

3. This term is found, for example, in F. Link, *Eclipse Phenomena in Astronomy* (New York: Springer Verlag, 1969).

CHAPTER 3

1. Bertrand Russell, *History of Western Philosophy*, 2nd ed. (London: George Allen and Unwin, 1961), p. 54, claims, "Pythagoras, as everyone knows, said that 'all things are number.'" Of course, what "everyone knows" is not always so, and the statement may first have been made by one of the Pythagoreans who followed him.

2. Camille Flammarion, *Dreams of an Astronomer*, trans. E. E. Fournier D'Albe (New York: D. Appleton, 1923), p. 95.

3. This identification of Venus as a male figure belies the common belief that the planet has been universally worshiped—because of its physical beauty, presumably—as a female divinity. Venus also served as a male divinity in India.

4. William Hickling Prescott, *History of the Conquest of Mexico and History of the Conquest of Peru* (New York: Cooper Square Press, 2000; reprints of 1843 and 1847 editions), p. 289.

5. Ibid., p. 429.

6. Ibid., pp. 434–36.

7. Two other books relevant to this chapter include J. L. E. Dreyer, *A History of Astronomy from Thales to Kepler*, 2nd ed. (New York: Dover, 1953; reprint, 1906 edition); and David Harry Grinspoon, *Venus Revealed: A New Look below the Clouds of Our Mysterious Twin Planet* (Reading, MA: Addison-Wesley, 1997).

CHAPTER 4

1. Jacob Burckhardt, *The Greeks and Greek Civilization*, trans. Sheila Stern (New York: St. Martin's Press, 1998), p. 322.

2. See Arthur A. Hoag, "Aristarchos Revisited," *Griffith Observer* 54, no. 8 (August 1990): 10–18.

3. Owen Gingerich, *The Eye of Heaven: Ptolemy, Copernicus, Kepler* (New York: American Institute of Physics, 1993), p. 8.

4. Nicolaus Copernicus, *Commentariolus*, in *Three Copernican Treatises*, trans. Edward Rosen (New York: Dover, 1959), p. 57.

5. Quoted in A. Koyré, *The Astronomical Revolution: Copernicus—Kepler—Borelli*, trans. R. E. W. Maddison (New York: Dover, 1973, reprint 1992), p. 65.

6. Andrew T. Young of San Diego State University has proposed that most likely the square sun was just that: he points out that sunsets sometimes do occur when the sun appears distorted by refraction into a rectangle; a rectangle of roughly equal proportions is usually referred to as "square."

7. Axel D. Willmann, *Catalog of Naked-Eye Sunspot Observations and Large Sunspots* (Göttingen, Germany: University Observatory, 1997).

8. Zhentao Xu, David W. Pankenier, and Yaotiao Jiang, *East Asian Archaeoastronomy: Historical Records of Astronomical Observations of China, Japan, and Korea* (Amsterdam: Gordon and Breach Science Library, 2000).

9. Richard McKim, *British Astronomical Association, Mars Section, Circular*, December 2001.

10. Stillman Drake, *Discoveries and Opinions of Galileo* (New York: Doubleday, 1957), p. 82.

11. Quoted in Drake, *Discoveries and Opinions of Galileo*, p. 117.

12. Quoted in Max Caspar, *Kepler*, trans. C. Doris Hellman (New York: Dover, 1993), p. 151.

13. Biographical note, in *Great Books of the Western World*, ed. Robert Maynard Hutchins (Chicago: Encyclopedia Britannica, 1952), 16:842.

14. Other sources on transits and sixteenth-century astronomy include Pierre Gassendi, *Mercurius in Sole Visa et Venus Invisa* (1632), in *Opera Omnia*, vol. 4 (Lyons, 1658); Robert Grant, *History of Physical Astronomy from the Earliest Ages to the Middle of the Nineteenth Century* (London: R. Baldwin, 1852); Hajo Holborn, *A History of Modern Germany*, vol. 1, *The Reformation* (Princeton, NJ: Princeton University Press, 1959); Jean Meeus, *Transits* (Richmond, VA: Willmann-Bell, 1989); Richard A. Proctor, *Transits of Venus: A Popular Account of Past and Coming Transits from the First Observed by Horrocks A.D. 1639 to the transit of A.D. 2012* (London: Longmans, Green, 1874); Albert van Helden, "The Importance of the Transit of Mercury of 1631," *Journal for the History of Astronomy* 7 (1976): 1–10.

CHAPTER 5

1. George Macaulay Trevelyan, *A Shortened History of England* (New York: Longmans, Green, 1942; reprint, Penguin, 1987), p. 284.

2. Winston Churchill, *History of the English-speaking Peoples*, arranged for one volume by Henry Steele Commager (New York: Barnes and Noble, 1995), p. 174.

3. Allan Chapman, "Jeremiah Horrocks, the Transit of Venus, and the 'New Astronomy' in Early Seventeenth-Century England," *Quarterly Journal of the Royal Astronomical Society* 31 (1990): 333–57.

4. As Joseph Edleston would later describe the similar position occupied by Isaac Newton as a sizar at Trinity College. See Frank E. Manuel, *A Portrait of Isaac Newton* (Cambridge, MA: Belknap Press, 1968), pp. 68–69.

5. Arundell Blount Whatton, *Memoir of the Life and Labours of the Rev. Jeremiah Horrox, Curate of Hoole, near Preston; to Which Is Appended a Translation of His Celebrated Discourse upon the Transit of Venus across the Sun* (London: Wertheim, Macintosh, and Hunt, 1859), p. 9.

6. Ibid., pp. 13–14.

7. Allan Chapman, *Three North Country Astronomers* (Manchester: N. Richardson, 1982), pp. 9–10.

8. Quoted in Whatton, *Memoir*, pp. 21–22.

9. Milton, *Paradise Lost*, bk. 8, lines 83–84.

10. J. E. Bailey, *The Palatine Note-books*, abridged in *The Observatory* 6 (1883): 318–28.

11. Quoted in Whatton, *Memoir*, p. 49.

12. Ibid., p. 43.

13. Ibid., p. 123.

14. Ibid., p. 124.

15. Ibid., p. 129.

16. Ibid., pp. 116–17.

17. Ibid., p. 134.

18. Trevelyan, *A Shortened History of England*, p. 294.

19. Quoted in Whatton, *Memoir*, p. 57.

20. Ibid., p. 58.

21. Oliver Cromwell, *Letters and Speeches*, ed. Thomas Carlyle (London: Methuen, 1904), 1:151.

22. Robert Grant, *History of Physical Astronomy from the Earliest Ages to the Middle of the Nineteenth Century* (London: R. Baldwin, 1852).

23. Information about Horrocks and his *Venus in Sole Visa* was assem-

bled in Whatton, *Memoir,* which remains an invaluable work. Another useful reference is W. F. Bushell, "Jeremiah Horrocks: The Keats of English Astronomy," *Mathematical Gazette* 43 (1959): 1–16.

CHAPTER 6

1. Betty Jo Teeter Dobbs and Margaret C. Jacob, *Newton and the Culture of Newtonianism* (Amherst, NY: Humanity Books, 1995), p. 62.

2. Richard Westfall, *The Life of Isaac Newton* (Cambridge: Cambridge University Press, 1993), p. 15. This is a very readable abridgment of the same author's massive *Never at Rest: A Biography of Isaac Newton* (Cambridge: Cambridge University Press, 1980).

3. William Stukeley, *Memoirs of Sir Isaac Newton's Life,* ed. A. Hastings White (London: Taylor and Francis, 1936), p. 45.

4. Westfall, *Never at Rest,* p. 61n.

5. Westfall, *Life of Isaac Newton,* p. 111.

6. Ibid., p. 19.

7. "Charles II," *Encyclopaedia Britannica,* 15th ed. (1980), pp. 54–56.

8. Quoted in Westfall, *Life of Isaac Newton,* p. 40.

9. Westfall, *Life of Isaac Newton,* p. 40.

10. D. T. Whiteside, "Before the *Principia,*" *Journal for the History of Astronomy* 1 (1970): 11n9.

11. Frank E. Manuel, *Portrait of Isaac Newton* (Cambridge, MA: Harvard University Press, Belknap Press, 1968), p. 105.

12. Alan Cook, *Edmond Halley: Charting the Heavens and the Seas* (Oxford: Clarendon Press, 1998), p. 41.

13. Ibid.

14. "Richelieu," *Encyclopaedia Britannica,* 15th ed. (1980), pp. 830–35.

15. See Richard Baum and William Sheehan, "G. D. Cassini and the Rotation Period of Venus: A Common Misconception," *Journal for the History of Astronomy* 23 (1992): 299–301.

16. For more on the longitude question, see Dava Sobel, *Longitude* (New York: Walker, 1995).

17. Manuel, *Portrait of Isaac Newton,* p. 294.

18. Quoted in Alan Cook, *Edmond Halley,* p. 52.

19. Manuel, *Portrait of Isaac Newton,* p. 318.

20. Carl Sagan and Ann Druyan, *Comet* (New York: Random House, 1985), p. 39.

21. Cited in Flamsteed to Moore, Greenwich, July 16, 1678, in *The Correspondence of John Flamsteed,* ed. E. G. Forbes, L. Murdin, and F. Willmoth (Bristol, UK, and Philadelphia: Institute of Physics Publishing, 1995), 1:643–46.

22. From James Ferguson's translation of the paper Halley submitted to the *Proceedings of the Royal Society,* 1716.

23. H. W. Turnbull, ed., *The Correspondence of Isaac Newton* (Cambridge: Cambridge University Press, 1960), 1:302.

24. Nicholas Kollerstrom, "The Path of Halley's Comet, and Newton's Late Apprehension of the Law of Gravity," *Annals of Science* 56 (1999): 331–56; see also Donald K. Yeomans, *Comets: A Chronological History of Observation, Science, Myth, and Folklore* (New York: John Wiley and Sons, 1991).

25. Kollerstrom, "The Path of Halley's Comet," p. 354.

26. Quoted in Westfall, *Never at Rest,* p. 403.

27. Abraham DeMoivre, quoted in Westfall, *Never at Rest,* p. 403.

28. Manuel, *Portrait of Isaac Newton,* p. 152.

29. Turnbull, ed., *The Correspondence of Isaac Newton,* 2:431.

30. Dana Densmore, *Newton's* Principia: *The Central Argument* (Santa Fe, NM: Green Lion Press, 1995), p. xxiii.

31. As in their title *Newton and the Culture of Newtonianism.*

32. Ibid., p. 64.

33. W. Stanley Jevons, *The Principles of Science: A Treatise on Logic and Scientific Method,* 3rd. ed. (London: Macmillan, 1879), pp. 294–95.

CHAPTER 7

1. Harry Woolf, *The Transits of Venus* (Princeton, NJ: Princeton University Press, 1959), p. 23.

2. Ibid.

3. Carl Sagan and Ann Druyan, *Comet* (New York: Random House, 1985), p. 54.

4. "Doctor Halley's Dissertation on the Method of Finding the Sun's Parallax and Distance from the Earth, by the Transit of Venus over the Sun's Disc, June the 6th, 1761," trans. James Ferguson, in *Astronomy Explained Upon Sir Isaac Newton's Principles, and Made Easy to Those Who Have Not Studied Mathematics,* 6th ed. (London, 1778); this has been conveniently reprinted in David Sellers, *The Transit of Venus: The Quest to Find the True Distance of the Sun* (Leeds, UK: MagaVelda Press, 2001), p. 204.

5. Quoted in Richard A. Proctor, *Transits of Venus: A Popular Account of Past and Coming Transits* (London: Longmans, Green, 1874), pp. 33–34.

6. Ibid., p. 36.

7. Quoted in Sellers, *The Transit of Venus*, p. 204.

8. Instead, he made his priority observing the moon, comparing its positions with those given in Isaac Newton's 1702 treatise, *Theory of the Moon's Motion*. For details, see Nicholas Kollerstrom's excellent book, *Newton's Forgotten Lunar Theory: His Contribution to the Quest for Longitude* (Santa Fe, NM: Green Lion Press, 2000), p. xx.

9. Christiaan Huygens, *Cosmotheros; sive de Terris Coelestis Earumque Ornatu Conjecturae* (The Hague, 1698); English translation, *Celestial Worlds Discover'd; or, Conjectures Concerning the Inhabitants, Plants, and Productions of the World in the Planets* (London: Frank Cass, 1968; reprint of 1698 ed.), p. 110.

10. Bianchini's book, *New Phenomena of Hesperus and Phosphorus, or Rather Observations Concerning the Planet Venus*, together with the correspondence of Briga and other contemporaries, was translated by Sally Beaumont and Peter Fay (Berlin: Springer Verlag, 1996), p. 32.

11. Ibid., p. 44.

12. Ibid., p. 149.

13. Woolf, *Transits of Venus*, p. 46.

14. César François Cassini, "Histoire abrégé de la parallaxe du Soleil" (Abridged History of the Solar Parallax), in Chappe, *Voyage en Californie pour l'observation du passage de Vénus sur le disque du Soleil* (Paris: A. Jombet, 1772), p. 114.

15. Quoted in Donald Fernie, *The Whisper and the Vision: The Voyages of the Astronomers* (Toronto: Clark, Irwin, 1976), p. 15.

16. Ibid., p. 16.

17. Woolf, *Transits of Venus*, p. 101.

18. Jean Chappe d'Auteroche, *A Journey into Siberia Made by Order of the King of France* (London, 1774), p. 38.

19. Woolf, *Transits of Venus*, p. 120.

20. Alexandre-Gui Pingré, *Relations de mon Voyage de Paris à l'ísle Rodrigue*, manuscript quoted, in French, in Woolf, *Transits of Venus*, p. 106.

21. Woolf, *Transits of Venus*, p. 126.

22. Quoted in Donald Fernie, *The Whisper and the Vision*, p. 42.

23. Chappe, *A Journey into Siberia*, p. 83.

24. As elucidated by Andy Young of San Diego State University, personal communication with W. Sheehan, March 8, 2002: "That ordinary

luminous ring is just the solar aureole caused by scattering in the haze at the cloud-tops. The particles are small and so not very forward-scattering; so the aureole is very dim compared to the surface of the sun itself—which is what is seen when Venus gets just to the edge of the sun and we see the solar surface refracted through the atmosphere of Venus. Sure, it begins as a tiny image near the point farthest from the center of the sun. But as Venus edges onto the sun's disk, this image spreads around the planet, forming a complete ring about the time Venus is bisected by the sun's limb." The details of what observers actually saw have been well described by F. Link, *Eclipse Phenomena in Astronomy* (New York: Springer Verlag, 1969), pp. 205–16, who even used them to produce quantitative estimates of the density of the atmosphere of Venus above the cloud-tops.

25. Mikhail V. Lomonosov, "The Appearance of Venus on the Sun, Observed in the St. Petersburg Academy of Sciences on May 26, 1761," quoted in Boris N. Menshutkin, *Russia's Lomonosov* (Princeton, NJ: Princeton University Press, 1952), p. 147.

26. Unfortunately, this paper, like many others by Lomonosov, was not printed during his lifetime, but it is the basis of Lomonosov's claim to be regarded as the discoverer of the atmosphere of Venus.

27. See Bradley E. Schaefer, "The Transit of Venus and the Notorious Black Drop Effect," *Journal for the History of Astronomy* 32 (2001): 325–36.

28. Ibid., pp. 327–28.

29. W. Hirst, "An Observation of the Same Transit of Venus over the Sun, June 6, 1761, at Madras," *Philosophical Transactions* 52 (1761): 396.

30. Schaefer, "The Transit of Venus and the Notorious Black Drop Effect," pp. 329–30.

31. Woolf, *Transits of Venus*, p. 193.

32. See, for instance, Albert van Helden, *Measuring the Universe* (Chicago: University of Chicago Press, 1985), p. 155.

33. Additional useful references dealing with this period include Angus Armitage, "Chappe d'Auteroche: A Pathfinder for Astronomy," *Annals of Science* 10, no. 4 (December 1954): 277–93; Angus Armitage, "The Pilgrimage of Pingré: An Astronomer-monk of Eighteenth-century France," *Annals of Science* 9, no. 1 (March 1953): 7–63; Thomas Pynchon, *Mason and Dixon* (New York: Henry Holt, 1997), a fictional account.

CHAPTER 8

1. Thomas Hornsby, "On the Transit of Venus in 1769," *Philosophical Transactions Abridged* 12 (1809): 265–74.

2. Harry Woolf, *The Transits of Venus* (Princeton, NJ: Princeton University Press, 1959), p. 164, citing Council Minutes for November 30, 1767.

3. Quoted in Richard Hough, *Captain James Cook: A Biography* (London: Hodder and Stroughton, 1994), p. 44.

4. Pink: "a ship with a narrow overhanging stern." *Merriam-Webster's Collegiate Dictionary*, 10th ed. (Springfield, MA: Merriam-Webster, 1998). We are indebted to Alan Gilmore of Mt. St. John University, New Zealand, for pointing this out.

5. Rex and Thea Rienits, *The Voyages of Captain Cook* (London: Paul Hamlyn, 1968), p. 14.

6. J. C. Beaglehole, *The Life of Captain James Cook* (Stanford, CA: Stanford University Press, 1974), pp. 148–49.

7. Ibid., p. 132. Beaglehole adds, "Perennial and brilliant visions!—those sunset continents on the vast Pacific horizon" (p. 133).

8. Samuel Johnson, *Dictionary* (1755).

9. Sir Everard Home, Hunterian Oration to the College of Surgeons, February 14, 1822; quoted in Hough, *Captain James Cook*, p. 58.

10. Quoted in Donald Fernie, *The Whisper and the Vision: The Voyages of the Astronomers* (Toronto: Clark, Irwin, 1976), p. 27.

11. Joseph Banks, journal entry, April 13, 1769.

12. Quoted in Beaglehole, *The Life of Captain James Cook*, p. 173.

13. Quoted in Lubert Stryer, *Principles of Biochemistry*, 2nd ed. (San Francisco: W. H. Freeman, 1981), p. 192.

14. Quoted in J. C. Beaglehole, *The Journals of Captain Cook*, selected and edited by Philip Edwards (London: Penguin, 1999), 1:74.

15. Sydney Parkinson, *A Journal of a Voyage to the South Seas in His Majesty's Ship, the* Endeavour (London, 1784), p. 21.

16. Quoted in Beaglehole, *The Journals of Captain Cook*, pp. 54–55.

17. Additional sources on Captain Cook and the transit of 1769 include J. C. Beaglehole, *The* Endeavour *Journal of Joseph Banks*, 2nd ed. (Sydney, Australia: Angus and Robertston, 1963); Charles Green and James Cook, "Observations Made, by Appointure of the Royal Society, at King George's Island in the South Seas," *Philosophical Transactions of the Royal Society* 61 (1771): 397–421; Wayne Orchiston, *James Cook and the 1769 Transit of Mercury*, Carter Observatory Information Sheet no. 3 (Wellington, New Zealand, 1994);

Wayne Orchiston, *Nautical Astronomy in New Zealand: The Voyages of James Cook* (Wellington, New Zealand: Carter Observatory, 1998).

CHAPTER 9

1. Thomas Hornsby, "On the Transit of Venus in 1769," *Philosophical Transactions Abridged* 12 (1809).

2. Quoted in Joseph Ashbrook, *The Astronomical Scrapbook: Skywatchers, Pioneers, and Seekers in Astronomy* (Cambridge, MA: Sky Publishing, 1984), p. 221.

3. Harry Woolf, *The Transits of Venus* (Princeton, NJ: Princeton University Press, 1959), p. 44.

4. A. H. Smyth, ed., *The Writings of Benjamin Franklin* (New York, 1905–1907), 5:137.

5. This and the following quotations are from William Wales, "Journal of a Voyage, Made by Order of the Royal Society, to Churchill River, on the North-west Coast of Hudson's Bay; of Thirteen Months Residence in the Country, and of the Voyage Back to England, in the Years 1768 and 1769," *Philosophical Transactions of the Royal Society of London* 60, no. 13 (1770): 100–36.

6. For Wales and Dymond, see Donald Fernie, *The Whisper and the Vision: The Voyages of the Astronomers* (Toronto: Clarke, Irwin, 1976), p. 21, and Wayne Orchiston and Derek Howse, "From Transit of Venus to Teaching Navigation: The Work of William Wales," *Astronomy and Geophysics* 39, no. 6 (December 1998): 21–24.

7. William Wales and James Dymond, "Astronomical Observations Made by Order of the Royal Society, at Prince of Wales' Fort, on the North-west Coast of Hudson's Bay," *Philosophical Transactions of the Royal Society* 59, no. 65 (1769): 467.

8. Ibid.

9. For details, see Brooke Hindle, *David Rittenhouse* (Princeton, NJ: Princeton University Press, 1964).

10. Benjamin Rush, *An Eulogium Intended to Perpetuate the Memory of David Rittenhouse, Late President of the American Geophysical Society* (Philadelphia: Ormrod and Conrad, 1796–97), p. 12.

11. Hindle, *David Rittenhouse*, p. 56.

12. D. B. Nunis, *The 1769 Transit of Venus* (Los Angeles: Natural History Museum of Los Angeles County, 1982), pp. 65–66.

13. G. Le Gentil, "Mémoire de M. Le Gentil . . . au subjet de l'observation qu'il va faire, par ordre du Roi, dans les Indes Orientale, du prochain passage de Venus pardevant le Soleil" (Memoir of M. Le Gentil . . . about the Observations He Will Make, by Order of the King, in the East Indies, of the Forthcoming Transit of Venus in Front of the Sun), *Journal des Sçavans* (March 1760): 132–42; translated by Helen S. Hogg, "Out of Old Books: Le Gentil and the Transits of Venus, 1761 and 1769," *Journal of the Royal Astronomical Society of Canada* 45 (1951): 90.

14. Ibid., p. 131.

15. Le Gentil, p. 35; quoted in Woolf, *The Transits of Venus*, p. 155; translated from the French by David Sellers, *The Transit of Venus: The Quest to Find the True Distance of the Sun* (Leeds, UK: MagaVelda Press, 2001), pp. 137–38.

16. Sellers, *The Transit of Venus*, p. 138.

17. Woolf, *Transits of Venus*, p. 130.

18. A good general reference on the 1769 transit and its observers is Willy Ley, *Watchers of the Skies* (New York: Viking Press, 1966). Some additional sources on specific observers include Académie Royale des Sciences (France), "Elogé de M. l'Abbé Chappe" (Eulogy for Abbot Chappe), *Histoire de l'Académie Royales des Sciences* (Paris, 1769), pp. 163–72; Thomas Hornsby, "The Quantity of the Sun's Parallax, as Deduced from the Observations of the Transit of Venus, on June 3, 1769," *Philosophical Transactions of the Royal Society* 61 (1771): 574–79; Simon Newcomb, "On Hell's Alleged Falsification of His Observations of the Transit of Venus in 1769," *Monthly Notices of the Royal Astronomical Society* 42 (1883): 371–81; Simon Newcomb, *The Reminiscences of an Astronomer* (Boston: Harper and Brothers, 1903); John Westfall, "The 1769 Transit of Venus Expedition to San José del Cabo," in *Research Amateur Astronomy*, ed. Stephen Edberg, Astronomical Society of the Pacific Conference Series 33 (San Francisco: Astronomical Society of the Pacific, 1992), pp. 234–42.

Chapter 10

1. William Blake, "Jerusalem" (London, 1804).

2. Quoted in Simon Schaffer, "Metrology, Metrication, and Victorian Values," in *Victorian Science in Context*, ed. Bernard Lightman (Chicago: University of Chicago Press, 1997), p. 438.

3. William Thomson, Lord Kelvin, *Popular Lectures and Addresses*

(London: Macmillan, 1891–94), 1:73. Ironically, it would be the more qualitative-thinking biologists and geologists—not the physicist, Kelvin—who would be closer to being right in the case of the burning question of the age of the earth. Kelvin relied on the theories of Kant and Laplace, which, according to a recent scholarly review of Kelvin's work on the age of the earth, "inevitably introduced into the resulting calculations the unquantifiable influence of accepted authority" (Joe D. Burchfield, *Lord Kelvin and the Age of the Earth* [London and Chicago: University of Chicago Press, 1990], p. 215).

4. Sir William Cecil Dampier, *A Shorter History of Science* (London: Scientific Book Club, 1946), p. 92.

5. Johann Franz Encke, *Der Venusdurchgange von 1769 als Fortsetzung der Abhandlung über die Entfernung der Sonne von der Erde* (The Transit of Venus of 1769 as a Continuation of the Discourse on the Distance of the Sun from the Earth) (Gotha, Germany: Becker'schen Buchhandlung, 1824), p. 108.

6. Quoted by Friedrich Engels, "The Chartist Movement: Meeting in Support of the National Petition," *La Réforme*, January 19, 1848. Undoubtedly the phrase was old even then; it has also been attributed to a statement made in 1773 by Sir George Macartney, as well as to Phillip II, the sixteenth-century king of Spain.

7. Michael G. Mulhall, *Balance-Sheet of the World for Ten Years, 1870–1880* (London: Edward Stanford, 1881), pp. 5, 22.

8. Ibid., p. 26.

9. Ibid.

10. Auguste Comte, *Course de Philosophie Positive* (Paris: Bachelier, 1835), 2:2.

11. Camille Flammarion, *Popular Astronomy*, trans. J. Ellard Gore (New York: Appleton, 1907), p. 371. (Originally published 1880.)

12. Gerald Merton, "Photography and the Amateur Astronomer," *Journal of the British Astronomical Association* 63 (1953): 8.

13. Frederick Scott Archer, "The Use of Collodion in Photography," *The Chemist*, no. 2 (1851): 87.

14. Abney, William de Wiveleslie, "Dry Plate Process for Solar Photography," *Monthly Notices, Royal Astronomical Society* 34 (1874): 275–78.

15. Two works on the topic of "personal equation" are Raynor Duncombe, "Personal Equation in Astronomy," *Popular Astronomy* 53 (1945): 2–13, 63–76, and 110–21; and Simon Schaffer, "Astronomers Mark Time: Discipline and the Personal Equation," *Science in Context* 2, no. 1 (1988): 115–45.

16. Nevil Maskelyne, *Astronomical Observations Made at the Royal Observatory at Greenwich from 1799 to 1810* (London: Nourse, 1811), 3:319.

17. Steven J. Dick, Wayne Orchiston, and Tom Love, "Simon Newcomb, William Harkness, and the Nineteenth-century American Transit of Venus Expeditions,"*Journal for the History of Astronomy* 29 (1998): 232–33.

18. Besides the references already cited, the following are useful works related to the nineteenth-century transits: William J. H. Andrewes, ed., *The Quest for Longitude* (Cambridge, MA: Collection of Historical Instruments, Harvard University, 1996); Jimena Canales, "The 'Cinematographic Turn' and Its Alternatives in Nineteenth-century France," *Isis* 93, no. 4 (2002): 585–613; Steven J. Dick, "The Naval Observatory and the American Transit of Venus Expeditions of 1874 and 1882," chapter 7 in *Sky and Ocean Joined: U.S. Naval Observatory, 1830–2000*, ed. Steven J. Dick (Cambridge: Cambridge University Press, 2003); J. L. E. Dreyer and H. H. Turner, eds., *History of the Royal Astronomical Society, vol. 1, 1820–1920* (London: Royal Astronomical Society, 1923); Peter J. Hughill, *Global Communications since 1844* (Baltimore: Johns Hopkins University Press, 1999); Tom Standage, *The Victorian Internet* (New York: Walker, 1998).

Chapter 11

1. Peter Andreas Hansen, "On the Value of the Solar Parallax Deduced from the Lunar Tables," *Monthly Notices of the Royal Astronomical Society* 23, no. 8 (1863): 242–43.

2. Urbain Jean Joseph Leverrier, "Parallaxe du Soleil tirée de l'equation parallactique de la Lune" (Solar Parallax Derived from the Parallactic Equation of the Moon), *Annales de l'Observatoire de Paris* 4 (1858): 101; Simon Newcomb, "Investigation of the Distance of the Sun, and the Elements Which Depend on It, from the Observation of Mars, Made during the Opposition of 1862, and from Other Sources," *Washington Observations, 1865* (Washington, DC: Government Printing Office, 1867), appendix H; Edward J. Stone, "A Rediscussion of the Observations of the Transit of Venus, 1769," *Monthly Notices of the Royal Astronomical Society* 28, no. 6 (1868): 255–66.

3. Hervé Faye, "Note sur les nouvelles tables des planètes intérieures" (Note on the New Tables of the Interior Planets), *Comptes Rendus des Séances de l'Académie des Sciences* 54 (1862): 630–39.

4. George Biddell Airy, "On the Means Which Will Be Available for

Correcting the Measure of the Sun's Distance, in the Next Twenty-Five Years," *Monthly Notices of the Royal Astronomical Society* 17, no. 7 (1857): 208–21.

5. Ibid., p. 209.

6. John Lankford, "Photography and the Nineteenth-century Transits of Venus," *Technology and Culture* 28, no. 3 (1987): 648–57.

7. Airy, "On the Means Which Will Be Available for Correcting the Measure of the Sun's Distance," p. 214.

8. George Biddell Airy, "On the Preparatory Arrangements Which Will Be Necessary for Efficient Observation of the Transits of Venus in 1874 and 1882," *Monthly Notices of the Royal Astronomical Society* 29, no. 2 (1868): 33–43.

9. Richard A. Proctor, *Transits of Venus: A Popular Account* (London: Longmans, Green, 1874) and *The Universe and the Coming Transits* (London: Longmans, Green, 1874).

10. Richard A. Proctor, *Old and New Astronomy* (London: Longmans, Green, 1892), p. 264.

11. George Biddell Airy, *Autobiography of Sir George Biddell Airy, K.C.B., M.A., L.L.D., D.C.L., F.R.S., F.R.A.S., Honorary Fellow of Trinity College, Cambridge, Astronomer Royal from 1836 to 1881*, ed. Wilfrid Airy (Cambridge: Cambridge University Press, 1896), pp. 302–303.

12. Quoted in A. J. Meadows, "The Transit of Venus in 1874," *Nature* 250 (August 1974): 751.

13. Proctor, *Old and New Astronomy*, pp. 302–303.

14. Henry Strommel, *Lost Islands* (Vancouver: University of British Columbia Press, 1984), pp. 73, 75, 105, 115.

15. Quoted in Norman White, *Gerard Manley Hopkins: A Literary Biography* (Oxford: Clarendon Press, 1992), p. 336.

16. Quoted in ibid., pp. 337–38.

17. Hermann A. Bruck, "Lord Crawford's Observatory at Dun Echt, 1872–1892," *Vistas in Astronomy* 35 (1992): 81–138. For an account setting Crawford's activities in a larger context, see Allan Chapman, *The Victorian Amateur Astronomer: Independent Astronomical Research in Britain, 1820–1920* (New York: John Wiley and Sons, 1998), pp. 133–35.

18. George Forbes, *The Transit of Venus* (London and New York: Macmillan Nature Series, 1874), p. 17.

19. Steven J. Dick, Wayne Orchiston and Tom Love, "Simon Newcomb, William Harkness, and the Nineteenth-century American Transit of Venus Expeditions," *Journal for the History of Astronomy* 29 (1998): 230.

20. Hervé Faye, "Rôle que pourra prendre la France dans l'observation du prochain passage de Vénus sur le Soleil: Passage d'un discours prononcé par M. Faye, à l'overture de la séance publique annuelle du 25 novembre 1872" (The Role That France Will Be Able to Take in the Observation of the Next Transit of Venus over the Sun: Passage from a Talk Given by M. Faye at the Commencement of the Annual Public Meeting of November 25, 1872), *Comptes Rendus des Séances de l'Académie des Sciences* 75 (1872): 1295–96.

21. Otto William Struve, "List of Stations Selected for Observation of the Transit of Venus by Russian Astronomers," *Monthly Notices of the Royal Astronomical Society* 33, no. 7 (1873): 415–17.

22. Quoted in George Biddell Airy, *Account of Observations of the Transit of Venus, 1874, December 8, Made under the Authority of the British Government and of the Reduction of the Observations* (London: Her Majesty's Stationery Office, 1881), p. 334.

23. George Lyon Tupman, "On the Mean Solar Parallax as Derived from the Observations of the Transit of Venus, 1874," *Monthly Notices of the Royal Astronomical Society* 38, no. 8 (1878): 443.

24. Hervé Faye, "Mémoire de M. Faye sur l'observation photographique des passages de Vénus, et sur un appareil de M. Laussedat" (Memoir by M. Faye on the Photographic Observation of Transits of Venus, and on an Apparatus of M. Laussedat), *Comptes Rendus des Séances de l'Académie des Sciences* 70 (1870): 541–48.

25. Hervé Faye, "Association française pour l'avancement des Sciences, Congrès de Lille, conférences publiques: Le prochain passage de Vénus sur Soleil" (French Association for the Advancement of the Sciences, Congress at Lille: The Next Transit of Venus over the Sun), *Revue Scientifique* 14, no. 16 (1874): 366.

26. Jimena Canales, "Photogenic Venus: The 'Cinematographic Turn' and Its Alternatives in Late Nineteenth-century Science," *Isis* 93, no. 4 (2002): 585–613. ("Immediately convince the spirit" is from C. Delaunay, "Note sur la distance du Soleil à la Terre" (Note on the Distance from the Sun to the Earth), extract from *L'annuaire pour l'an 1816* (The Almanac for the Year 1816), published by the Bureau des Longitudes, p. 94).

27. Pierre Jules César Janssen, "Passage de Vénus, méthode pour obtenir photographiquement l'instant des contacts, avec les circonstances physiques qu'ils présentent" (Transit of Venus: A Method for Photographically Obtaining the Instant of the Contacts, along with the Physical Circumstances They Present), *Comptes Rendus des Séances de l'Académie des Sciences* 76 (1873): 677–79.

28. The management and operations of transit expeditions were similar to those for solar eclipse expeditions of the same period. For the latter, see Alex Soojung-Kim Pang, *Empire and the Sun: Victorian Solar Eclipse Expeditions* (Stanford, CA: Stanford University Press, 2000).

CHAPTER 12

1. Michael E. Chauvin, "Astronomy in the Sandwich Islands: The 1874 Transit of Venus," *The Hawaiian Journal of History* 27 (1993), p. 207.

2. Jules Janssen, "Note sur l'observation du passage de la planète Vénus sur le Soleil" (Note on the Observation of the Transit of the Planet Venus upon the Sun), *Comptes Rendus des Séances de l'Académie des Sciences* 96 (January 1883): 288–92.

3. Quoted in Wayne Orchiston, Tom Love, and Steven J. Dick, "Refining the Astronomical Unit," *Journal of Astronomical History and Heritage* 3, no. 1 (2000): 23–44.

4. George Lyon Tupman, "On the Mean Solar Parallax as Derived from the Observations of the Transit of Venus, 1874," *Monthly Notices of the Royal Astronomical Society* 38, no. 8 (1878): 434.

5. Ibid., p. 432.

6. Ibid., pp. 432–33.

7. Ibid., p. 433.

8. Agnes Mary Clerke, *A Popular History of Astronomy during the Nineteenth Century*, 4th ed. (London: Adam and Charles Black, 1902), p. 234.

9. Quoted in Joseph Ashbrook, "Father Perry's Expedition to Kerguélen Island," in *The Astronomical Scrapbook: Skywatchers, Pioneers, and Seekers in Astronomy* (Cambridge, MA: Sky Publishing, 1984), pp. 221–27.

10. Lord Lindsay, "Observations of the Transit of Venus at Mauritius," *Monthly Notices of the Royal Astronomical Society* 35 (January 1875): 231.

11. Tupman, "On the Mean Solar Parallax," p. 431.

12. Ibid., p. 509.

13. Jimena Canales, "Photogenic Venus: The 'Cinematographic Turn' and Its Alternatives in Nineteenth-century France," *Isis* 93, no. 4 (2002): 588.

14. George Forbes, *David Gill, Man and Astronomer: Memoirs of Sir David Gill, K.C.B., H.M. Astronomer (1879–1907) at the Cape of Good Hope* (London: Watts, 1909).

15. Isobel Gill, *Six Months in Ascension* (London: John Murray, 1878), p. 63.

16. Besides works already cited in chapters 11 and 12, some useful sources for the 1874 transit operations are Gerard de Vaucouleurs, *Astronomical Photography: From the Daguerreotype to the Electron Camera* (New York: Macmillan, 1961); Hervé Faye, "Sur les passages de Vénus et la parallaxe du soleil" (On the Transits of Venus and the Solar Parallax), *Comptes Rendus des Séances de l'Académie des Sciences* 68 (1869): 42–49; Académie des Sciences (France), *Passage de Vénus sur le Soleil du 9 décembre 1874: Mission de l'île Campbell* (Transit of Venus upon the Sun of December 9, 1874: Mission to Campbell Island) (n.p., 1875); Académie des Sciences (France), *Recueil de Mémoires, Rapports et Documents Relatifs à l'Observation du Passage de Vénus sur le Soleil* (Compilation of Memoirs, Reports, and documents Relating to the Observation of the Transit of Venus upon the Sun), Memoires de l'Académie des Sciences 41 (Paris: Librairie de Firmin Didot Frères, Fils et Cie, 1874); Paul M. Janiczek, "Remarks on the Transit of Venus Expedition of 1874," in *Sky with Ocean Joined: Proceedings of the Sesquicentennial Symposia of the U.S. Naval Observatory, December 5 and 8, 1980*, ed. Steven J. Dick and LeRoy E. Doggett (Washington, DC: U.S. Naval Observatory, 1983), pp. 52–73; Paul M. Janiczek and L. Houchins, "Transits of Venus and the American Expedition of 1874," *Sky and Telescope* 48 (December 1974): 366–71; Wayne Orchiston and A. Buchanan, "Illuminating Incidents in Antipodean Astronomy: Campbell Town and the 1874 Transit of Venus," *Australian Journal of Astronomy* 5 (1993): 11–31; L. Pigatto and V. Zanini, "Spectroscopic Observations of the 1874 Transit of Venus: The Italian Party at Muddapur, Eastern India," *Journal of Astronomical History and Heritage* 4 (2001): 43–58; Henry Norris Russell, Raymond Smith Dugan, and John Quincy Stewart, *Astronomy: A Revision of Young's Manual of Astronomy* (New York: Ginn, 1926). In addition to the specific sources above, various reports of varying length provided information at numerous points in our text, appearing in the following publications: *Astronomical Register*, vols. 12–17 (1874–79); *Astronomische Nachrichten*, vol. 85 (1875); *Comptes Rendus des Séances de l'Académie des Sciences*, vols. 78-85 (1874–77); *Monthly Notices of the Royal Astronomical Society*, vols. 35–38 (1874–78).

Chapter 13

1. Ministére de l'Instruction Publique et des Beaux-Arts (France), *Conférence internationale du passage de Vénus: Procès-verbaux* (International

Conference on the Transit of Venus: Verbal Proceedings) (Paris: Imprimerie Nationale, 1881), October 5 meeting, pp. 6, 8.

2. George Biddell Airy, *Account of Observations of the Transit of Venus, 1874, December 8, Made under the Authority of the British Government and of the Reductions of the Observations* (London: Her Majesty's Stationery Office, 1881), appendix p. 19.

3. Simon Newcomb, "Investigation of the Distance of the Sun, and the Elements Which Depend on It, from the Observation of Mars, Made during the Opposition of 1862, and from Other Sources," *Washington Observations, 1865* (Washington, DC: Government Printing Office, 1867), p. 29.

4. Nautical Almanac Office of Great Britain, *Explanatory Supplement to the Astronomical Ephemeris and the American Ephemeris and Nautical Almanac* (London: Her Majesty's Stationery Office, 1961), pp. 181, 188.

5. John Krom Rees, "Observations of the Transit of Venus, December 6, 1882," *Annals of the New York Academy of Sciences* 2 (1882): 384–90.

6. Previously the U.S. Coast Survey; see Thomas G. Manning, *U.S. Coast Survey vs. Naval Hydrographic Office: A Nineteenth-century Rivalry in Science and Politics* (Tuscaloosa: University of Alabama Press, 1988).

7. Vincent Ponko Jr., "Nineteenth-century Science in New Mexico: The 1882 Transit of Venus Observations at Cerro Roblero," *Journal of the West* 33 (1994): 44–51.

8. Frederick Brodie, "Observation of the Transit of Venus 1882, December 6, made at Fernhill, Wootton Bridge, Isle of Wight," *Monthly Notices of the Royal Astronomical Society* 43 (1883): 76.

9. Charles Leeson Prince, "Note on the Transit of Venus, 1882, Dec. 6," *Monthly Notices of the Royal Astronomical Society* 43 (1883): 278.

10. Simon Newcomb, *The Reminiscences of an Astronomer* (Boston: Harper and Brothers, 1903), p. 28.

11. Quoted in Jean Baptiste Dumas, "Passage de Vénus du 6 décembre 1882: Rapports préliminaires" (Transit of Venus of December 6, 1882: Preliminary Reports), *Comptes Rendus des Séances de l'Académie des Sciences* 97 (1883): 403. (Dumas was the president of the French Transit of Venus Commission. The entire article [pp. 353–443] is a convenient summary of the government-sponsored French expeditions in 1882–83.)

12. For a firsthand account, see David Todd, "Professor Todd's Own Story of the Mars Expedition," *The Cosmopolitan*, January 1908.

13. Quoted in Polly Longsworth, *Austin and Mabel: The Amherst Affair and Love Letters of Austin Dickinson and Mabel Loomis Todd* (Amherst: University of Massachusetts Press, 1984), p. 5.

14. For the history of the observatory, see Helen Wright, *James Lick's Monument: The Saga of Captain Richard Floyd and the Building of the Lick Observatory* (Cambridge: Cambridge University Press, 1987), and Donald E. Osterbrock, John E. Gustafson, and W. J. Shiloh Unruh, *Eye on the Sky: Lick Observatory's First Century* (Berkeley and Los Angeles: University of California Press, 1988).

15. As noted later in Taliesin Evans, "A Californian's Gift to Science," *The Century Illustrated Monthly Magazine* 30, no. 1 (May 1886): 72.

16. Wright, *James Lick's Monument,* p. 102.

17. Ibid., p. 101.

18. Richard B. Sewall, preface to Longsworth, *Austin and Mabel,* p. xiii.

19. David P. Todd, "On the Observations of the Transit of Venus, 1882, December 5-6, Made at the Lick Observatory, Mount Hamilton, California," *Monthly Notices of the Royal Astronomical Society* 43 (1883): 273-76.

20. Quoted in Wright, *James Lick's Monument,* p. 101.

21. For details of these efforts, see Steven J. Dick, *The Biological Universe* (Cambridge: Cambridge University Press, 1996), pp. 406-408.

22. Evans, "A Californian's Gift to Science," p. 72.

23. In addition to the specific references already cited, a large number of contemporary reports and several modern articles were consulted for this chapter, many of the former appearing in late December 1882 and the first months of 1883 in such scientific periodicals as the *Astronomical Register, Astronomische Nachrichten, Comptes Rendus des Séances de l'Académie des Sciences,* and the *Monthly Notices of the Royal Astronomical Society.* Some especially useful sources include Ronaldo Rogério de Freitas Mourão, *Os Eclipses, da Superstição à Previsão Mathmática* (Eclipses: From Superstition to Mathematical Forecast) (São Leopoldo, Brazil: Ed. Unisinos, 1993), esp. pp. 125-91; S. P. Langley, "Observation of the Transit of Venus, 1882, December 6, Made at the Alleghany Observatory," *Monthly Notices of the Royal Astronomical Society* 43 (1883): 72-73; John Lankford, "The Impact of Photography on Astronomy," chapter 2 in *The General History of Astronomy, vol. 4, Astrophysics and Twentieth Century Astronomy to 1950: Part A.* (Cambridge: Cambridge University Press, 1984); Vincent Ponko Jr., "Cedar Key, Florida, and the Transit of Venus: The 1882 Site Observations," *Gulf Coast Historical Review* 10, no. 2 (1995): 47-66; Royal Astronomical Society, "Report of the Council to the Sixty-third Annual General Meeting," *Monthly Notices of the Royal Astronomical Society* 43 (1883): 189-202 (Reports of Observatories), 211-15 (The Transit of Venus, 1882); Royal Astronomical

Society, "Report of the Council to the Sixty-fourth Annual General Meeting," *Monthly Notices of the Royal Astronomical Society* 44 (1883): 182–84; United States Coast and Geodetic Survey, *Report of the Superintendent of the U.S. Coast and Geodetic Survey Showing the Progress of the Work during the Fiscal Year Ending with June 1883* (Washington, DC: Government Printing Office, 1884); United States Transit of Venus Commission, *Instructions for Observing the Transit of Venus, December 6, 1882, Prepared by the Commission Authorized by Congress and Printed for the Use of the Observing Parties by Authority of the Hon. Secretary of the Navy* (Washington, DC: Government Printing Office, 1882); Hermann W. Vogel, "Beobachtungen des Venusdurchgangs am 6. December 1882, angestellt auf dem Astrophysikalischen Observatorium zu Potsdam" (Observations of the Transit of Venus on December 6, 1882, Made at the Astrophysical Observatory of Potsdam), *Astronomische Nachrichten* 104 (1883): 257–62.

CHAPTER 14

1. Harold Spencer Jones, "The Distance of the Sun," *Endeavour* 1, no. 1 (1942): 17.

2. P. Kenneth Seidelman, ed., *Explanatory Supplement to the* Astronomical Almanac (Mill Valley, CA: University Science Books, 1992), p. 298.

3. See "Astrodynamic Constraints and Parameters," JPL Solar System Dynamics, http://ssd.jpl.nasa.gov/astro_constants.html (accessed December 1, 2003).

4. See Alan W. Hirshfeld, *Parallax: The Race to Measure the Cosmos* (New York: W. H. Freeman, 2001).

5. William Sheehan and Thomas Dobbins, "The Spokes of Venus: An Illusion Explained," *Journal for the History of Astronomy* 34 (2003): 53–63.

6. There are a number of books on the modern spacecraft exploration of Venus, such as David Henry Grinspoon, *Venus Revealed* (Reading, MA: Addison-Wesley, 1997), an eloquent account that includes discussions of the Magellan spacecraft results.

7. A. T. Young and L. G. Young, "Observing Venus near the Sun," *Sky & Telescope* 43, no. 3 (1972): 140–44.

8. See also B. Ralph Chou, "Solar Filter Safety," *Sky & Telescope* 95, no. 2 (1998): 36–40.

9. L. Eduardo Vega, "Homemade Mylar Filters," *Sky & Telescope* 95, no. 2 (1998): 39.

10. See also Michael Davis and David Staup, "Shooting the Planets with Webcams," *Sky & Telescope* 105, no. 6 (2003): 117–22.

11. European Southern Observatory Educational Office, "The Venus Transit 2004: Exoplanets and the Size of the World!" http://www.eso.org/outreach/eduoff/vt-2004 (accessed July 25, 2003). Three similar projects are described at http://www.transitofvenus.co.za/index.html, http://didaktik.physik.uni-essen.de/~backhaus/VenusProject.htm, and http://eclipse.astroinfo.org/transit/venus/project2004.

12. George F. Chambers, *A Handbook of Descriptive and Practical Astonomy* (Oxford: Clarendon Press, 1889), 1:349.

13. For details about similar illusions and a physiological explanation, see Floyd Ratliff, *Mach Bands: Quantitative Studies on Neural Networks in the Retina* (San Francisco: Holden-Day, 1965).

14. F. Link, *Eclipse Phenomena in Astronomy* (New York: Springer Verlag, 1969), pp. 205–25.

15. Herbert George Wells, "Atom: Boon or Doom?" *Daily Herald*, August 9, 1945.

16. Additional references that may help in planning for the 2004 and 2012 transits of Venus are William R. Corliss, *The Moon and the Planets: A Catalog of Astronomical Anomalies* (Glen Arm, MD: Sourcebook Project, 1985); F. Link, "Allongement des cornes de Vénus" (Extension of the Horns of Venus), *Bulletin of the Astronomical Institutes of Czechoslovakia* 1, no. 6 (1949): 77–81; Eli Maor, *June 8, 2004: Venus in Transit* (Princeton, NJ: Princeton University Press, 2000); Michael Maunder and Patrick Moore, *Transit: When Planets Cross the Sun* (London: Springer Verlag, 2000); Willy Rudloff, *World-Climates: With Tables of Climatic Data and Practical Suggestions* (Stuttgart: Wissenschaftliche Verlagsgesellschaft, 1981); James A. Ruffner and Frank E. Bair, eds., *The Weather Almanac*, 5th ed. (Detroit: Gale Research, 1987); Henry Norris Russell, "The Atmosphere of Venus," *Astrophysical Journal* 9 (1899): 284–99; United States Naval Observatory, *Nautical Almanac* Office, *The Astronomical Almanac for the Year 2004* (Washington, DC: U.S. Government Printing Office, 2002), pp. A88–A95.

Bibliography

Abney, William de Wiveleslie. "Dry Plate Process for Solar Photography." *Monthly Notices of the Royal Astronomical Society* 34, no. 6 (1874): 275–78.

Académie royale des sciences (France). "Éloge de M. l'Abbé Chappe" (Eulogy for Abbot Chappe). *Histoire de l'Académie Royales des Sciences* (1769): 163–72.

Académie des sciences (France). *Recueil de mémoires, rapports et documents relatifs à l'observation du passage de Vénus sur le Soleil* (Compilation of Memoirs, Reports, and Documents Relating to the Observation of the Transit of Venus upon the Sun). Mémoires de l'Académie des Sciences 41. Paris: Librairie de Firmin Didot Frères, Fils, 1874.

———. *Passage de Vénus sur le Soleil du 9 décembre 1874: Mission de l'île Campbell* (Transit of Venus upon the Sun of December 9, 1874: The Campbell Island Mission). N.p., 1875.

Airy, George Biddell. "On the Means Which Will Be Available for Correcting the Measure of the Sun's Distance, in the Next Twenty-five Years." *Monthly Notices of the Royal Astronomical Society* 17, no. 7 (1857): 208–21.

———. "On the Preparatory Arrangements Which Will Be Necessary for Efficient Observation of the Transits of Venus in 1874 and 1882." *Monthly Notices of the Royal Astronomical Society* 29, no. 2 (1868): 33–43.

———. *Account of Observations of the Transit of Venus, 1874, December 8, Made under the Authority of the British Government and of the Reduction of the Observations*. London: Her Majesty's Stationery Office, 1881.

———. *Autobiography of Sir George Biddell Airy, K.C.B., M.A., L.L.D., D.C.L., F.R.S., F.R.A.S., Honorary Fellow of Trinity College, Cambridge, Astronomer Royal from 1836 to 1881*. Edited by Wilfrid Airy. Cambridge: Cambridge University Press, 1896.

Andrewes, William J. H., ed. *The Quest for Longitude*. Cambridge, MA: Collection of Historical Instruments, Harvard University, 1996.

Archer, Frederick Scott. "The Use of Collodion in Photography." *The Chemist*, no. 2 (1851): 257.

Armitage, Angus. "The Pilgrimage of Pingré. An Astronomer-monk of Eighteenth-century France." *Annals of Science* 9, no. 1 (1953): 47–63.

———. "Chappe d'Auteroche: A Pathfinder for Astronomy." *Annals of Science* 10, no. 4 (1954): 277–93.

Ashbrook, Joseph. "Father Perry's Expedition to Kerguélen Island." In *The Astronomical Scrapbook: Skywatchers, Pioneers, and Seekers in Astronomy*, pp. 221–27. Cambridge, MA: Sky Publishing, 1984.

Bailey, J. E. *The Palatine Note-books*. Abridged in *The Observatory* 6 (1883): 318–28.

Baum, Richard, and William Sheehan. "G. D. Cassini and the Rotation Period of Venus: A Common Misconception." *Journal for the History of Astronomy* 23 (1992): 299–301.

Beaglehole, J. C. *The* Endeavour *Journal of Joseph Banks*. 2nd ed. Sydney: Angus and Robertston, 1963.

———. *The Life of Captain James Cook*. Stanford, CA: Stanford University Press, 1974.

Bianchini, Francesco. *Hesperi et Phosphori Nova Phaenomena sive Observationes circe Planetam Veneris*. Rome: Ioannem Mariam Salvioni, 1728. Translated into English as *New Phenomena of Hesperus and Phosphorus, or Rather Observations Concerning the Planet Venus*. Translated by Sally Beaumont and Peter Fay. Berlin: Springer Verlag, 1996.

———. *Observations Concerning the Planet Venus*. Translated by Sally Beaumont, assisted by Peter Fay. Berlin: Springer Verlag, 1996.

Breasted, James H. *A History of Egypt from the Earliest Times to the Persian Conquest*. New York: Scribner, 1909.

Brodie, Frederick. "Observation of the Transit of Venus 1882, December 6, Made at Fernhill, Wootton Bridge, Isle of Wight." *Monthly Notices of the Royal Astronomical Society* 43 (1883): 76.

Bruck, Hermann A. "Lord Crawford's Observatory at Dun Echt, 1872–1892." *Vistas in Astronomy* 35 (1992): 81–138.

Burchfield, Joe D. *Lord Kelvin and the Age of the Earth*. London and Chicago: University of Chicago Press, 1990.

Burckhardt, Jacob. *The Greeks and Greek Civilization*. Translated by Sheila Stern. New York: St. Martin's Press, 1998.

Bushell, W. F. "Jeremiah Horrocks: The Keats of English Astronomy." *Mathematical Gazette* 43 (1959): 1–16.

Campbell, James, and Edmund Neison [Nevil]. "On the Determination of the Solar Parallax by Means of the Parallactic Inequality in the Motion of the Moon." Parts 1 and 2. *Monthly Notices of the Royal Astronomical Society* 30, no. 7 (May 14, 1880): 366–411; no. 8 (June 11, 1880): 441–72.

Canales, Jimena. "The 'Cinematographic Turn' and Its Alternatives in Nineteenth-century France." *Isis* 93, no. 4 (2002): 585–613.

Carlyle, Thomas, ed. *Letters and Speeches of Oliver Cromwell*. 3 vols. London: Methuen, 1904.

Caspar, Max. *Kepler*. Translated by C. Doris Hellman. New York: Dover, 1993.

Cassini, C. F. "Histoire abrégé de la parallaxe du Soleil" (Abridged History of the Solar Parallax). In Chappe, *Voyage en Californie pour l'observation du Passage de Vénus sur le disque du Soleil, le 3 juin 1769 . . .*, p. 114. Paris: Chez Charles-Antoine Jombet, 1772.

Chambers, George F. *A Handbook of Descriptive and Practical Astronomy*. 3 vols. Oxford: Clarendon Press, 1889.

Chapman, Allan. *Three North Country Astronomers*. Manchester: N. Richardson, 1982.

———. "Jeremiah Horrocks, the Transit of Venus, and the 'New Astronomy' in Early Seventeenth-century England." *Quarterly Journal of the Royal Astronomical Society* 31 (1990): 333–57.

———. *The Victorian Amateur Astronomer: Independent Astronomical Research in Britain, 1820–1920*. New York: John Wiley and Sons, 1998.

Chappe d'Auteroche, Jean-Baptiste. *Voyage en Californie pour l'observation du passage de Vénus sur le disque du Soleil, le 3 juin 1769; contenant les observations de ce phénomène, et la description historique de la route de l'auteur à travers le Mexique* (Journey to California for the Observation of the

Transit of Venus upon the Disk of the Sun, June 3, 1769, with the Observations of this Phenomenon and a Historical Description of the Author's Route across Mexico). Paris: Chez Charles-Antoine Jombet, 1772.

———. *A Journey into Siberia Made by Order of the King of France*. London, 1774.

Chauvin, Michael E. "Astronomy in the Sandwich Islands: The 1874 Transit of Venus." *Hawaiian Journal of History* 27 (1993): 185–225.

Chou, B. Ralph. "Solar Filter Safety." *Sky & Telescope* 95, no. 2 (1998): 36–40.

Churchill, Winston. *History of the English-speaking Peoples*. Arranged for one volume by Henry Steele Commager. New York: Barnes and Noble, 1995.

Clerke, Agnes Mary. *A Popular History of Astronomy during the Nineteenth Century*. 4th ed. London: Adam and Charles Black, 1902.

Comte, Auguste. *Cours de philosophie positive* (Course on positive philosophy). Vol. 2. Paris: Bachelier, 1835.

Cook, Alan. *Edmond Halley: Charting the Heavens and the Seas*. Oxford: Clarendon Press, 1998.

Cook, James. *The Journals*. Prepared from the original manuscripts by J. C. Beaglehole for the Haklug Society, 1955–67; selected and edited by Philip Edwards. London and New York: Penguin, 1999.

Copernicus, Nicolaus. *The Commentarius of Copernicus*. In *Three Copernican Treatises*, translated by Edward Rosen. New York: Dover, 1959.

Corliss, William R. *The Moon and the Planets: A Catalog of Astronomical Anomalies*. Glen Arm, MD: The Sourcebook Project, 1985.

Dampier, William Cecil. *A Shorter History of Science*. London: Scientific Book Club, 1946.

Davis, Michael, and David Staup. "Shooting the Planets with Webcams." *Sky & Telescope* 105, no. 6 (2003): 117–22.

Delaunay, Charles Eugène. "Note sur la distance du Soleil à la Terre" (Note on the Distance from the Sun to the Earth). In *L'annuaire pour l'an 1816* (Almanac for the Year 1816), p. 94. Paris: Bureau des Longitudes, 1816.

Densmore, Dana. *Newton's Principia: The Central Argument*. Santa Fe, NM: Green Lion Press, 1995.

Dick, Steven J. *The Biological Universe*. Cambridge, UK: Cambridge University Press, 1996.

———. "The Naval Observatory and the American Transit of Venus Expeditions of 1874 and 1882." Chapter 7 in *Sky and Ocean Joined: U.S. Naval*

Observatory, 1830–2000, edited by Steven J. Dick. Cambridge: Cambridge University Press, 2003.

Dick, Steven J., Wayne Orchiston, and Tom Love. "Simon Newcomb, William Harkness, and the Nineteenth-century American Transit of Venus Expeditions." *Journal for the History of Astronomy* 29 (1998): 232–33.

Dobbs, Betty Jo Teeter, and Margaret C. Jacob. *Newton and the Culture of Newtonianism*. Amherst, NY: Humanity Books, 1995.

Drake, Stillman. *Discoveries and Opinions of Galileo*. New York: Doubleday, 1957.

Dreyer, J. L. E. *A History of Astronomy from Thales to Kepler*. 2nd ed. New York: Dover, 1906. Reprint 1953.

Dreyer, J. L. E., and H. H. Turner, eds. *History of the Royal Astronomical Society*. Vol. 1. *1820–1920*. London: Royal Astronomical Society, 1923.

Dumas, Jean Baptiste. "Passage de Vénus du 6 décembre 1882: Rapports préliminaires" (The Transit of Venus of December 6, 1882: Preliminary Reports). *Comptes Rendus des Séances de l'Académie des Sciences* 97 (1883): 353–443.

Duncombe, Raynor. "Personal Equation in Astronomy." *Popular Astronomy* 53 (1945): 2–13, 63–76, 110–21.

Encke, Johann Franz. *Die Entfernung der Sonne von der Erde aus dem Venusdurchgange von 1761* (The Distance from the Sun to the Earth Based on the Transit of Venus of 1761). Gotha: Becker'schen Buchhandlung, 1822.

———. *Der Venusdurchgange von 1769 als Forsetzung der Abhandlung über die Entfernung der Sonne von der Erde* (The Transit of Venus of 1769 as a Continuation of the Discussion of the Distance from the Sun to the Earth). Gotha: Becker'schen Buchhandlung, 1824.

Evans, Taliesin. "A Californian's Gift to Science." *Century Illustrated Monthly Magazine* 32, no. 1 (1886): 62.

Faye, Hervé. "Note sur les nouvelles tables des planètes intéreures" (Note on the New Tables of the Interior Planets). *Comptes Rendus des Séances de l'Académie des Sciences* 54 (1862): 630–39.

———. "Sur les passages de Vénus et la parallaxe du Soleil" (On the Transits of Venus and the Solar Parallax). *Comptes Rendus des Séances de l'Académie des Sciences* 68 (1869): 42–49.

———. "Mémoire de M. Faye sur l'observation photographique des passages de Vénus, et sur un appareil de M. Laussedat" (Memoir by M. Faye on the Photographic Observation of the Transits of Venus, and on an Apparatus of M. Laussedat). *Comptes Rendus des Séances de l'Académie des Sciences* 70 (1870): 541–48.

———. "Rôle que pourra prendre la France dans l'observation du prochain passage de Vénus sur le Soleil: Passage d'un discours prononcé par M. Faye, à l'overture de la séance publique annuelle du 25 novembre 1872" (The Role That France Will Be Able to Take in the Observation of the Next Transit of Venus upon the Sun: Passage of a Talk Given by M. Faye at the Commencement of the Annual Public Meeting of November 25, 1872). *Comptes Rendus des Séances de l'Académie des Sciences* 75 (1872): 1295–96.

———. "Association française pour l'avancement des sciences, Congrès de Lille, conférences publiques: Le prochain passage de Vénus sur Soleil" (French Association for the Advancement of the Sciences, Congress at Lille, Public Conferences: The Next Transit of Venus upon the Sun). *Revue Scientifique* 14, no. 16 (1874): 366.

Ferguson, James. *Astronomy Explained upon Sir Isaac Newton's Principles, and Made Easy to Those Who Have Not Studied Mathematics*. 6th ed. London, 1778.

Fernie, Donald. *The Whisper and the Vision: The Voyages of the Astronomers*. Toronto: Clarke, Irwin, 1976.

Flammarion, Camille. *Les terres du ciel: Description astronomique, physique, climatologique, géographique des planètes qui gravitent avec la terre autour du soleil et de l'état probable de la vie à leur surface* (The Worlds of the Sky: An Astronomical, Physical, Climatological, and Geographic Description of the Planets That Revolve with the Earth around the Sun and the Probable State of Life on Their Surface). Paris: Didier, 1877.

———. *Popular Astronomy*. 1880. Translated by J. Ellard Gore. New York: Appleton, 1907.

———. *Dreams of an Astronomer*. Translated by E. E. Fournier D'Albe. New York: D. Appleton, 1923.

Forbes, E. G., L. Murdin, and F. Willmoth, eds. *The Correspondence of John Flamsteed*. Vol. 1. Bristol, UK, and Philadelphia: Institute of Physics Publishing, 1995.

Forbes, George. *The Transit of Venus*. London and New York: Macmillan Nature Series, 1874.

———. *David Gill, Man and Astronomer: Memoirs of Sir David Gill, K.C.B., H.M. Astronomer (1879–1907) at the Cape of Good Hope*. London: Watts, 1909.

Galle, Johann Gottfried. "On the Observations of Flora, Made with a View of Determining the Solar Parallax." *Monthly Notices of the Royal Astronomical Society* 35, no. 1 (1874): 11–12.

Gassendi, Pierre. *Mercurius in Sole Visa et Venus Invisa* (Mercury Seen on the Sun, and Venus Unseen). 1632. In *Opera Omnia*. Vol. 4. Lyons, 1658.

Gill, Isobel. *Six Months in Ascension*. London: John Murray, 1878.

Gingerich, Owen. *The Eye of Heaven: Ptolemy, Copernicus, Kepler*. New York: American Institute of Physics, 1993.

Grant, Robert. *History of Physical Astronomy from the Earliest Ages to the Middle of the Nineteenth Century: Comprehending a Detailed Account of the Establishment of the Theory of Gravitation by Newton, and Its Development by His Successors; with an Exposition of the Progress of Research on All the Other Subjects of Celestial Physics*. London: R. Baldwin, 1852.

Green, Charles, and James Cook. "Observations Made, by Appointure of the Royal Society, at King George's Island in the South Seas." *Philosophical Transactions of the Royal Society* 61 (1771): 397–421.

Grinspoon, David Harry. *Venus Revealed: A New Look below the Clouds of Our Mysterious Twin Planet*. Reading, MA: Addison-Wesley, 1997.

Halley, Edmond. "Doctor Halley's Dissertation on the Method of Finding the Sun's Parallax and Distance from the Earth, by the Transit of Venus over the Sun's Disc, June the 6th, 1761." Translated by James Ferguson. In David Sellers, *The Transit of Venus: The Quest to Find the True Distance of the Sun*, pp. 201–17. Leeds, UK: MagaVelda Press, 2001.

Hansen, Peter Andreas. "On the Value of the Solar Parallax Deduced from the Lunar Tables." *Monthly Notices of the Royal Astronomical Society* 23, no. 8 (1863): 242–43.

———. "Calculation of the Sun's Parallax from the Lunar Theory." *Monthly Notices of the Royal Astronomical Society* 24, no. 1 (1863): 8–12.

Harkness, William. "The Solar Parallax and Its Related Constants, Including the Figure and Density of the Earth." Appendix III in *Washington Observations for 1885*. Washington, DC: Government Printing Office, 1891.

Hindle, Brooke. *David Rittenhouse*. Princeton, NJ: Princeton University Press, 1964.

Hirshfeld, Alan W. *Parallax: The Race to Measure the Cosmos*. New York: W. H. Freeman, 2001.

Hirst, W. "An Observation of the Same Transit of Venus over the Sun, June 6, 1761, at Madras." *Philosophical Transactions of the Royal Society* 52, no. 62 (1761): 396.

Hoag, Arthur A. "Aristarchos Revisited." *Griffith Observer* 54, no. 8 (1990): 10–18.

Hogg, Helen S. "Out of Old Books: Le Gentil and the Transits of Venus, 1761 and 1769." *Journal of the Royal Astronomical Society of Canada* 45 (1951): 37–44, 89–93, 127–34, 173–78.

Holborn, Hajo. *A History of Modern Germany*. Vol. 1, *The Reformation*. Princeton, NJ: Princeton University Press, 1959.

Hooijberg, Maarten. *Practical Geodesy Using Computers*. Berlin: Springer Verlag, 1997.

Hornsby, T. "The Quantity of the Sun's Parallax, as Deduced from the Observations of the Transit of Venus, on June 3, 1769." *Philosophical Transactions of the Royal Society* 61 (1771): 574–79.

———. "On the Transit of Venus in 1769." *The Philosophical Transactions of the Royal Society of London, from Their Commencement in 1665, to the Year 1800* 12 (1809): 265–74.

Horrocks, Jeremiah. *Venus in Sole Visa* (Venus in the Face of the Sun). Appended to Arundell Blount Whatton, *Memoirs of the Life and Labours of the Rev. Jeremiah Horrox*. London, 1859.

Hough, Richard. *Captain James Cook: A Biography*. London: Hodder and Stroughton, 1994.

Hughill, Peter J. *Global Communications Since 1844*. Baltimore: Johns Hopkins University Press, 1999.

Huygens, Christiaan. *Cosmotheoros: Sive de Terris Coelestis Earumque Ornatu Conjecturae*. The Hague, 1698. English translation, Celestial Worlds Discover'd; or, Conjectures Concerning the Inhabitants, Plants and Productions of the Worlds in the Planets. Translated by Timothy Childe. 1698. Reprint, London: Cass, 1968.

Janiczek, Paul M. "Remarks on the Transit of Venus Expedition of 1874." In *Sky with Ocean Joined: Proceedings of the Sesquicentennial Symposia of the U.S. Naval Observatory, December 5 and 8, 1980*, edited by Steven J. Dick and LeRoy E. Doggett, pp. 52–73. Washington, DC: U.S. Naval Observatory, 1983.

Janiczek, Paul M., and LeRoy E. Houchins. "Transits of Venus and the American Expedition of 1874." *Sky & Telescope* 48 (December 1974): 366–71.

Janssen, Pierre Jules César. "Passage de Vénus: Méthode pour obtenir photographiquement l'instant des contacts, avec les circonstances physiques qu'ils présentent" (Transit of Venus: A Method for Photographically Obtaining the Instant of the Contacts, along with the Physical Circumstances They Present). *Comptes Rendus des Séances de l'Académie des Sciences* 76 (1873): 677–79.

———. "Note sur l'observation du passage de la planète Vénus sur le Soleil" (Note on the Observation of the Transit of the Planet Venus upon the Sun). *Comptes Rendus des Séances de l'Académie des Sciences* 96 (1883): 288–92.

Jeans, James. *Physics and Philosophy*. London: Cambridge University Press, 1943.

Jevons, W. Stanley. *The Principles of Science: A Treatise on Logic and Scientific Method*. 3rd ed. London: Macmillan, 1879.

Jones, Harold Spencer. "The Distance of the Sun." *Endeavour* 1, no. 1 (1942): 9–17.

Kelvin, Lord. *See* Thomson, William.

Kollerstrom, Nicholas. *Newton's Forgotten Lunar Theory: His Contribution to the Quest for Longitude*. Santa Fe, NM: Green Lion Press, 2000.

———. "The Path of Halley's Comet, and Newton's Late Apprehension of the Law of Gravity." *Annals of Science* 56 (1999): 331–56.

Koyré, A. *The Astronomical Revolution: Copernicus–Kepler–Borelli*. Translated by R. E. W. Maddison. New York: Dover, 1992 reprint of 1973 ed.

Langley, S. P. "Observation of the Transit of Venus, 1882, December 6, Made at the Alleghany Observatory." *Monthly Notices of the Royal Astronomical Society* 43 (1883): 72–73.

Lankford, John. "The Impact of Photography on Astronomy." Chap. 2 in *The General History of Astronomy*, edited by O. Gingerich, Vol. 4, *Astrophysics and Twentieth Century Astronomy to 1950: Part A*, pp. 16–39. Cambridge: Cambridge University Press, 1984.

———. "Photography and the Nineteenth-century Transits of Venus." *Technology and Culture* 28, no. 3 (1987): 648–57.

Le Gentil, G. "Memoire de M. Le Gentil . . . au subjet de l'Observation qu'il va faire, par ordre du Roi, dans les Indes Orientale, du prochain passage de Venus pardevant le Soleil" (Memoir of M. Le Gentil . . . about the Observation He Will Make, by Order of the King, in the East Indies, of the Next Transit of Venus in Front of the Sun). *Journal des Sçavans* (March 1760): 132–42. English translation, ". . . Le Gentil and the Transits of Venus. . . ." Translated in part by Helen S. Hogg (1951).

Leverrier, Urbain Jean Joseph. "Parallaxe du Soleil tirée de l'equation parallactique de la Lune" (The Solar Parallax Derived from the Parallactic Equation of the Moon). *Annales de l'Observatoire de Paris* 4 (1858): 101.

Ley, Willy. *Watchers of the Skies*. New York: Viking Press, 1966.

Lightman, Bernard, ed. *Victorian Science in Context*. Chicago: University of Chicago Press, 1997.

Lindsay, Lord. "Observations of the Transit of Venus at Mauritius." *Monthly Notices of the Royal Astronomical Society* 35 (1875): 131–32.

Link, F. "Allongement des cornes de Vénus" (Extension of the Horns of Venus). *Bulletin of the Astronomical Institutes of Czechoslovakia* 1, no. 6 (1949): 77–81.

———. *Eclipse Phenomena in Astronomy*. New York: Springer Verlag, 1969.

Longsworth, Polly. *Austin and Mabel: The Amherst Love Affair and Love Letters of Austin Dickinson and Mabel Loomis Todd*. Amherst: University of Massachusetts Press, 1984.

Manning, Thomas G. *U.S. Coast Survey vs. Naval Hydrographic Office: A Nineteenth-century Rivalry in Science and Politics*. Tuscaloosa: University of Alabama Press, 1988.

Manuel, Frank E. *A Portrait of Isaac Newton*. Cambridge, MA: Harvard University Press, Belknap Press, 1968.

Maor, Eli. *June 8, 2004: Venus in Transit*. Princeton, NJ: Princeton University Press, 2000.

Marshall, Lorna J. *Nyae Nyae !Kung Beliefs and Rites*. Cambridge, MA: Peabody Museum of Archaeology and Ethnology, Harvard University, 1999.

Maskelyne, Nevil. *Astronomical Observations Made at the Royal Observatory at Greenwich*. London, 1799–1800.

Maunder, Michael, and Patrick Moore. *Transit: When Planets Cross the Sun*. London: Springer Verlag, 2000.

McKim, Richard. *British Astronomical Association Mars Section Circular*. British Astronomical Association Mars Section, December 2001.

Meadows, A. J. "The Transit of Venus in 1874." *Nature* 250 (August 30, 1974): 749–52.

Meeus, Jean. *Transits*. Richmond, VA: Willmann-Bell, 1989.

Menshutkin, Boris N. *Russia's Lomonosov*. Princeton, NJ: Princeton University Press, 1952.

Merton, Gerald. "Photography and the Amateur Astronomer." *Journal of the British Astronomical Association* 63 (1953): 7–30.

Ministère de l'instruction publique et des beaux-arts (France). *Conférence internationale du passage de Vénus: Procès verbaux* (International Conference on the Transit of Venus: Verbal Proceedings). Paris: Imprimerie Nationale, 1881.

Mourão, Ronaldo Rogério de Freitas. *Os eclipses, da superstição à previsão mathmática* (Eclipses: From Superstition to Mathematical Forecast). São Leopoldo, Brazil: Ed. Unisinos, 1993.

Mulhall, Michael G. *Balance-sheet of the World for Ten Years, 1870–1880.* London: Edward Stanford, 1881.

Nautical Almanac Office (Great Britain). *Explanatory Supplement to the Astronomical Ephemeris and the American Ephemeris and Nautical Almanac.* London: Her Majesty's Stationery Office, 1961.

Newcomb, Simon. "Investigation of the Distance of the Sun, and the Elements Which Depend on It, from the Observation of Mars, Made during the Opposition of 1862, and from Other Sources." Appendix H in *Washington Observations, 1865*. Washington, DC: Government Printing Office, 1867.

———. "On Hell's Alleged Falsification of His Observations of the Transit of Venus in 1769." *Monthly Notices of the Royal Astronomical Society* 42 (1883): 371–81.

———. *The Reminiscences of an Astronomer*. Boston: Harper and Brothers, 1903.

Nunis, Doyce B., Jr., ed. *The 1769 Transit of Venus: The Baja California Observations of Jean-Bapiste Chappe d'Auteroche, Vicente de Doz, and Joaquín Velázquez Cárdenas de León*. Los Angeles: Natural History Museum of Los Angeles County, 1982.

Orchiston, Wayne. *James Cook and the 1769 Transit of Mercury*. Carter Observatory Information Sheet 3. Wellington, New Zealand: Carter Observatory, 1994.

———. *Nautical Astronomy in New Zealand: The Voyages of James Cook.* Wellington, New Zealand: Carter Observatory, 1998.

Orchiston, Wayne, and A. Buchanan. "Illuminating Incidents in Antipodean Astronomy: Campbell Town and the 1874 Transit of Venus." *Australian Journal of Astronomy* 5 (1993): 11–31.

Orchiston, Wayne, and Derek Howse. "From Transit of Venus to Teaching Navigation: The Work of William Wales." *Astronomy and Geophysics* 39, no. 6 (1998): 21–24.

Orchiston, Wayne, Tom Love, and Steven J. Dick. "Refining the Astronomical Unit: Queenstown and the 1874 Transit of Venus." *Journal of Astronomical History and Heritage* 3, no. 1 (2000): 23–44.

Osterbrock, Donald E., John E. Gustafson, and W. J. Shiloh Unruh. *Eye on the Sky: Lick Observatory's First Century*. Berkeley and Los Angeles: University of California Press, 1988.

Pang, Alex Soojung-Kim. *Empire and the Sun: Victorian Solar Eclipse Expeditions*. Stanford, CA: Stanford University Press, 2000.

Pannekoek, Anton. *History of Astronomy*. New York: Dover, 1961, reprint 1989.

Parkinson, Sydney. *A Journal of a Voyage to the South Seas in His Majesty's Ship, the* Endeavour. London, 1784.

Pigatto, L., and V. Zanini. "Spectroscopic Observations of the 1874 Transit of Venus: The Italian Party at Muddapur, Eastern India." *Journal of Astronomical History and Heritage* 4 (2001): 43–58.

Ponko, Vincent, Jr. "Nineteenth-century Science in New Mexico: The 1882 Transit of Venus Observations at Cerro Roblero." *Journal of the West* 33 (1994): 44–51.

———. "Cedar Key, Florida, and the Transit of Venus: The 1882 Site Observations." *Gulf Coast Historical Review* 10, no. 2 (1995): 47–66.

Powalky, C. "Beiträge zu einer vollständigeren Beurtheilung der Venusdurchgänge und Ermittelung einiger genauerer Resultate aus denselben" (Contributions to a More Complete Analysis of the Transits of Venus and a Determination of More Exact Results from Them). *Astronomische Nachrichten* 76, no. 1811–12 (1870): 161–92.

Prescott, William Hickley. *History of the Conquest of Mexico and History of the Conquest of Peru*. New York: Cooper Square Press, 2000. Unabridged reprints of 1843 and 1847 editions.

Prince, Charles Leeson. "Note on the Transit of Venus, 1882, Dec. 6." *Monthly Notices of the Royal Astronomical Society* 43 (1883): 278.

Proctor, Richard A. *Transits of Venus: A Popular Account of Past and Coming Transits from the First Observed by Horrocks* A.D. *1639 to the Transit of* A.D. *2012*. London: Longmans, Green, 1874.

———. *The Universe and the Coming Transits*. London: Longmans, Green, 1874.

———. *Old and New Astronomy*. London: Longmans, Green, 1892.

Pynchon, Thomas. *Mason and Dixon*. New York: Henry Holt, 1997.

Ratliff, Floyd. *Mach Bands: Quantitative Studies on Neural Networks in the Retina*. San Francisco: Holden-Day, 1965.

Rawlins, Dennis. "Astronomy vs. Astrology: The Ancient Conflict." *Queen's Quarterly* 91, no. 4 (1984): 969–89.

Rees, John Krom. "Observations of the Transit of Venus, December 6, 1882." *Annals of the New York Academy of Sciences* 2 (1882): 384–90.

Rienits, Rex, and Thea Rienits. *The Voyages of Captain Cook*. London: Paul Hamlyn, 1968.

Royal Astronomical Society. "Report of the Council to the Sixty-third Annual General Meeting." Parts 1 and 2. *Monthly Notices of the Royal Astronomical Society* 43 (1883): 189–202, 211–15.

———. "Report of the Council to the Sixty-fourth Annual General

Meeting." *Monthly Notices of the Royal Astronomical Society* 44 (1883): 182–84.

Rudloff, Willy. *World-Climates: With Tables of Climatic Data and Practical Suggestions*. Stuttgart: Wissenschaftliche Verlagsgesellschaft, 1981.

Ruffner, James A., and Frank E. Bair, eds. *The Weather Almanac*. 5th ed. Detroit: Gale Research, 1987.

Rush, Benjamin. *An Eulogium Intended to Perpetuate the Memory of David Rittenhouse, Late President of the American Geophysical Society*. Philadelphia: Ormrod and Conrad, 1796/97.

Russell, Bertrand. *History of Western Philosophy*. 2nd ed. London: George Allen and Unwin, 1961.

Russell, Henry Norris. "The Atmosphere of Venus." *Astrophysical Journal* 9 (1899): 284–99.

Russell, Henry Norris, Raymond Smith Dugan, and John Quincy Stewart. *Astronomy*. A revision of Young's Manual of Astronomy. New York: Ginn, 1926.

Sagan, Carl, and Ann Druyan. *Comet*. New York: Random House, 1985.

Sarton, George. *A History of Science*. 2 vols. New York: W. W. Norton, 1970.

Schaefer, Bradley E. "The Transit of Venus and the Notorious Black Drop Effect." *Journal for the History of Astronomy* 32 (2001): 325–36.

Schaffer, Simon. "Astronomers Mark Time: Discipline and the Personal Equation." *Science in Context* 2, no. 1 (1988): 115–45.

———. "Metrology, Metrication, and Victorian Values." In *Victorian Science in Context*, edited by Bernard Lightman, pp. 438–74. Chicago: University of Chicago Press, 1997.

Schroeter, Johann Hieronymus. *Selenotopographische Fragmente*. 2 vols. Helmstedt: Johann Georg Rosenbusch, 1791.

Schweiger-Lerchenfeld, Armand von. *Atlas der Himmelskunde auf Grundlage der Ergebnisse der coelestichen Photographie* (Astronomical Atlas Based on the Results of Celestial Photography). Vienna, Pest, and Leipzig: A. Hartlebens' Verlag, 1898.

Seidelman, P. Kenneth, ed. *Explanatory Supplement to the* Astronomical Almanac. Mill Valley, CA: University Science Books, 1992.

Sellers, David. *The Transit of Venus: The Quest to Find the True Distance of the Sun*. Leeds, UK: MagaVelda Press, 2001.

Sheehan, William, and Thomas Dobbins. "The Spokes of Venus: An Illusion Explained." *Journal for the History of Astronomy* 34 (2003): 53–63.

Smith, James R. *Introduction to Geodesy: The History and Concepts of Modern Geodesy*. New York: Wiley, 1997.

Smyth, A. H., ed. *The Writings of Benjamin Franklin*. 10 vols. New York: Macmillan, 1905–1907.

Sobel, Dava. *Longitude*. New York: Walker, 1995.

Standage, Tom. *The Victorian Internet*. New York: Walker, 1998.

Stone, Edward James. "A Rediscussion of the Observations of the Transit of Venus, 1769." *Monthly Notices of the Royal Astronomical Society* 28, no. 6 (1868): 255–66.

———. "On the Telescopic Observations of the Transit of Venus 1874, and the Conclusions to be Deduced from These Observations." *Monthly Notices of the Royal Astronomical Society* 38, no. 5 (1878): 279–95.

Strommel, Henry. *Lost Islands*. Vancouver: University of British Columbia Press, 1984.

Struve, Otto William. "List of Stations Selected for Observation of the Transit of Venus by Russian Astronomers." *Monthly Notices of the Royal Astronomical Society* 33, no. 7 (1873): 415–17.

Stryer, Lubert. *Principles of Biochemistry*. 2nd ed. San Francisco: W. H. Freeman, 1981.

Stukeley, William. *Memoirs of Sir Isaac Newton's Life*. Edited by A. Hastings White. London: Taylor and Francis, 1936.

Sydney Observatory. *Observations of the Transit of Venus, 9 December, 1874: Made at Stations in New South Wales, Illustrated with Photographs and Drawings*. Sydney: C. Potter, Government Printer, 1892.

Thomson, William (Lord Kelvin). *Popular Lectures and Addresses*. 3 vols. London: Macmillan, 1891–94.

Todd, David P. "On the Observations of the Transit of Venus, 1882, December 5–6, Made at the Lick Observatory, Mount Hamilton, California." *Monthly Notices of the Royal Astronomical Society* 43 (1883): 273–76.

———. "Professor Todd's Own Story of the Mars Expedition." *The Cosmopolitan*, January 1908.

Todhunter, I. *A History of Mathematical Theories of Attraction and the Figure of the Earth*. 2 vols. New York: Dover, 1962.

Trevelyan, George Macaulay. *A Shortened History of England*. New York: Longmans, Green, 1942; reprint, Penguin, 1987.

Tupman, George Lyon. "On the Mean Solar Parallax as Derived from the Observations of the Transit of Venus, 1874." *Monthly Notices of the Royal Astronomical Society* 38, no. 8 (1878): 429–57, 513.

Turnbull, H. W., ed. *The Correspondence of Isaac Newton*. 2 vols. Cambridge: Cambridge University Press, 1960.

United States Coast and Geodetic Survey. *Report of the Superintendent of the U.S. Coast and Geodetic Survey Showing the Progress of the Work during the Fiscal Year Ending with June, 1883*. Washington, DC: Government Printing Office, 1884.

United States Naval Observatory, *Nautical Almanac* Office. *The Astronomical Almanac for the Year 2004*. Washington, DC: U.S. Government Printing Office, 2002.

United States Transit of Venus Commission. *Instructions for Observing the Transit of Venus, December 6, 1882, Prepared by the Commission, Authorized by Congress, and Printed for the Use of the Observing Parties by Authority of the Hon. Secretary of the Navy*. Washington, DC: Government Printing Office, 1882.

Van Helden, Albert. "The Importance of the Transit of Mercury of 1631." *Journal for the History of Astronomy* 7 (1976): 1–10.

———. *Measuring the Universe: Cosmic Dimensions from Aristarchus to Halley*. Chicago: University of Chicago Press, 1985.

Vanícek, Petr, and Edward J. Krakiwsky. *Geodesy: The Concepts*. Amsterdam: North-Holland, 1986.

Vaucouleurs, Gérard Henri de. *Astronomical Photography: From the Daguerreotype to the Electron Camera*. New York: Macmillan, 1961.

Vega, L. Eduardo. "Homemade Mylar Filters." *Sky & Telescope* 95, no. 2 (1998): 39.

Vogel, Hermann W. "Beobachtungen des Venusdurchgangs am 6. December 1882, angestellt auf dem Astrophysikalischen Observatorium zu Potsdam" (Observations of the Transit of Venus on December 6, 1882, Made at the Astrophysical Observatory of Potsdam). *Astronomische Nachrichten* 104 (1883): 257–62.

Wales, William. "Journal of a Voyage, Made by Order of the Royal Society, to Churchill River, on the North-west Coast of Hudson's Bay; of Thirteen Months' Residence in That Country, and of the Voyage Back to England, in the Years 1768 and 1769." *Philosophical Transactions of the Royal Society* 60, no. 13 (1770): 100–36.

Wales, William, and Joseph Dymond. "Astronomical Observations Made by Order of the Royal Society, at Prince of Wales' Fort, on the North-west Coast of Hudson's Bay." *Philosophical Transactions of the Royal Society* 59, no. 65 (1769): 467–488.

Weinberg, Boris. "Über den wahrscheinlichsten Wert der Sonnenparallaxe nach den bisherigen astronomischen Bestimmungen" (On the Most Probable Value of the Solar Parallax Based on the Previous Astronomical Determinations). *Astronomische Nachrichten* 163, no. 3866 (1903): 17–30.

———. "Endgültige Ausgleichung der wahrcheinlichsten Werte der Sonnenparallaxe, der Aberrationskonstante, der Lichtgleichung, der Verbreitungsgeschwindigkeit der Störungen im Äther nach den bisherigen Messungen" (Final Adjustment of the Most Probable Value of the Solar Parallax, the Constant of Aberration, the Equation of Light, the Speed of Propagation of Disturbances in the Ether, Based on Previous Measurements). *Astronomische Nachrichten* 165, no. 3945 (1904): 133–42.

Wells, Herbert George. "Atom: Boon or Doom?" *Daily Herald*, August 9, 1945.

Westfall, John. "The 1769 Transit of Venus Expedition to San José del Cabo." In *Research Amateur Astronomy*, edited by Stephen Edberg, pp. 234–42. Astronomical Society of the Pacific Conference Series 33. San Francisco: Astronomical Society of the Pacific, 1992.

Westfall, Richard. *The Life of Isaac Newton*. Cambridge: Cambridge University Press, 1993.

———. *Never at Rest: A Biography of Isaac Newton*. Cambridge: Cambridge University Press, 1980.

Whatton, Arundell Blount. *Memoir of the Life and Labours of the Rev. Jeremiah Horrox. . . .* London, 1859.

White, Norman. *Gerard Manley Hopkins: A Literary Biography*. Oxford: Clarendon Press, 1992.

Whiteside, D. T. "Before the *Principia*." *Journal for the History of Astronomy* 1 (1970): 5–19.

Willmann, Axel D. *Catalog of Naked-Eye Sunspot Observations and Large Sunspots*. Göttingen, Germany: University Observatory, 1997.

Woolf, Harry. *The Transits of Venus*. Princeton, NJ: Princeton University Press, 1959.

Wright, Helen. *James Lick's Monument: The Saga of Captain Richard Floyd and the Building of the Lick Observatory*. Cambridge: Cambridge University Press, 1987.

Xu, Zhentao, David W. Pankenier, and Yaotiao Jiang. *East Asian Archaeoastronomy: Historical Records of Astronomical Observations of China, Japan, and Korea*. Amsterdam: Gordon and Breach Science Library, 2000.

Yeomans, Donald K. *Comets: A Chronological History of Observation, Science, Myth, and Folklore*. New York: John Wiley and Sons, 1991.

Young, A. T., and L. G. Young. "Observing Venus near the Sun." *Sky & Telescope* 43, no. 3 (1972): 140–44.

INDEX

Note: References to figures or tables are given in italics.

Abney, William de Wiveleslie, 216
Adams, John Couch, 228
Airy, Sir George Biddell
  appointed Astronomer Royal, 225
  as director of Cambridge Observatory, 225
  initially favors method of Delisle for 1874 transit of Venus, 228, 230
  and Neptune controversy, 228
  orders telescopes for 1874 British parties, 243
  plans for 1874 transit of Venus, 225, 235
  portrait, *228*
  at Trinity College, 225
  views attacked by Proctor, 232
alignments, celestial
  conjunctions, 35
  definition, 35, *36*
  eclipses, 35
  eclipses, lunar, 38–39, *39*
  eclipses, planetary satellites, 40, *40*
  eclipses, solar, 37–38, *38*, *39*, 269, *269*, 270
  occultations, 35
  transit contacts, 131, *132*
  transits, 35. *See also* Venus, transits of
  triple conjunction of 7–6 BCE, 36–37

Ammisaduqua (Babylonian king), 44
André, Charles, 274–75
Apollonius of Perga, 52
Archer, Frederick Scott, 216
Aristarchus of Samos, 28
  distance to moon, 51, *51*
  heliocentric system, 50–53
Aristotle, 26
Ascension Island, 264–65
astronomical unit (distance to sun)
  definition, 14, 29
  "derived constant," definition of, 299–300
  estimated by Horrocks, 86
  measured by Encke, 204
  measured by Eros approaches, 298–99
  measured by Gill, by observing Mars, 265, 276
  measured by timing Galilean satellite eclipses, *227*, 245
  measured by Venus radio ranging, 299
Auwers, Georg Friedrich Julius Arthur von, 240

Banks, Joseph, 171, 173
Barnard, Edward Emerson, 279
Baume-Pluvinel, Comte Aymar de la, 284
Bayley, William, 163–64, 186
Bergmann, Torbern Olaf, 153
Bernardières, Octave Marie Gabriel Joachim de, 284–86
Berthoud, Ferdinand, 217
Bevis, John, 163
Bianchini, Francesco, 138–40
  observations of Venus, 138–40, *140*
Bigg-Wither, Archibald Cuthbert, 260
Bougainville, Louis Antoine de, 169–70
Brahe, Tycho, 59–60, 62
Brodie, Frederick, 281
bubonic plague, London (1665–66), 95
Buchan, Alexander, 171
Bunsen, Robert, 212
Burton, Charles Edward, 263

Callandreau, Pierre Jean Octave, 284
Cassegrain, Guillaume, 148
Cassini, Giovanni Domenico
  first director, Royal Observatory (Paris), 101, *101*, 136–37
  observes Venus, 101–102, *102*
  measures solar parallax, 102–104, *103*, *104*
Cassini, Jacques, 102, 136–37, 140
Cassini, Jacques Dominique de, 136–37
Cassini Thury, César François de, 136–37, 141
  observes 1761 transit of Venus, 146–47, 153
Chappe d'Auteroche, Abbé Jean-Baptiste, 145–46, *146*
  death of, 196–97, *197*
  determines position of San José del Cabo, 196, 197
  encounters epidemic at San José del Cabo, 196
  observatory at San José del Cabo, 195–96
  observes 1761 transit of Venus, 153, 154

observes 1769 transit of Venus, 196
travels to San José del Cabo for 1769 transit of Venus, 195
travels to Tobolsk for 1761 transit of Venus, 145–46, 149–50
Chapuis, Eugène, 284, *285*
Charles I (king of England), 76, 87–88, 95
Charles II (king of England, Charles Stuart), 95, 99, 120
founds Royal Observatory at Greenwich, 105
Civil War, English, 90, 95
Clarke, Alexander Ross, 218, 224
comets, 112–13, 117, 135, 141–42, 269, *269*
Compte, Auguste, 212
Cook, James
early maritime experience, 166
portrait, *167*
summary of voyages, 162
Cook, James, first voyage of (1768–71)
anchors at Matavai Bay, Tahiti, 173–74
arrives back at England 1771, 183
chosen to lead expedition, 166
crew suffers dysentery epidemic, 183
Fort Venus, *177*, 179
given command of the *Endeavour*, 170
instruments for observation, 179, *180*
observes 1769 Nov 09 transit of Mercury, 183
relations with Tahitians, 176–79
scurvy prevention, 174–75
seeks Great Southern Continent, 169, 183
transit of Venus 1769, contacts timed by Cook and Green, 181
transit of Venus 1769, drawings by Cook and Green, *182*
transit of Venus 1769, observation stations at Moorea and Taaupiri, 180
voyage from Plymouth to Tahiti, 1768–69, 171–73
weather conditions for transit of Venus, 1769, 180–81
Copeland, Ralph, 261
Copernicus, Nicolaus
Copernican (heliocentric) System, 48, 49–50, *53*, 56
De Revolutionibus Orbium Caelestium, 49, 56, 72–73
Cortés, Hernando, 46–47
Counterreformation, Roman Catholic, 60–61
Crabtree, William, 78, 80, 82, 83, 89, 90
Cysat, Jean-Baptiste (Cysatus), 70

D'Abbadie, Antoine Thompson, 284
Daguerre, Louis Jacques Mandé, 215–16
Dalrymple, Alexander
experience in East Indies, 165
recommended for South Pacific expedition for 1769 transit of Venus, 165–66
dating convention, 16
Davidson, George, 255, 280, *281*
de la Rue, Warren, 216

Delisle, Joseph-Nicolas, 136–38, *136*
  advocates observation of transit of Mercury 1753 May 06, 141
  appointed *Astronome de la Marine*, 141
  appointed chair of mathematics, *Collège Royal*, 137
  Delislean Poles, 138
  method of, 137–38, 228
  moves to Russia, 138
  prepares *mappemonde* for transit of Venus of 1761, 144, *144*
Descartes, René, 97, 111
Diaz Covarrubias, Don Francisco, 255
Dickinson, Austin, 287, 291
Dickinson, Emily, 49, 267, 287, 293
Dixon, Jeremiah
  attempts to cancel 1761 transit expedition, 149
  joins C. Mason to observe 1761 transit of Venus, 146
  observes 1761 transit of Venus, 151, 153
  travels to Norway to observe 1769 transit of Venus, 165, 186
Dollond, John, 148, 208, 211
Dom Pedro II (Brazilian emperor), 284
Draper, John William, 216
Dumas, Jean Baptiste, 270
Dymond, James. *See also* Wales, William
  accompanies W. Wales to Hudson Bay to observe 1769 transit of Venus, 189
  assigned to Hudson Bay to observe 1769 transit of Venus, 165
Dynamic Time (DT, "atomic time"), 337

earth
  geoid (spheroid, ellipsoid) model, 218
  radius, measurements of, *27*
  size measurement, 25–27
  transits sun, as seen from Mars, 302, *302*
eclipse phenomena. *See* alignments, celestial
Egypt, ancient
  Akhenaten (pharaoh), 32–34, *33*
  Akhetaten (Tell el-Amarna, capital), 33
  sun worship, 32–35
  Tutankhamen (pharaoh), 32–33, *33*
Ellery, Robert Lewis John, 258
Encke, Johann Franz, 203–204, *204*
  recomputes 1761/69 transit of Venus parallaxes, 154, 204
ephemerides, national
  American Ephemeris, 223, 276
  Nautical Almanac (British), 223
Eratosthenes of Cyrene, 26
Europe, overseas expansion, 120–21
  extent of colonial empires, 142, 205–206

Fabricius, Johann (David Goldschmidt), 65–66
Faye, Hervé, 225, 241, 245
Ferdinand II (Ferdinand of Styria), 67–68
Ferry, Jules, 274
Fizeau, Armand Hippolyte Louis, 216, 229

Flammarion, Camille
  Les Terres du Ciel, 213, *214*
  life on Venus, speculations, 213–14
Flamsteed, John
  biography by Francis Baily, 118
  comet of 1680, 112
  feud with Hooke, 100, 105
  as first Astronomer Royal, 100, 105, *105*
  Historia Coelestis Britannicae (star catalog), 105–106, 127
  measures solar parallax, 104
Floyd, Richard S., 289, 290
Fontana, Francesco, 65
Forbes, George, 254
Foucault, Léon, 216, 224, 245
Fouchy, Grandjean de, 154
Franklin, Benjamin, 189
Fraser, Thomas, 289
Fraunhofer, Joseph von, 211

galaxies, distances to, 301
Galilei, Galileo, 64–65, *65*
  arrest of, 72–73, 82
  sunspot observations, 66–67
Galle, Johann Gottfried, 224, 228, 276
Gallet, Jean Charles, 109–110
Gascoigne, William, 90, 211
Gassendi, Pierre, 70–72, *71*, 87
  observes 1631 transit of Mercury, 70–71, *71*
Gellibrand, Henry, 79
George III (king of England), 143
Gill, Isobel, 264–65
Gill, Sir David
  astronomer on 1874 transit of Venus Mauritius expedition, 240, 261
  Astronomer Royal at Cape of Good Hope, 240
  determines solar parallax using Mars observations, 264–65, 278
  portrait, *241*
Gordon, Charles George, 240–41, 268
great fire, London (1666), 95
Great Southern Continent
  Cook instructed to seek, 169, 182–83
  legend of, 165, 167
Green, Charles, 170–71, 180, 181
Gregory, James
  designs Gregorian reflecting telescope, 147
  proposes use of planetary transits to determine solar parallax, 127–28
Gyre, Bouquet de la, 257

Hall, Asaph, 255
Halley, Edmond, *126*, *128*
  Astronomer Royal, 135
  birth in Hackney, 106
  childhood, 106–107
  comet Halley (Halley's Comet), 112–13, 117, 135, 141–42
  education at Queens College, 107
  Halleyan and Cishalleyan Poles, 130–31
  method of, 110–11, 122, 126–27, 128–35, *130*, 135
  observes transit of Mercury 1677 Nov 07, 110–11, 126
  relations with Flamsteed, 107, 128
  relations with Newton, 113–14, 116–17

secretary of Royal Society, 116, 127
voyage in *Paramour*, 127, 166
Hansen, Peter Andreas, 224
Hardy, Thomas, 269–70
Two on a Tower, 269, 270
Harkness, William, 257–58
member, United States Transit of Venus Commission, 229
Harriot, Thomas, 64
Harrison, John, 217
Hell, Father Maximilian
controversy over 1769 observations, 187–88
observes 1761 transit of Venus, 146–47, 153
observes 1769 transit of Venus, 186–87
portrait, *187*
Henry, Joseph, 229, 269
Heraclides of Pontus, 50
Herschel, William, 212, 213
Hevelius (Johann Hewelcke, Hevel), 90
Hipparchus of Rhodes
distance to moon, 28
geocentric system, 52
Hirst, William, 153
Hooke, Robert, 99, 100
theory of gravitation, 111–12, 113
Hornsby, Thomas, 162
Horrocks (Horrox), Jeremiah, 77–87, 88–91, *89*, 93
birth near Toxteth, 77, *77*
computes motion of Venus, 81–84
education at Cambridge, 78–79
measures apparent diameter of Venus, 85–86
observes 1639 transit of Venus, 84–87, *85*
studies Kepler's writings, 80–82
Venus in Sole Visa, *85*, 86–87, 90
Huygens, Christiaan, 90, 139
observes 1661 transit of Mercury, 109

James II (king of England), 120
Janssen, Pierre Jules César, 215, *247*
leads 1874 French expedition to Japan, 255–56
observes 1882 transit of Venus from Oran, 283
photographic revolver (*revolver photographique*), 220–21, 245, 247, 247–49, *248*, 255, 256, *256*, 258, 260, 264, 272

Kepler, Johannes, 59–64, *59*, 68
death of, 69
ellipticity of planetary orbits, 63, 80
harmonic law, 80–81, 224
Harmonice Mundi, 81
horoscope, 61, *62*
Keplerian eyepiece, 65
Phaenomenon Singulare seu Mercurius in Sole, 64. *See also* Mercury, transits of, supposed transit of 1607
prediction of transit of Mercury 1631 Nov 07, 69
prediction of transit of Venus 1631 Dec 06, 69
Rudolphine Tables, 69
Kirchoff, Gustav, 212
!Kung, 32

Lacaille, Nicholas-Louis, 218
Lalande, Joseph-Jérôme François de
 constructs *mappemonde* for 1769 transit of Venus, 186, *186*
 explains black drop effect, 156
Langley, Samuel Pierpont, 280, *280*
Lansberg, Philippe van, 79
Le Gentil de la Galasiere, Guillaume Joseph Hyacinthe Jean Baptiste
 abortive attempt to observe 1761 transit of Venus from Pondicherry, 145, 151–52
 at sea for 1761 transit of Venus, 152
 clouded out for 1769 transit of Venus, 198–99
 conducts research in East Indies between transits, 198, 200
 returns to France, 199–200
 travels to Manila and Pondicherry, 198
Leibniz, Gottfried Wilhelm, 119
Lemonnier, Pierre-Charles, 147, 154
Le Roy, Pierre, 217
Leverrier, Urbain Jean Joseph, 224, 228, 229
Lick, James, 288–89
Lindsay, James Ludovic (fourth Earl of Crawford and Balcarres), 240, 261
Lippershey, Hans, 64
Littrow, Carl Ludwig von, 188
Lomonosov, Mikhail V., 154–55
longitude
 difficulty in determining, 105–106, 154, 162
 found by timing eclipses of Galilean satellites, 106, 196, 217
 found with marine chronometers, 217
 method of lunar culminations, 217–18
 method of lunar distances ("lunars"), 106, 177, 217
 telegraphic determination, 208, 218
Louis XIV (king of France), 100, 120
Lovell, John L., 289, 290
Lumière, Auguste Marie Louis Nicolas, 264
Lumière, Louis Jean, 264

Marey, Étienne Jules, 264
Martial, Louis Ferdinand, 286, *286*
Maskelyne, Nevil
 encourages observation of 1769 transit of Venus from North America, 189
 unsuccessful observation of 1769 transit of Venus from St. Helena, 144, 152
Mason, Charles
 abandoned destinations to observe 1761 transit of Venus, 144, 146
 attempts to cancel 1761 transit expedition, 149
 observes 1761 transit of Venus, 151, 153
 observes 1769 transit of Venus, 186
Mayans, observatories, 45–46
Meldrum, Charles, 284
Mercury, transits of
 black drop effect, 156–57, *157*
 supposed transit of 1607 May 29, 63

transit of 1631 Nov 07, observed by Gassendi, 70–71, *71*
transit of 1631 Nov 07, predicted by Kepler, 69
transit of 1651 Nov 03, 109
transit of 1661 May 03, 109
transit of 1677 Nov 07, observed by Halley and Gallet, 109–11
transit of 1753 May 06, 141
transit of 1769 Nov 09, 183
transit of 1878 May 06, 270
transit of 1881 Nov 07/08, 270, 290
transit of 1999 Nov 15, 131, *157*
transit of 2003 May 07, *110*
Messier, Charles, 141–42
Milton, John, 82
Misch, Anthony, *292*, 293, *294*
Mitchel, Ormsby Macknight, 220
Mitchell, Maria, 279
Montezuma (Aztec ruler), 47
moon
distance, 27–28
moon illusion, 24
motion pictures, 264

Newcomb, Simon
employs David Peck Todd, 287–88, 289, 291
investigates lunar equation, 224
leads 1882 expedition to South Africa, 274, 283
member, United States Transit of Venus Commission, 229, 276–77
portrait, *274*
recomputes 1761/69 transit of Venus parallaxes, 154
reviews solar parallax values up to 1870, 276
vindicates Hell's observations of 1769 transit of Venus, 188
Newton, Isaac
annus mirabilis (1665–66), 96–97
appearance of, 115, 118, *119*
biography by Francis Baily, 118
born at Woolsthorpe, 93
childhood, 94
"Culture of Newtonianism," 118
De Motu Corporum in Gyrum, 114
education at Trinity College, 94–95
invents reflecting telescope, 98, *99*, 147
inverse-square law of gravitation, 113–14, 116–17, 123
as Lucasian professor, 98
as Master of the Mint, 118
Principia (The Mathematical Principles of Natural Philosophy), 114, 115–16, 116–18
Principia publication aided by Halley, 116–17
relations with Flamsteed, 118–19
relations with Halley, 113–14, 116–17
relations with Hooke, 111–12, 118
relations with Royal Society, 98, 99, 118
religious interests, 97, 119–20
self-absorption, 97–98, 99–100, 116
studies comet orbits, 112–13, 117
Nichol, J. Walter, 243
Noël, Alexander-Jean, 195, 197, *197*
Nursing Row (Narsinga Rao), Ankitam Venkata, 258–59

observatories
  status in 1760s, 148
  status in 1874/82, 208, *210*, 235
orreries, 122, 192

Palitzsch, Johann, 142
parallax
  definition, 23
  horizontal equatorial, 28
  lunar, 27, 28
  Mars, 102–104, *103*, *104*, 224, 276, *277*
  solar. *See* solar parallax
  Venus, 28
Paris, Royal Observatory, 101
Parkinson, Sydney, 171
Peiresc, Nicolas-Claude Fabri de (Peirescius), 71, 87
Perry, Father Stephen Joseph, 239–40, 260, 283
personal equation, 219
Peter I (Peter the Great, czar of Russia), 107
Peters, Christian Heinrich Friederich, 257, *257*
photography
  daguerreotype, 215–16
  daguerreotype applied to moon and sun, 216
  daguerreotype applied to transits of Venus, 1874 and 1882, 216, 244–45
  dry plates, 216, 272, 274
  objectivity, supposed, 220
  revolver-camera. *See* Janssen, Pierre Jules César, photographic revolver
  significance in astrophysics, 211–12
  wet plates, 216
  wet plates applied to transits of Venus, 1874 and 1882, 216, 272, 274
Pigott, George, 153
Pingré, Alexandre-Gui, 145
  clouds interfere with observation of 1761 transit of Venus, 152
  travels to Rodrigues Island to observe 1761 transit of Venus, 145, 149, 150–51, 152
planets
  apparent sizes, supposed, 63, 70
  distances, 28–29
  recognition of, 31–32
Pogson, Norman Robert, 215
Polynesians, 46
Prague (capital of Bohemia), 59, 60, 61, 62, 68
precision in science, quest for, 202–203
Prince, Charles Leeson, 281–82
Proctor, Richard Anthony
  advocates antarctic/subantarctic stations for 1874 transit of Venus, 232–35, *234*
  emigrates to Florida, 233
  portrait, *232*
  role in Royal Astronomical Society, 232, 233
  Transits of Venus: A Popular Account, 232
  Universe and the Coming Transits, The, 232
projection, eyepiece
  description, 311–12
  invented by Benedetto Castelli, 66–67
  used by Gassendi, 70–71

used by Horrocks, 84
used for 1874 transit of Venus, 243
Ptolemy, Claudius (Ptolemaeus, Ptolemy of Alexandria)
Mathemetike syntaxis (Almagest), 55
Planetary Hypotheses, 57
Ptolemaic (geocentric) System, 29, 48, 52, 54–55, *54*
Puiseux, Victor-Alexandre, 275

Rayet, Georges A. P., 215
Rees, John Krom, 279–80
Reformation, Protestant, 60
Remus Quietanus (Johannes Remus), 70
Rittenhouse, David
ability in mathematics and clock-making, 191–92
collapses during 1769 transit of Venus, 193–94
observatory, 192
observes 1769 transit of Venus, 193–95, *196*
Rømer, Ole, 103, 245
Royal Astronomical Society (of London), 232, 233
Royal Society (Royal Society for the Improvement of Natural Knowledge), 98–99, 144, 146, 149
Rudolf II (Holy Roman emperor), 59–60, 61, 67–68
Russell, Henry Chamberlain, 258, *259*
Russell, Henry Norris, 278
Rutherfurd, Lewis Morris, 279

Sands, Benjamin F., 229
Schaefer, Bradley E., 156–57
Scheiner, Christoph, 66
Schroeter, Johann Hieronymus
measures rotation period of Venus, 213
observes Venus cusp extension, 154, 212
suspects Venusian mountains ("Himalayas of Venus"), 213, *213*, *214*
Seven Years' War, 121, 143
Shakerley, Jeremiah, 110
Smith, Edwin, 257, 287
Solander, Daniel Carl, 171
solar parallax, 28, 29, 126
annual (stellar) aberration results, 224, *227*, *277*
asteroid (minor planet) parallax results, *226*
early proposals for use of transits to determine, 127–28
Eros approach, 1900–1901, 298
Eros approach, 1931, 298–99
found by Gill by observing Mars, 265, 276
lunar equation results, 224, *227*
"official" values, 223, 224, 276
parallactic inequality results, 224, *277*
parallax of Mars results, *226*, 265, 276, *277*
planetary equation results, *226*, *277*
results from 1761 and 1769 transits of Venus, 204, *226*
results from 1761 transit of Venus, 157–58
results from 1874 transit of Venus, 274–75, *275*, *277*

values found 1771–1874, 224, *226–27*
values found 1874–1882, 276, *277*
Venus parallax result, *226*
Solar System Dynamics Group, Jet Propulsion Laboratory, 300
Sousa, John Philip
Transit of Venus, The, 269
"Transit of Venus March," 269
spectroscopy
applied to Mars, 247
applied to sun, 215, 216, 247
applied to Venus, 212, 213, 255–56, *256*, 258
significance in astrophysics, 211–12
Spöring, Herman, 171
stars, distances to, 300–301
Stone, Edward James, 224
Strahan, George, 260
Struve, Otto Wilhelm von, 242
sun
limb darkening, 219
sunspots, early telescopic observations, 66–67
sunspots, naked-eye visibility, 34–35, 57–59

Tahiti
chosen by Transit Committee as 1769 transit of Venus station, 170
description, 161–62, 168, *169*, 173–74
discovery by Samuel Wallis, 167–168
visited by Bougainville, 169–70
Tarde, Jean, 66
telescopes and accessories
achromatic refracting telescope, 148, 208, 211
Cassegrain reflecting telescope, 147–48
early, discoveries, 64
early, drawbacks, 64–66
early, long focal ratios, 66, 102
equatorial mounting, clock-driven, 211
eyepiece filters, 243–44
Gregorian reflecting telescope, 147, *180*
heliometers, 211, 244, 261, 265
heliostats, 211, 263
invention, 64
invention of reflecting telescope by Newton, 98, *99*
micrometers, 65, 90, 211, 244
negative (Galilean) eyepiece, 64–65
Newtonian reflecting telescope, 98, *99*, 147
photoheliograph, 245, *246*
photoheliostats used in 1874 and 1882 transits, 244–45, *246*, 284, *285*, 289
silver-on-glass mirrors, 211
smoked glass filters, 243
solar prisms, 243
Tennant, James Francis, 215, 260
Thirty Years' War, 68–69, 72–73, 76, 87, 100
Tisserand, François Felix, 255
Todd, David Peck
early career, 287–88
late in life, 291, 293
photographs 1882 transit of Venus, 290–91, *292*, 293, *294*

photoheliostat used, 289
relationship with Austin Dickinson and Mabel Loomis Todd, 287, 291
travels to Lick Observatory, 289
Todd, Mabel Loomis, 287, 288, 291, 293
Trépied, Charles, 283
Tupman, George, 243, 263, 287

Ueno, Hikoma, 255
units of measure, 16
stade, 25

Venus
apparent diameter, 85–86
atmosphere, modern findings, 303
cusp extension seen near inferior conjunction, 154, *155*
names, selected cultures, 41, 42, 42–43, 44–45, 45, 46, 46–47
observed by F. Bianchini, 138–39, *140*
observed by G. D. Cassini, 101–102, *102*
period, synodic, 43
ring of light seen at transit ingress and egress, 154, 154–155
rotation period, 102, 139, 140, 213, 303
spectroscopy, 212, 215
surface, nature of, 303, *304*
Venus, apparitions of, 2003/04
telescopic appearance, *307*
telescopic appearance near inferior conjunction, 305
visibility in sky, 305, *306*, *307*
Venus, apparitions of, 2012/13, *44*
Venus, transit of, 1631 Dec 06, 69, 71–72
Venus, transit of, 1639 Dec 04
observed by Crabtree, 86
observed by Horrocks, 84–87
predicted by Horrocks, 83
track across sun's disk, *336*
Venus, transit of, 1761 Jun 06, 126
black drop effect, 156, 157
British expeditions, 144, 146
Delislean Poles, *134*, 138
disappointing results, 153–54, 158–59, 162
French expeditions, 144, 144–46
general data, *337–38*
Halleyan Poles, 133, *134*
international cooperation, 142–43
resulting solar parallaxes, 157–58
results recomputed, 154
ring of light (aureole) observed, 154, 154–56
track across sun's disk, *336*
visibility zones, 133, *134*
Venus, transit of, 1769 Jun 03, 126
black drop effect, 181, *182*, 191
British observing plans, 163–65
British South Pacific plans, 165–66
British Transit Committee, 163, 165, 166
general data, *337–38*
international efforts, 162
observed by Cook and Green, 181, *182*
observed by Wales and Dymond, 191
ring of light (aureole) observed, 191

track across sun's disk, 185, *336*
visibility zones, 163, *164*, 186, *186*
Venus, transit of, 1874 Dec 09
American plans, 241
American standard telescope, 243, *244*
American stations, 250
Australian Woodford station, 251, *251*
black drop effect, 258, *259*, 261, *262*
British controversy over station locations, 231–35, *234*
British "districts" for stations, 235
British Egyptian stations, 240, 251
British Hawaiian stations, 250–51
British Kerguélen stations, 239–40, 260–61
British Mauritius station, 240
Dutch expedition to Réunion, 242
French *Commission du passage de Vénus de l'Académie des sciences*, 229, 241
French Île St.-Paul station, 251–52, *252*
general data, *337–38*
German commission, 229
German expeditions, 241–42
German Kerguélen station, *261*
international expeditions, 235, *236–39*
Italian expedition to Madhepur, 242
Mexican expedition to Yokohama, 242
observations, American, 255, 257–58
observations, Australian, 256, 258, *259*
observations, British, 253–54, 256, 257, 260–61, *262*
observations, French, 255–56, *256*, 257, 260
observations, German, 256–57
observations, Indian, 258–59
observations, Italian, *256*, 258
observations, Mexican, 255
observations, Russian, 254–55, 261
observing equipment, 243
photographic results, 263–64, 271–72
planning, 225, 229–30, 235
resulting solar parallaxes, *275*
ring of light (aureole) observed, 258, *259*, 260
Russian stations, 242
track across sun's disk, *336*
United States Transit of Venus Commission, 229, 241, *242*, 274
visibility zones, 229–30, *231*
Venus, transit of, 1882 Dec 06
black drop effect, 282, *283*, 284–86
British stations, 283–84, *284*, 286, 287
Coast and Geodetic Survey stations, 280–81, *281*, 287
and comets of 1882, 269, *269*
conditions, general, 277
general data, *337–38*
international conference, 1881, 279–72, *271–72*

national participation, 270–72, *273*
observations, American, 279–80, *280*
observations, British, 281–82
observations, French, 282–83, *282*, 284, 284–86, *285*
observations, German, 283, *283*
observing stations, *273*, 277–78
photography, 272, 274
public observing, 278–79, *279*
ring of light (aureole) observed, 279–80, *280*, 281–82, *282*, 285–86
track across sun's disk, *336*
United States Naval Observatory stations, 280, 283
visibility zones, 268, 272, *273*, 274, 277

Venus, transit of, 2004 Jun 08
earth-Venus distance, 300
general data, 307, *337–38*
local circumstances, 305–306, 339–43
track across sun's disk, *336*
viewing site selection, 308, 328–31
visibility zones, 304–305, 307–308, *309*, *310*, 327, 328

Venus, transit of 2012, Jun 05–06
earth-Venus distance, 300
local circumstances, 345–50
observing technology, possible new, 323–24
track across sun's disk, *336*
viewing site selection, 320, 322–23
visibility zones, 304–305, 320, *321*, 322–23, *323*

Venus, transit of, 2117 Dec 11, 324–26

Venus, transits of (general)
black drop effect, 156–57, *157*, *158*, 159, 181, 258, *259*, 261, *262*, 282, *283*, 285–86
chronology 2971 BCE–7464 CE, 333–34
geometrical circumstances, 1761–2012, *337–38*
naked-eye/pretelescopic visibility, 57
ring of light (aureole) observed, 154, 154–56, 191, 258, *259*, 260, 279–80, *280*, 281–82, *282*, 285–86
role of women, 15, 249–50, 279
series, 334
sunshine probability, December, 308, *353*
sunshine probability, June, 308, 339–43, 345–50, *352*
tracks across sun's disk 1639–2012, 335–36, *336*
transit simulators, 163, 219–20
transits, partial, 334

Venus, transits of, 1874 and 1882
observing sites, establishing, 250–52
organization of personnel, 249–50
reaching destinations, 250

Venus, transits of, 2004 and 2012
atmospheric turbulence, 312
CCD cameras, 315
documenting observations, 317
eyepiece projection, 311–12, 313
eyes, dangers to, 308
filters, hydrogen-alpha, 313

filters, safe, 310–11
finder, covering, 312
image enlargement, 315, 316
image processing, 316–17
infrared imaging, 319, 324
observing projects, 317–19
photography and imaging, 312–17
webcams, 316
Web sites, 319
Vogel, Hermann Karl, 215, 283, *283*
Voltaire (François-Marie Arouet), 121–22

Wales, William
conditions experienced at Hudson Bay, 189–90
observations of 1769 transit of Venus, 190–91
observatory and equipment at Hudson Bay for 1769 transit of Venus, 189
travels to Hudson Bay to observe 1769 transit of Venus, 165, 189, *190*
Wallis, John, 78
Wallis, Samuel, 167–69
Wargentin, Pehr Wilhelm, 153
wars between England and Holland, 95, 104–105
West, Benjamin, 191
William III (William of Orange, king of England), 120
Wilson, William Parkinson, 258
Winthrop, John, 147, 153, 188
Wolf, Charles, 219–20
world
wars circa 1874–82, 206
changes 1761/69–1874/82, 201–202, 205, 207–208, *209*, 229, 268–69, *286*
Worthington, John, 78
Wren, Christopher, 98–99

Yanagi, Naranoyoshi, 255

DATE DUE
